AF677685

Integrating Spatial Analysis and Regional Science to Guide Urban Planning

Integrating Spatial Analysis and Regional Science to Guide Urban Planning

Guest Editors

Apostolos Lagarias
Poulicos Prastacos
Despina Dimelli
Alexandra Delgado-Jiménez

MDPI

Basel • Beijing • Wuhan • Barcelona • Belgrade • Novi Sad • Cluj • Manchester

Guest Editors
Apostolos Lagarias
Department of Geography
and Regional Planning
National Technical University
of Athens
Athens
Greece

Poulicos Prastacos
Division of Spatial
Analysis-Geographic
Information Systems &
Remote Sensing
Institute of Applied and
Computational Mathematics
of the Foundation of Research
and Technology-Hellas
(FORTH)
Heraklion
Greece

Despina Dimelli
School of Architecture
Technical University of Crete
Chanai
Greece

Alexandra Delgado-Jiménez
Department of Architecture
Escuela Politécnica Superior
University
Nebrija
Spain

Editorial Office
MDPI AG
Grosspeteranlage 5
4052 Basel, Switzerland

This is a reprint of the Special Issue, published open access by the journal *Land* (ISSN 2073-445X), freely accessible at: https://www.mdpi.com/journal/land/special_issues/M22E9AG36C.

For citation purposes, cite each article independently as indicated on the article page online and as indicated below:

Lastname, A.A.; Lastname, B.B. Article Title. *Journal Name* **Year**, *Volume Number*, Page Range.

ISBN 978-3-7258-5827-9 (Hbk)
ISBN 978-3-7258-5828-6 (PDF)
https://doi.org/10.3390/books978-3-7258-5828-6

© 2025 by the authors. Articles in this book are Open Access and distributed under the Creative Commons Attribution (CC BY) license. The book as a whole is distributed by MDPI under the terms and conditions of the Creative Commons Attribution-NonCommercial-NoDerivs (CC BY-NC-ND) license (https://creativecommons.org/licenses/by-nc-nd/4.0/).

Contents

About the Editors

Apostolos Lagarias

Apostolos Lagarias is an Assistant Professor in the Department of Planning and Regional Development at the University of Thessaly. As of 2013, he holds a PhD from the Department of Architecture at Aristotle University of Thessaloniki and has worked as a postdoctoral researcher at the Foundation for Research and Technology–Hellas (Institute of Applied and Computational Mathematics) and at the National Technical University of Athens (School of Rural and Surveying Engineering–Geoinformatics Engineers). His research focuses on the analysis of urban transformations through contemporary methodological approaches that employ geospatial methods to support decision-making in urban and spatial planning. His research interests include urban form models, land-use analysis through Geographic Information Systems (GISs), the relationship between spatial planning and Climate Change, and the application of spatial analysis methods in planning. He has 26 papers published in Scopus-indexed international journals (h-index 10) and currently participates in national research programs on Climate Resilience (CLIMASPIN project).

Poulicos Prastacos

Poulicos Prastacos is Researcher Emeritus at the Institute of Applied and Computational Mathematics (IACM) of the Foundation for Research and Technology-Hellas (FORTH), Heraklion, Greece. He holds a Ph.D. (1981) in Transportation Systems from the department of Civil Engineering, University of Illinois at Urbana-Champaign, a Master of Science in Construction Management (1976) from the same university and a B.S. in Civil Engineering (1976) from Columbia University. His research interests include the development of spatial decision support systems, statistical analysis of spatial data in the areas of urban and regional analysis, Geographic information systems (GIS) applications, remote sensing methodologies and others. From 1990-2020 he was in charge of the Regional Analysis Division, an R&D group involved with the analysis of urban and regional issues and the development of Geographic information systems (GIS) applications. He has participated in about 30 projects funded by EU and various Greek organizations and has published about 100 papers in scientific journals and conference proceedings.

Despina Dimelli

Despina Dimelli is a Professor in urban and regional planning, and environmental management. Her work integrates spatial analysis, participatory governance, and data-driven approaches to advance resilient and sustainable urban development. She has contributed to EU-funded and national research projects addressing climate-resilient planning and heritage preservation. Her expertise encompasses qualitative and quantitative methods, stakeholder engagement, and the facilitation of knowledge-exchange workshops. Her research outputs include peer-reviewed publications and presentations on sustainable tourism, urban resilience, and innovative planning practices, reflecting a commitment to evidence-based solutions that reconcile environmental, social, and economic objectives in urban and regional contexts.

Alexandra Delgado-Jiménez

Alexandra Delgado-Jiménez is an Architect and Doctor of Urban Planning with a distinguished international profile in urban design related to sustainability, climate change, and energy transition. She is the Principal Investigator and founder of the At-the-oUTSET group, Architecture and Urban

Planning in the Face of Social, Economic, and Territorial Transformations, at Nebrija University, Spain. She is the Director of the Advanced Urban Planning Workshop for the master's degree in Architecture at Nebrija University and Adjunct Professor of Urban Planning at the undergraduate level. She has also been a visiting professor at La Sapienza University. She has also been a visiting professor at La Sapienza University in Rome, IAUV University in Venice, the United Nations Economic Commission for Latin America and the Caribbean (ECLAC) in Santiago, Chile, and in academic programmes organised by the Harbin Institute of Technology and TEC Monterrey. She has contributed to more than 90 chapters and articles in national and international journals.

Preface

From its early foundations in quantitative human geography to today's advanced applications of big data and remote sensing, spatial analysis has continuously evolved as both a scientific discipline and an analytical tool for urban and regional planning. This Special Issue, *Integrating Spatial Analysis and Regional Science to Guide Urban Planning*, contributes to this direction, demonstrating how methodological innovation and applied research can converge to address today's pressing planning and environmental challenges.

The contributions collected reflect the diversity of approaches in spatial analysis. Some focus on secondary geospatial data, offering refined insights into urban and regional dynamics, while others employ cutting-edge remote sensing technologies, simulations, and spatial–environmental models to investigate climate impacts, land-use change, or ecosystem degradation. Together, these perspectives illustrate not only the power of spatial analysis to reveal hidden patterns and relationships, but also its potential to inform forward-looking planning decisions at multiple scales, from the urban neighborhood to the wider regional landscape.

Overall, this Special Issue has an interdisciplinary and international scope, bringing together scholars from different regions who address topics ranging from urban sprawl and mobility to participatory mapping and climate resilience. Their collective work underscores the importance of integrated spatial approaches, ensuring that planning is data-informed and forward-looking. We hope that readers will find in this Reprint inspiration for how spatial analysis can be embedded into planning practice worldwide.

Apostolos Lagarias, Poulicos Prastacos, Despina Dimelli, and Alexandra Delgado-Jiménez
Guest Editors

land

Editorial

Integrating Spatial Analysis and Regional Science to Guide Urban Planning (Editorial for Special Issue Reprint)

Apostolos Lagarias [1,*], Poulicos Prastacos [2], Despoina Dimelli [3] and Alexandra Delgado-Jiménez [4]

1 Department of Planning and Regional Development, University of Thessaly, 38334 Volos, Greece
2 Institute of Applied and Computational Mathematics of the Foundation of Research and Technology-Hellas (FORTH), 70013 Heraklion, Greece; poulicos@yahoo.com
3 School of Architecture, Technical University of Crete, 73100 Chania, Greece; ddimelli@tuc.gr
4 School of Architecture and Building, Higher Polytechnic School, University Nebrija, 28015 Madrid, Spain; adelgado@nebrija.es
* Correspondence: lagarias@uth.gr

1. Introduction

1.1. Advances in Spatial Analysis

Spatial analysis is a research paradigm that employs specialized techniques and models to analyze and model spatial data, focusing on the variation over space and helping to reveal hidden patterns and model relationships [1]. Since the 1950s, spatial analysis has been established as a core discipline in quantitative geography [2], with numerous applications in the scientific fields in which the location of activities/events must be explicitly considered. Areas of application include, among others, urban analysis, environmental planning, transportation analysis, health planning, etc. In the 1990s, with the wide availability of microcomputers, spatial data, and commercial/open software, spatial analysis was integrated into the geographic information systems (GISs) that offered a powerful computing environment for managing and analyzing spatial data. During the past two decades a new paradigm has been gradually emerging, related to the massive use of big data in the analysis of space at several spatial scales [3], from the most local to the regional or even the global one.

Spatial analysis emerged historically from simple models that estimated the interaction/relationship of two objects/events as a function of the distance that separated them. The simple gravity function, proposed in the 1950s, was gradually transformed to an elaborate set of equations and mathematical constructs that described the impact of the geographic location on the environment and our lives. In city planning, cities were initially recognized as static systems governed by regular patterns. Classical approaches, such as Christaller's central place theory or the rank–size law, described settlement hierarchies in abstract isotropic space. Early urban economics introduced density gradients and bid–rent models that explained how population densities and land values decline with distance from a center, followed by large-scale interaction models, such as simulating the flows of people from residences to employment using spatial interaction functions.

Since the 1980s spatial analysis has been transformed, and three major axes can be mentioned:

First, GPS and remote sensing have resulted in a revolution in the availability of spatial data. The GPS and the mobile devices permitted the on-demand precise identification of the location of any event. Volunteered geographic information (VGI), geotagged social media, and sensor networks introduced unprecedented spatial detail and update frequency. With remote sensing it was possible to observe the Earth's surface at high resolution and at regular time intervals, thus providing consistent data for urbanization, land-use change, natural resources monitoring, climate change, etc. Landsat and other sensors provided a continuously updated open repository of satellite images. In the EU, the Copernicus program and the Sentinel constellation resulted in a wealth of free satellite images at detailed resolutions, as well as processed data for European countries ready to be used for spatial analysis. The CORINE Land Cover program—alongside other land use/land cover geospatial databases of the Copernicus Land Monitoring Service and the European Environmental Agency—is another pioneering case study in urban and territorial transformation, allowing comparative studies at a pan-European level. This "datafication" of geography has shifted spatial analysis from coarse zones to detailed, dynamic observations of space and society.

Second, advanced modeling techniques such as agent-based models, spatial statistics, raster analysis, and others were introduced, and nowadays they are enhanced by machine learning and artificial intelligence (AI). Since the 1990s, cellular automata and agent-based models have been widely used to reproduce urban growth, sprawl, or segregation as emergent processes [4]. providing a bridge between micro-behavior and system dynamics [5]. These methodologies can be used to analyze spatial-temporal time series of environmental data, to simulate traffic patterns and urban growth, etc.

Third, advances in computing facilities permitted the introduction of platforms that can process and analyze data using available software [6,7]. The Google Earth Engine provides an integrated environment for analyzing satellite images, producing vast geo-referenced datasets that require explicit spatial-statistical handling [8]. The R software includes ready-to-use spatial statistics algorithms, whereas for users who want to develop their own code, Python libraries for spatial data are widely available. Complete modeling environments are readily and freely available, for example, models for simulating urban growth like UrbanSim [9] and SLEUTH [10], models of land surface temperature like SURFEX [11], models for flood modeling like HEC-RAS [12], and models for microclimate and environmental simulations (ENVI-met-type models) at the urban canopy level. Such models can be used for environmental simulations that link human activities, housing and employment location, topology, and existing infrastructure to climate and ecosystem performance [13] or to test various scenarios for flood or other hazard mitigation.

1.2. Spatial Analysis in Planning

Planning at the local/metropolitan/regional scale is revolutionized by advances in spatial analysis and the availability of extremely large datasets (big data) and methods for processing, managing, and viewing those data, either in the form of remote sensing analysis [14], advanced software analysis [15,16], or generally, in the form of complex algorithms, including machine-learning methods and, more recently, artificial intelligence (AI).

The integration of large-scale data and advanced analytical methods is reshaping current decision-making processes at multiple spatial scales. The rapid expansion of data sources provides planners with better ways of understanding urban dynamics. For instance, combining satellite imagery, mobile phone records, and geotagged social media with machine learning techniques enables predictive modeling of urban growth, mobility patterns, and environmental risks [17].

Remote sensing and GIS-based spatial analyses remain central to these developments, offering the mapping and visualization tools needed to interpret complex datasets. Those systems enable detailed assessments of land use, vegetation cover, and infrastructural conditions, supporting both strategic and operational planning; while when paired with machine learning, such datasets can reveal spatio-temporal patterns and scenario testing, offering planners the ability to evaluate multiple interventions before their implementation.

Nevertheless, researchers and practitioners increasingly recognize that purely technical approaches cannot fully capture the lived realities of urban spaces and pinpoint the need to integrate traditional planning knowledge, local expertise, and participatory methods into data-driven processes. Therefore, participatory GIS, citizen sensing, and co-creation workshops are increasingly employed to complement computational methods, ensuring grounding analyses in local expertise and lived experience [18]. This combination of advanced analytics with participatory approaches represents a broader trend toward "hybrid planning," where technology augments, rather than replaces, human judgment.

With the integration of real-time data streams, cities can now implement dynamic policies—in the sense of the "real-time city", as described by Kitchin [19]—highlighting a shift from static, masterplan approaches toward flexible, evidence-informed decision-making. Overall, today's trends reveal a planning paradigm in which technological innovation, big data, and AI coexist with traditional spatial methods and participatory engagement, producing more responsive, evidence-based, and socially attuned planning processes.

This Special Issue provides a repository of novel approaches in spatial analysis that follow the trends discussed above and that could be potentially integrated into planning methodology at different scales, from the urban to the regional. The significance of the value of the research discussed in these papers cannot be underestimated since only planning policy and practice can address the existing challenges of quality of life and the new and evolving challenges of climate change and its impacts on cities and rural areas (urban heat, floods, drought, degradation of ecosystems, increasing urbanization, environmental degradation, etc.).

Issues like urban sprawl and inclusive sustainable growth underline the need for innovative approaches on how to handle the environmental consequences of urban growth at a metropolitan or even regional scale [20] Climate change introduces additional challenges for urban and regional systems, including increased frequency of extreme weather events, rising temperatures, flooding, and pressures on water and energy systems. These dynamics necessitate planning strategies that not only accommodate growth but also enhance the resilience of urban and peri-urban areas [21].

As shown by the selection of papers included in this Special Issue, spatial analysis facilitates the integration of diverse datasets, helping planners understand the interdependencies among different urban processes. For example, spatial analysis and modeling can help quantify urban sprawl (List of Contributions, 1,2) identify climate-related high-risk areas like urban heat island hotspots (List of Contributions, 3) or other environmental issues in the urban context like noise perception (List of Contributions, 4) optimize the placement of green infrastructure (List of Contributions, 5), and evaluate the cumulative impact of new developments on regional ecosystems (List of Contributions, 6) Such methods usually enable scenario-based planning, allowing decision-makers to foresee the potential outcomes of different urban growth strategies and assess trade-offs between economic development, social equity, and environmental protection. When combined with traditional spatial analysis methods, new spatial planning methods and technologies allow a more holistic approach to planning, where policies are not only reactive but anticipatory, promoting resilience and sustainability across regional systems.

2. Research Papers and Thematic Areas

The aim of this Special Issue is to contribute to the challenges discussed earlier and to explore novel approaches connecting spatial analysis and spatial planning elements. The

Special Issue includes papers that provide methodological advances and/or have practical planning implications. The Special Issue includes eight research papers, covering two basic thematic areas: (a) Spatial analysis using census and secondary geospatial data at the urban or regional level. (b) Spatio-environmental analysis using remote-sensing data and simulations at the urban level.

Additionally, a research contribution (List of Contributions, 7) that discusses participatory mapping is included. In this perspective, integrated planning brings together spatial science with participatory governance, with this paper being a complement to the more data-driven contributions in the SI. Although not exhaustive in their selection of topics and methods, the eight papers provide an insight into the current research interests of international academics, including scholars from Europe, USA, and North Africa (Table 1).

Table 1. Case study areas and research issues.

Case Study Area	Topic
Southern Europe	Equity in health planning (Rome); SDG urban metrics (Thessaloniki); ecosystem services and land capitals (inland Italy); impervious land expansion (Greece); gender-inclusive planning (Serbia)
North Africa	Urban heat and microclimate impacts (Morocco, Cairo)
USA	Urban morphology and noise modeling (NYC)

Table 2 tabulates the papers in terms of the study region, the methodology used, and its application and relevance in planning. A variety of models are noted, including spatial indicators estimation, land use/land cover (LULC) changes, simulation models, spatial autocorrelation and spatiotemporal heterogeneity, geographically weighted regression (GWR) modeling, demand and supply methods, land surface model (LSM), etc.

Table 2. List of published works, key methods used, and relevance to planning.

List of Contributions	Author (Year)	Study Region	Methods	Application in Planning
1	Lagarias (2023)	Greece, coastal zone	Spatial analysis based on land use detection, spatial indicators	Urban density/sprawl
2	Yiannakou & Zografos (2025)	Thessaloniki, Greece	Spatial indicators	Urban density/sprawl
3	Lachkham et al. (2025)	Morocco	Biophysically based land surface model	Land surface temperature estimation in urban areas
4	Chen et al. (2025)	New York City, USA	Multi-scale spatial analysis: spatial autocorrelation and spatiotemporal heterogeneity, geographically weighted regression (GWR) modeling	Noise perception in urban environments
5	Abdelmejeed et al. (2023)	Cairo, Egypt	ENVI-met analysis (urban geometry and tree scenarios)	Microclimate in urban environments
6	Marino et al. (2024)	Italy, inland areas	LULC changes, demand and supply methods	Spatial dynamics and ecosystem services
7	Čolić Marković & Danilović Hristić (2025)	Belgrade, Serbia	Participatory planning and other participatory methods	Inclusion and public spaces
8	Mazzalai, et al. (2025)	Rome, Italy	Socio-demographic indicators	Definition of health districts in urban areas

A brief description of the research objectives, the methodology used, and the results/major findings obtained in each paper is presented below:

Lagarias (List of Contributions, 1) applies geospatial methodologies on high-resolution imperviousness raster data to identify urban sprawl patterns in Greece's coastal zones between 2006 and 2018. The analysis shows that although during the economic crisis there was a significant slowdown in construction, some areas continued to be developed, even within areas of ecosystemic priority (Natura 2000) and climate-change sensitivity (flood-prone areas). Methodologically, the study highlights imperviousness as a more effective indicator than building density or urban land cover for monitoring and managing land-use change and calls for stronger spatial planning instruments in Greece and across the Mediterranean to explicitly regulate impervious expansion and safeguard climate resilience.

Yiannakou & Zografos (List of Contributions, 2) address peri-urban sprawl in Thessaloniki and estimate the ratio of land consumption to population growth, using GIS-based land-use analysis at different time periods. The methodological innovation lies in estimating the global indicator of Sustainable Development Goals' metrics at a fine-grained local scale, which allows evaluation of its precision and relevance. The results reveal that land consumption has outpaced population growth, indicating inefficient sprawl. The study's significance rests on its ability to bridge global sustainability frameworks with site-specific spatial analysis, reinforcing the operational value of SDGs in urban planning practice.

Lachkham et al. (List of Contributions, 3) explore the link between urbanization and land surface temperature (LST) in multiple cities in Morocco, applying remote sensing integrated with spatial statistical techniques. The methodological strength is the comparative, multi-city analysis that identifies common and divergent patterns across diverse urban morphologies. Results confirm a strong relationship between urban expansion and surface heat and reveal intra-urban variations affected by the population density and urban form. This work underscores the need for climate-sensitive planning in rapidly urbanizing contexts of the Global South.

Chen et al. (List of Contributions, 4) explore the relationship between urban morphology and noise perception, employing a multi-scale spatiotemporal modeling approach that integrates morphological metrics with acoustic data and perception surveys. Methodologically, it links physical urban form to subjective sensory experiences, combining quantitative spatial data with qualitative perceptions. Results show scale dependent relationships: while dense environments intensify noise annoyance at certain scales, they also mitigate perception in others by diffusing sound or supporting adaptive behavior.

To optimize microclimate conditions Abdelmejeed et al. (List of Contributions, 5) apply the ENVI-Met simulation model in Cairo, Egypt, to test different urban morphology and vegetation configurations. Methodologically, it relies on design-oriented scenario modeling, allowing urban form and greenery placement to be virtually explored. Results indicate that depending on the morphology of the urban streets (aspect ratio) and the presence of urban canyons, different strategies regarding placement and density of trees should be applied to mitigate heat stress. The results confirm ENVI-Met's value as a useful practical planning tool, enabling forward-looking design choices and showing how simulation modeling can be used for implementing adaptation strategies for climate-sensitive design in urban environments.

Marino et al. (List of Contributions, 6) analyze territorial change in Italian inland territories through an ecosystem services framework that integrates ecological, social, and economic indicators. Methodologically, it is noteworthy because it uses an assessment model that considers how land transformations affect natural resources, economic viability, and social well-being. Results show that degradation of natural capital reduces ecosystem services and undermines the economic potential and the social resilience of these regions.

Čolić Marković and Danilović Hristić (List of Contributions, 7) address the inclusion of gender perspectives in urban planning, using qualitative, participatory, and policy analysis methods rather than quantitative modeling. The study uncovers persistent gender gaps in planning participation and decision-making, showing how planning cultures reproduce inequalities. Its major finding is that procedural inclusivity is as crucial as spatial equity, since technical solutions alone cannot address systemic exclusion. The significance lies in highlighting that integrated planning must combine spatial science with participatory governance.

Finally, Mazzalai et al. (List of Contributions, 8) propose a method to redefine urban boundaries to improve equity in healthcare access in Rome. They apply a socio-demographic spatial model, combining census data, socio-economic variables, and geospatial mapping of health services. This methodological approach incorporates equity lenses, highlighting disparities between administrative divisions and populated urban geographies. Results show that official boundaries do not align with actual demographic and health needs, producing systematic inequities in service provision.

3. Summary and Perspectives

Spatial analysis has moved from stylized, equilibrium-based theories to data-intensive, dynamic approaches. It now integrates satellite monitoring, big-data analytics, complex simulation procedures, machine learning, and AI-driven tools, providing a much richer basis for understanding and guiding urban and regional change and related spatial policies. Integrating spatial analysis methods into planning processes has become particularly critical in managing the spatial complexity of urban development, as spatial analysis tools allow planners to visualize and simultaneously analyze land-use patterns and environmental constraints at multiple scales.

Furthermore, contemporary planning increasingly emphasizes the multi-dimensional nature of urban systems. Social, economic, and environmental factors interact in complex ways. Spatial analysis facilitates the integration of diverse datasets—from demographic and economic statistics to mobility patterns and environmental indicators—helping planners understand the interdependencies among different urban processes.

This Special Issue demonstrates how spatial analysis methods are instrumental in dealing with urban-environmental challenges. Cross-regional case studies illustrate how context-specific problems (e.g., urban heat islands, ecosystem services) can be addressed through a spatial–regional science lens. The focus on tools such as ENVI-Met (a leader in holistic 3D modeling software) and Sustainable Development Goals (SDG) indicators underscores a commitment to actionable research and informed policymaking. Overall, this Special Issue brings a convergence of theoretical advancement and practical application.

Author Contributions: Conceptualization, A.L. and P.P.; methodology, A.L., P.P., D.D.; writing—original draft preparation, A.L., P.P., D.D.; writing—review and editing, A.L., P.P., D.D. and A.D.-J.; visualization, A.L., P.P., D.D. and A.D.-J.; supervision A.L. All authors have read and agreed to the published version of the manuscript.

Funding: This research received no external funding.

Conflicts of Interest: The authors declare no conflicts of interest.

List of Contributions:

1. Lagarias, A. Impervious land expansion as a control parameter for climate-resilient planning on the Mediterranean coast: Evidence from Greece. *Land* **2023**, *12*, 1844.

2. Yiannakou, A.; Zografos, G. Spatial patterns of land take in a Mediterranean city: An assessment of the SDG indicator 11.3.1 in the peri-urban area of Thessaloniki. *Land* **2025**, *14*, 965.
3. Lachkham, M.A.; Bounoua, L.; Ed-dahmany, N.; Yacoubi Khebiza, M. Impact of urbanization on surface temperature in Morocco: A multi-city comparative study. *Land* **2025**, *14*, 1280.
4. Chen, S.; Yu, B.; Shi, G.; Cai, Y.; Wang, Y.; He, P. Scale-dependent relationships between urban morphology and noise perception: A multi-scale spatiotemporal analysis in New York City. *Land* **2025**, *14*, 476.
5. Abdelmejeed, A.Y.; Gruehn, D. Optimization of microclimate conditions considering urban morphology and trees using ENVI-Met: A case study of Cairo City. *Land* **2023**, *12*, 2145.
6. Marino, D.; Barone, A.; Marucci, A.; Pili, S.; Palmieri, M. The integrated analysis of territorial transformations in inland areas of Italy: The link between natural, social, and economic capitals using the ecosystem service approach. *Land* **2024**, *13*, 1455.
7. Čolić Marković, N.; Danilović Hristić, N. Integrating gender perspectives in participation to guide changes in urban planning in Serbia. *Land* **2025**, *14*, 258.
8. Mazzalai, E.; Caminada, S.; Paglione, L.; Salvatori, L.M. Redefining urban boundaries for health planning through an equity lens: A socio-demographic spatial analysis model in the city of Rome. *Land* **2025**, *14*, 1574.

References

1. Fischer, M.M. Spatial Analysis in Geography. In *International Encyclopedia of the Social & Behavioral Sciences*; Smelser, N.J., Baltes, P.B., Eds.; Pergamon: Oxford, UK, 2001; pp. 14752–14758.
2. Liu, P. Spatial analysis. In *The Encyclopedia of Human Geography*; Warf, B., Ed.; Springer: Cham, Switzerland, 2024.
3. Batty, M. Big data and the city. *Built Environ.* **2016**, *42*, 321–337. [CrossRef]
4. White, R.; Engelen, G. High-resolution integrated modelling of the spatial dynamics of urban and regional systems. *Comput. Environ. Urban Syst.* **2000**, *24*, 383–400. [CrossRef]
5. Batty, M. *Cities and Complexity*; MIT Press: Cambridge, MA, USA, 2005.
6. Gorelick, N.; Hancher, M.; Dixon, M.; Ilyushchenko, S.; Thau, D.; Moore, R. Google Earth Engine: Planetary-scale geospatial analysis for everyone. *Remote Sens. Environ.* **2017**, *202*, 18–27. [CrossRef]
7. Pham-Duc, B.; Nguyen, H.; Phan, H.; Tran-Anh, Q. Trends and applications of Google Earth Engine in remote sensing and earth science research: A bibliometric analysis using Scopus database. *Earth Sci. Inform.* **2023**, *16*, 2355–2371. [CrossRef]
8. Van der Meer, F. Remote-sensing image analysis and geostatistics. *Int. J. Remote Sens.* **2012**, *33*, 5644–5676. [CrossRef]
9. Waddell, P. UrbanSim: Modeling urban development for land use, transportation, and environmental planning. *J. Am. Plan. Assoc.* **2002**, *68*, 297–314. [CrossRef]
10. Silva, E.A.; Clarke, K.C. Calibration of the SLEUTH urban growth model for Lisbon and Porto, Portugal. *Comput. Environ. Urban Syst.* **2002**, *26*, 525–552. [CrossRef]
11. Masson, V.; Le Moigne, P.; Martin, E.; Faroux, S.; Alias, A.; Alkama, R.; Belamari, S.; Barbu, A.; Boone, A.; Bouyssel, F.; et al. The SURFEX v7.2 land and ocean surface platform for coupled or offline simulation of earth surface variables and fluxes. *Geosci. Model Dev.* **2013**, *6*, 929–960. [CrossRef]
12. Peker, İ.B.; Gülbaz, S.; Demir, V.; Orhan, O.; Beden, N. Integration of HEC-RAS and HEC-HMS with GIS in flood modeling and flood hazard mapping. *Sustainability* **2024**, *16*, 1226. [CrossRef]
13. Yang, X.; Zhao, L.; Bruse, M.; Meng, Q. Evaluation of a microclimate model for predicting the thermal behavior of different ground surfaces. *Build. Environ.* **2013**, *60*, 93–104. [CrossRef]
14. Yu, D.; Fang, C. Urban Remote Sensing with Spatial Big Data: A Review and Renewed Perspective of Urban Studies in Recent Decades. *Remote Sens.* **2023**, *15*, 1307. [CrossRef]
15. Rey, S.J.; Anselin, L. Recent advances in software for spatial analysis in the social sciences. *Geogr. Anal.* **2006**, *38*, 1–23. [CrossRef]
16. Anselin, L. From SpaceStat to CyberGIS: Twenty years of spatial data analysis software. *Int. Reg. Sci. Rev.* **2012**, *35*, 131–157. [CrossRef]
17. Arribas-Bel, D. Accidental, open and everywhere: Emerging data sources for the understanding of cities. *Cities* **2014**, *47*, 51–57. [CrossRef]
18. Goodchild, M.F. Citizens as sensors: The world of volunteered geography. *GeoJournal* **2007**, *69*, 211–221. [CrossRef]
19. Kitchin, R. The real-time city? Big data and smart urbanism. *GeoJournal* **2014**, *79*, 1–14.

20. Lagarias, A. Exploring land use policy scenarios with the use of a cellular automata-based model: Urban sprawl containment and sustainable development in Thessaloniki. *Geocarto Int.* **2015**, *30*, 1033–1051. [CrossRef]
21. Sellberg, M.M.; Ryan, P.; Borgström, S.T.; Norström, A.V.; Peterson, G.D. From resilience thinking to Resilience Planning: Lessons from practice. *J. Environ. Manag.* **2018**, *217*, 906–918. [CrossRef] [PubMed]

Disclaimer/Publisher's Note: The statements, opinions and data contained in all publications are solely those of the individual author(s) and contributor(s) and not of MDPI and/or the editor(s). MDPI and/or the editor(s) disclaim responsibility for any injury to people or property resulting from any ideas, methods, instructions or products referred to in the content.

land

MDPI

Article

Impervious Land Expansion as a Control Parameter for Climate-Resilient Planning on the Mediterranean Coast: Evidence from Greece

Apostolos Lagarias

Department of Planning and Regional Development, School of Engineering, University of Thessaly, 38334 Volos, Greece; lagarias@iacm.forth.gr

Abstract: Impervious land expansion is linked to ecosystem fragmentation and degradation, posing threats to nature conservation and multiplying climate change impacts. This is particularly true on the Mediterranean coast, where persistent urbanization is coupled with tourism development, further intensifying coastal erosion, flood risk, heat stress and biodiversity loss, while decreasing carbon sequestration. In this research, high-resolution imperviousness data were analyzed through a geospatial methodology to detect patterns and processes in a heavily burdened Mediterranean coastal area, namely Greece. The methodology was structured on a set of GIS tools, analyzing the distribution of new impervious cover between 2006 and 2018, to evaluate pressures exerted on coastal territories and on the environmental protection network. The results revealed relatively slow rates of impervious land expansion at a nationwide scale, mostly attributed to the economic recession period in Greece. However, certain locations exhibited continuing artificialization of land even within Natura 2000 areas, while future pressures on coastal territories are expected to increase due to the restarting of construction activity and the continuing dynamic of the mass tourism sector. The conclusions imply that controlling for imperviousness is important in order to develop spatial planning policies for climate resilience, which should be decisively enforced in the Mediterranean to prevent a business-as-usual scenario.

Keywords: impervious land; coastal zone; climate resilience; climate change; spatial planning; geospatial analysis; economic crisis; mass tourism development; Mediterranean region; Greece

1. Introduction

Based on current projections and estimations, coastal European areas will be confronted with important threats and environmental challenges up to 2050, emanating from land use changes [1] related to urbanization, tourism development, and agricultural intensification. Due to intense land take, ecosystems and biodiversity will be heavily affected [2] by deforestation, ecosystem fragmentation and degradation [3], with a loss of open habitats, natural vegetation, and agricultural land [2]. Simultaneously, higher built-up pressure can be related to overexploitation of natural resources [1]; air, water, and land pollution; increasing vulnerability to coastal erosion [4]; and coastal inundation.

These threats are expected to further worsen due to climate change (CC) impacts [5], with highly urbanized coastal areas usually coinciding with high-risk prone areas due to significant population and infrastructure concentration in relation to hazard-amplifying conditions [2]. Simultaneously, increased artificialization of land is limiting the amount of carbon sinks and the ability of natural ecosystems to capture carbon dioxide; it is also related to increased water demand and a loss of fertile soils due to poor land management [6,7]. This is in line with IPCC directives [8], where controlling urban sprawl is considered a major element of climate change adaptation and a sustainable use of land.

The present paper handles key research questions regarding urban encroachment in coastal areas, setting a target of exploring a characteristic case study from the Mediterranean context and providing links between the analysis of impervious land expansion and climate change adaptation. In the EU-Mediterranean region, intense land use exploitation in coastal areas is currently reported [9], heavily transforming the traditional Mediterranean landscape and its spatial and ecological structure. Generally, it is accepted that in the coastal Mediterranean zone, complex ecological and resource systems (in both the marine and land seashore parts) coexist and interact with human communities and relevant socioeconomic activities [10].

However, the restructuring of historically compactly built cities towards more dispersed periurban patterns [11,12] was further intensified by tourism and second housing developments in coastal zones [13–15], with the Mediterranean region being characterized as an exemplary case of a heavily burdened territory due to incoming tourist flows [16,17]. As a result, the dominant tourism development model unfolds as a land- and resource-intensive activity, taking up space along the coastline to support a growing demand for facilities and related infrastructure, posing pressure on natural resources, coastal communities, and ecosystems. Moreover, due to long-term deficiencies in spatial planning, urbanization in coastal Mediterranean areas is also linked to low-density/dispersed patterns, with increasing CO_2 emissions and energy demands, under a car-dependent model of mobility that is prevailing in exurban and island areas [18].

Characteristic hotspots of coastal urban development related to mass tourism are identified on the Adriatic coast of Italy [19], on the Cote d'Azur in France [20], in Costa Blanca and Costal del Sol along the Spanish Mediterranean coast [21–24], in island states like Malta [25] and Cyprus [26], and also in emerging tourism destinations along the Dalmatian coast [27,28]. In Greece, signs of overtourism have already been identified [29–32], especially affecting highly reputable destinations like the South Aegean islands, Ionian Islands, and Crete, while intense coastalization is also affecting mainland coastal areas like Chalkidiki and Pieria in the Central Macedonia region, Attica, and certain parts of the Peloponnese [33], among other areas. This model of coastal urban development usually coincides with areas of unique ecological value [34,35], and is related to a long-term tradition of a largely unplanned or poorly regulated characteristic of settlement expansion [11], illegal construction [36,37], and an insufficient implementation of environmental protection policies, prioritizing economic profit over sustainability concerns while generating local and regional environmental conflicts [38].

The dramatic impact of these processes in coastal areas and ecosystems has been largely reported in the literature [39–42], while it is estimated that in certain parts of the coastline zone, and more notably in island territories, carrying capacity thresholds have already been reached [32,43]. At the same time, according to current scenarios, the Mediterranean Basin is a CC "hotspot" region with an increased number of multiple climatic hazards, including coastal flooding, heat extremes and fire risk, the frequency and severity of meteorological droughts, a decrease in precipitation and river flow, and increasing water demand for agriculture [44]. Overall, it is assessed that the Mediterranean coastal region currently lies at a critical standpoint, undergoing intensive socioeconomic and environmental changes [45], which, coupled with climate change risks and vulnerabilities, could result in further environmental degradation and threats to sustainability achievements [46].

This comes at a time where the EU seems determined to put forward more strict and concrete policies for nature preservation and protection and no-net-land-take policies [47,48], such as those included in the Biodiversity Strategy for 2030 [7] as a core part of the European Green Deal and the new Nature Restoration Law [49]. Coastal ecosystems in particular are considered a major asset to combat climate change, providing benefits including storm surge mitigation, protection against coastal erosion and floods, water filtration, and drought protection, while also acting as carbon sinks and carbon storage systems. To this end, strategies regarding climate resilience are marked by a considerable turn towards nature-based solutions (NbSs), providing links and synergies between CC adaptation and biodiversity [50].

Impervious land expansion is defined as the covering of the soil surface with impermeable materials (i.e., any surface which water cannot infiltrate) because of urban development and infrastructure construction, resulting in the substitution of the original natural/agricultural land cover or water surface with an artificial one [51]. According to the literature, the amount of impervious surface in a landscape is an important indicator [52] directly related to land take [47] and to the presence of residential, commercial, industrial complexes including paved ways, roads [53], while the analysis and mapping of impervious surface area is essential for improving urban environmental quality toward ecological, livable, and sustainable goals [54]. Researchers like Torbick and Corbiere [55], Zhou and Wang [56], Xian et al. [57], Ghazaryan et al. [58], and Li et al. [59], among others, have used imperviousness data to evaluate urbanization processes, while, specifically in the Mediterranean, relevant recent research includes the study of urbanization and peri-urbanization processes [60–63], built-up pattern analysis [15,64], polycentricity [65], and landscape conservation [66]. Imperviousness is directly related to a set of climate change impacts [67–69], changing microclimate conditions [70] and determining the strength of the urban heat island effect, while the sealing of the surface diminishes rainwater infiltration, reduces aquifer recharge [71], and increases water runoff [53] and subsequent soil erosion and flood risk [44].

In respect to vulnerability related to climate change, imperviousness can be regarded as a better control parameter than built-up density. This is because not only buildings, but also related structures and paved surfaces (like parking lots, pool areas, canopy roofs, sealed parts of sport and recreation areas, etc.), have an environmental impact, altering the use of land as well as climatic behavior at the local level. Therefore, these structures should be co-estimated in the assessment of pressures exerted on natural, seminatural, and agricultural areas and ecosystems; this is also the case for road structures that have been linked with ecosystem fragmentation and habitat loss, even beyond their immediate physical footprint, especially when they penetrate natural ecosystems or wilderness areas [72].

Based on the above remarks, we can deduce that the analysis and monitoring of impervious land expansion can be a crucial control parameter, supporting climate-resilient strategies for Mediterranean coastal zones at the national, regional, and local levels. This is line with the European Environmental Agency, which states that reducing imperviousness is an important factor for climate-resilient planning and is regarded as a "green" adaptation tool [44], while also supporting EE initiatives for green infrastructure and resource efficiency [73,74]. Our research hypothesis is that the continuing expansion of impervious land is exacerbating CC-related pressures on coastal territories. Towards this direction, specific methodologies and research studies are still needed, especially in the field of integrating a spatial analysis and spatial planning perspective into climate resilience issues. This appears to be crucial in the Mediterranean context, and more specifically, in Greece. Greece can be regarded as a representative example in light of this research due to (a) the significance of its extensive coastline zone including 6000 islands and islets, (b) the fact that coastal zones (both land and marine ones) contain a dense network of ecosystems particularly vulnerable to environmental and climatic pressures, and (c) the fact that climate change adaptation is still in its infancy in the Greek context, especially in the field of land use planning and environmental protection; this is related to severe problems in coping with increasing challenges and vulnerabilities related to the built environment in the coastal zone.

Imperviousness is not yet incorporated as a control parameter in national and regional CC adaptation plans, while consistent relevant studies taking advantage of high-resolution spatial data are still sparse. To fill this gap, a GIS-enabled methodology was adopted, combining imperviousness data from the imperviousness density high-resolution layer, as provided by the Copernicus Land Monitoring Service, with spatial information related to the network of protected areas, according to three subsequent zones near the coast: the frontline zone, buffer zone, and transitional zone. The research aim is to provide a full-scale assessment of new impervious cover in the period 2006–2018 (i.e., the time period of imperviousness data availability), downscaling results to the regional and local levels and providing considerable inputs for planners and decision makers regarding climate resilience.

This is particularly important as, currently, the country is emerging from the long-term economic crisis of the 2009–2018 period, which was followed by the COVID-19 pandemic. Our main hypothesis is that after a slowdown in impervious land expansion during the years of the economic recession, a "rebound" must be expected, especially in coastal locations, depending on large-scale investments in the tourism industry in combination with emerging spatial and developmental practices and policies in Greece. Therefore, the spatial analysis's results are discussed in light of the general context regarding the spatial and developmental repercussions of the crisis period in Greece and their impact on coastal territories; this is a crucial viewpoint in order to properly assess the current "state of the play" and to provide insights on future environmental pressures regarding coastal territories.

The structure of the paper contains five sections. Section 2 describes the methodology and data, including a brief presentation of the case study area and expected climate change impacts. Section 3 presents the results regarding the estimation and evaluation of impervious land expansion in the coastal zones at a regional and local level, while Section 4 puts forward a discussion relating the major research findings and results to the general context of the economic crisis period and its spatial and developmental repercussions. Finally, Section 5 highlights the conclusions and pathways for future research towards climate-resilient planning in the coastal region.

2. Materials and Methods

2.1. Data and Methodological Considerations

Methodological issues regarding land use analysis in coastal zones can be important, as the magnitude of changes related to urban expansion, as well the distributional characteristics of the built-up patterns (e.g., compact, dispersed forms of development), are largely dependent on the accuracy of the data used. Despite the current availability of high-resolution data, many works regarding land use analysis within coastal zones are still largely dependent on Corine Land Cover (CLC) data [1,21,75], where a minimum mapping unit (MMU) of 25 hectares (ha) for areal phenomena and a minimum width of 100 m for linear phenomena are used. However, using such a crude-resolution landcover zoning approach can lead to a significant underestimation or overestimation of urban development, depending on the pattern and geometry of development (i.e., compact/dispersed). This is important as information regarding impervious areas might be "lost", especially in exurban areas. Specifically, CLC data cannot capture a large portion of low-density built-up areas scattered within zones identified as agricultural seminatural/natural areas or wetlands. Moreover, the 25 ha MMU and the condition that only changes when $\geq$5 ha is mapped impede a sound comparison of changes between different versions of CLC [18].

To handle this problem, the EEA has recently released new sets of data (https://land.copernicus.eu/en/products accessed on 24 September 2023) focusing on different hotspots, i.e., areas that are prone to specific environmental challenges, with a spatial resolution equivalent to the one of Urban Atlas (MMU 0.25–1 ha). These databases include a subset of the Natura 2000 Network, Riparian Zones, and a database specifically targeted to the coastal zone covering a 10 km wide buffer zone from the coastline (https://land.copernicus.eu/local/coastal-zones accessed on 24 September 2023). However, the above-mentioned Coastal Zones database is not yet validated, and relevant data have not yet been incorporated in applied research. High-resolution national land cover databases have been developed in several Mediterranean countries, such as the ISPRA database (Institute for Environmental Protection and Research) in Rome, Italy, the BD TOPO®/RGE Geodatabase in France [3], and SIOSE2005 and HR SIOSE2017 in Spain [76]. However, national data are not comparable across different Mediterranean countries; this fact hampers a harmonized analysis of the EU-Mediterranean zone, which is an important issue in order to handle inter-regional environmental problems and better co-ordinate relevant policies.

In Greece, the classification of multitemporal satellite images has been extensively used by individual studies to obtain high-resolution data regarding impervious areas [77], while in other cases, researchers have depended on the manual processing and digitization of satellite images or orthophotograph interpretation [78], a process that can be time-consuming. Moreover, land use analysis, as specified in the official technical specifications of urban planning and regional planning studies in Greece [79,80], still largely depends on aggregated Corine data, while the same holds for recently accomplished Regional Adaptation Action Plans (RAAPs) and the currently developed Special Environmental Studies (SEP) for the Natura 2000 network, this posing severe constraints in accurately monitoring and evaluating urban patterns to support planning decisions.

2.2. The Imperviousness High-Resolution Dataset

The imperviousness dataset (IMP) is part of the Copernicus High-Resolution Layers, provided on a pan-European scale and based on the processing of satellite data obtained from the Copernicus Sentinel-1 and Sentinel-2 sensors, considering spatially detailed Google Earth imagery as a supplementary data source for validation [81]. These layers include specific land use/land cover products, such as on impervious surfaces, forests, grasslands, water bodies, woody features, and wetness conditions [82]. The IMP dataset captures the spatial distribution of artificially sealed areas. Sealed pixels are classified based on the level of soil sealing (imperviousness (or sealing) degree, IMD), produced using a semiautomated classification based on calibrated NDVI (Normalized Difference Vegetation Index). Non-sealed areas are characterized by a value of IMD = 0%, while built-up areas are characterized by IMD values in the range of 1–100%.

According to the technical specifications of the dataset [83], impervious surfaces include a large set of structures like housing areas and isolated buildings, industrial/commercial areas, traffic areas, roads, railways inside built-up areas, greenhouses, solar panel parks, construction sites with discernible evolving built-up structures, sealed surfaces associated with sport and recreation/tourist areas, and other paved surfaces. On the other hand, they do not include mines, quarries, peat extraction areas, sand and sand pits, dump sites, natural and cultivated vegetated areas, unvegetated or sparsely vegetated areas, arable land, vineyards, fruit plantations, grass surfaces used for sports, green roofs, etc. The IMP dataset spans through time covering the 2016–2018 period, at a spatial resolution of 20 m for the years 2006, 2009, 2012, and 2015 and 10 m for 2018, while aggregated rasters are also delivered at a 100 m $\times$ 100 m spatial resolution.

In this work, to assess pressure on coastal areas, the imperviousness change layers were used, as delivered for 3-year periods (IMCC_06: 2006–2009, IMCC_09: 2009–2012, IMCC_12: 2012–2015, and IMCC_15: 2015–2018) at a comparable resolution of 20 m. These layers include a raster dataset mapping the percentage of sealing increase or decrease in the period covered (IMC layer), and a classified change product (IMCC layer) that maps the most relevant categories of sealing change, namely: (a) unchanged no sealing (class = 0), (b) new cover (from unsealed to sealed, class = 1), (c) loss of cover (from sealed to unsealed, class = 2), (d) unchanged sealed (class = 10), (e) increased sealing (class = 11), (f) decreased sealing (class = 12). In the IMCC layer, a binary approach was used for sealed areas, as all land with IMD $\geq$ 1% was considered as "sealed". To handle the issue of varying sealing values caused by the variance in input data quality for the processing of the historical layers, differences below a dynamic threshold for sealing increase (set to 20%) were considered as stable, as a lower threshold would cause false detected changes [83].

As already noted in the introductory part, imperviousness is an important control parameter in the estimation of climate-related impacts in coastal areas. This is particularly important in coastal areas where hotels, commercial spaces, and vacation houses usually include extensive paved surfaces like parking lots, pool areas, marines, etc., which result in an area many times larger being affected (Figure 1). However, building regulations usually focus on permitted built-up surfaces and the volume of new construction, not accounting for the accompanying environmental impact and land take of other artificial structures.

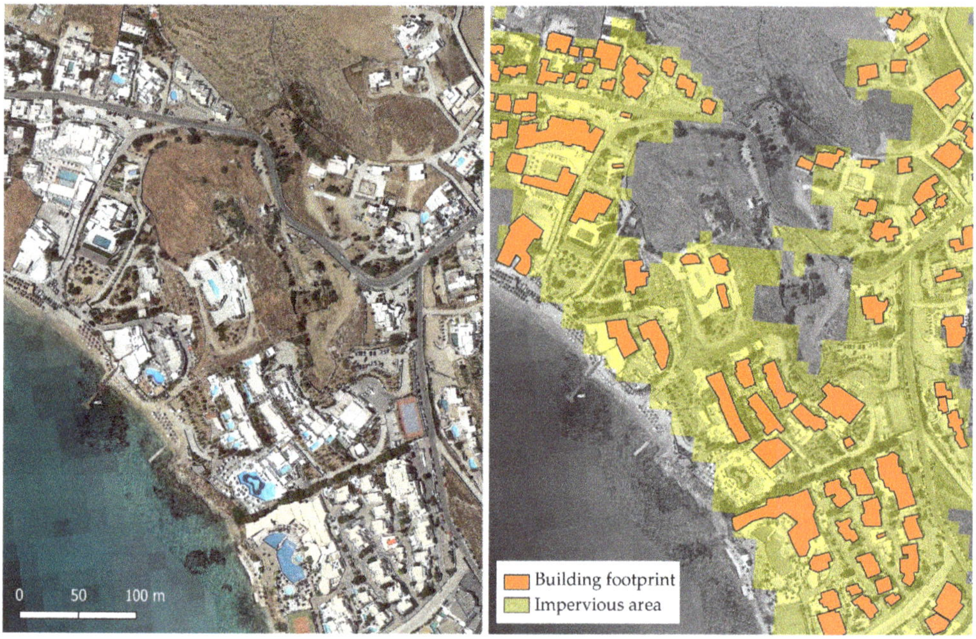

Figure 1. Mapping of tourism development on Mykonos Island (Greece): the imperviousness dataset (2018) captures the land take more accurately than the building footprint.

In order to apply the IMP dataset in the analysis of coastal areas, several methodological considerations needed to be taken into account. First of all, as the spatial resolution improved from 20 m to 10 m in the 2018 dataset, the identification of sealed areas became more detailed, capturing more buildings and structures in low-density areas than previous versions. This fact resolved several omissions and underestimations of sealed areas in previous releases and does not correspond to "new cover", meaning that newly identified cells due to technical reasons are classified as "unchanged sealed" in the IMCC layers. Therefore, in this work, we chose to use IMCC for the period of 2015–2018 as a reference, combining this more accurate product with "new cover" cells identified in the previous periods (2006–2009, 2009–2012, and 2012–2015).

Secondly, according to the imperviousness dataset methodology [83], not all the elements in the list of impervious/nonimpervious data may be separated clearly with automated classification methods. This was handled by the product developers with the use of in situ information to enhance accuracy levels. However, inspection of data in coastline zones revealed several misclassifications. For example, land attributed to sand/beach areas or bare rocks was, in several cases, incorrectly identified as "highly-sealed", especially in areas where coastal settlements were developed (Figure 2). This can be a disadvantage of using the IMP dataset to analyze coastal areas; however, applying the change layers (IMC and IMCC) to examine new cover allowed minimizing the effect classification inaccuracies could have on our results, as normally, the beach zone remains stable through time.

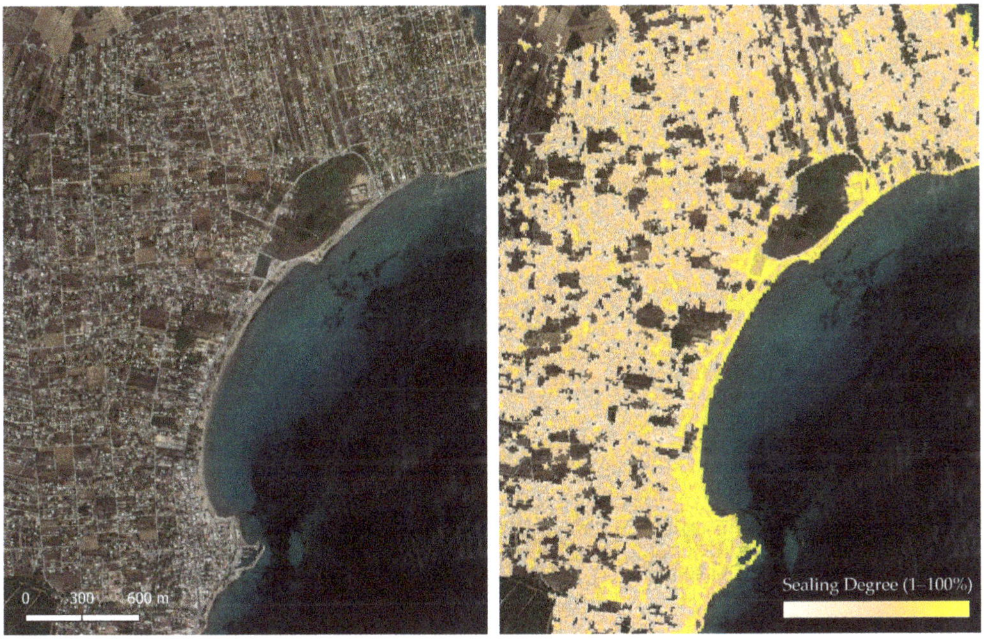

Figure 2. Incorrect classification in the eastern Attica zone (Greece), with the beach area classified as highly sealed impervious area (sealing degree close to 100%).

Inspection of data also showed that narrow roads were captured as "sealed areas" only if they were located between buildings. However, this also did not largely affect the current research target, as the focus was placed on new roads that were constructed in relation to residential, commercial, or tourism infrastructure and not on roads crossing between agricultural parcels and natural land. On the other hand, critical transport infrastructure like highways was included in the imperviousness database; therefore, in the present research, we applied a control of how new highway construction might affect the research results.

2.3. Methodological Steps

The basic workflow of the methodology adopted in the research is presented in Figure 3. The first step was to define the critical coastal zone. According to the literature, this may vary depending on the research focus and also on the geographical characteristics of the area examined [18]. In certain studies, the coastal zone is defined by a combination of the distance-to-coast and elevation dimensions [84,85], while distance thresholds could span from a belt zone of a few hundred meters [19] to a zone of several kilometers. For example, in the pan-European study by Lavalle et al. [1], the geographical delimitation of coastal zones was based on a 10 km buffer from the coastline, with an extra 2 km buffer from coastal wetlands, salt marshes, and salines, so as to include all transitional waters areas that were under the direct influence of maritime environments.

To construct a research-specific definition of the coastal zone, in this study, a combination of elevation and distance criteria was applied by taking into account the specific particularities of coastal development and geomorphology in the case study area. A coastal subzoning method was used, with the first zone including areas with high accessibility to the shore (including the seashore and the beach) and secondary (transitional) zones demarked towards the hinterland. A proximity raster calculating Euclidean distance to the coastline was combined with the Digital Elevation Model (DEM) of Europe to produce a

25 m raster dataset with cell values defined as $A_i = D_i + 5 \cdot E_i$, where A_i is coastal accessibility, D_i is the distance from the coast, and E_i is the terrain altitude of cell i. By setting different threshold values for this raster, the following set of coastal subzones was defined:

- Frontline coastal zone: $A_i \leq 500$ (direct accessibility to the sea, within what can be perceived as "walking distance"—up to 500 m; the elevation is up to 100 m).
- Buffer coastal zone: $500 \leq A_i \leq 1000$ (high accessibility to the sea, usually in continuity with the frontline zone; this zone could include higher-elevation areas between 100 and 200 m).
- Transitional coastal zone: $1000 < A_i \leq 2000$ (lower accessibility to the sea; this zone could include higher-elevation areas up to 200–400 m).

The specified zones were converted to vectors by considering polygon holes with a relatively small area ($\leq$10 ha) as an integral part of each zone (cartographic generalization procedure). Areas with $A_i > 2000$ were considered as "hinterland" and left out of this study, despite the fact that they can also be related to several coastal phenomena and influenced by maritime climatic conditions. This choice was mostly based on evidence that, in Greece, development in exurban areas is usually "squeezed" within a considerably narrow belt from the coast [86] due to increasing demand for immediate access to the sea for tourism and second housing development.

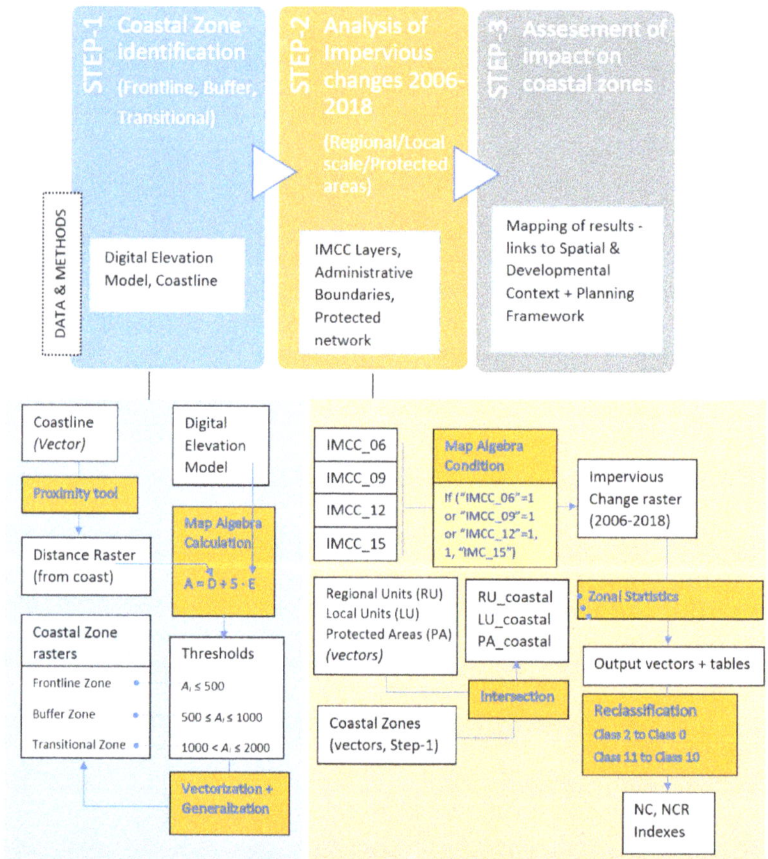

Figure 3. Methodological steps and GIS methods.

The second step was the processing of imperviousness data. Map algebra methods were used to combine the IMCC datasets for different periods in order to produce a raster dataset identifying new impervious cover (class = 1) and unchanged sealed land (class = 10) for the period of 2006–2018. The following condition was used: If (IMCC_06 = 1 or IMCC_09 = 1 or IMCC_12 = 1, true: 1, false: IMCC_15). Loss of cover (class = 2) and increased sealing (Class = 11) for the period of 2015–2018 were reclassified as Class 0 (unsealed) and as Class 10 (unchanged sealed), respectively. Class 12 pixels (decreased sealing) were not identified in the case study area. To evaluate impervious land changes, two main indicators were specified, namely:

(a) *NC* (new cover) index: the number of cells in Class 1 with new cover divided by the total zone area (Class 0 + Class 1 + Class 10).

(b) *NCR* (new cover rate) index: the number of cells in Class 1 with new cover divided by the sealed areas in 2006 (Class 10 cells) and multiplied by 100.

The third step was to combine the produced IMCC 2006–2018 raster with data of the protected network, i.e., Natura 2000 areas (obtained from the Copernicus Land Monitoring Service, https://land.copernicus.eu/local/natura/n2k-20181 accessed on 24 September 2023), Natural Parks and Wildlife Sanctuaries (obtained from the Geodata Geoportal of Greece, https://geodata.gov.gr accessed on 24 September 2023), and Important Bird Areas [87]. Zonal statistics methods were used in order to quantify the raster cell values per zone, while geoprocessing tools (intersection, spatial join, and dissolve) were applied to process the vector files of the specified coastal zones at the national, regional, and local levels. Geospatial analysis was processed using the Quantum GIS version 3.28.3 open source software.

2.4. Case Study Area

Despite its relatively small size, Greece is the country with the most extensive coastline in Europe, estimated to be close to 16 thousand km, including 6000 islands and islets (112 of which are inhabited) and including the Aegean Archipelago, the Ionian Archipelago, and the Libyan Archipelago (Figure 4). The coastal zone of Greece has been particularly important since ancient times, with major port cities and a large share of economic activities dependent on the sea, resulting in a dense network of archaeological spaces and coastal monuments. According to data from the 2021 Census [88], the population in Greece has reached 10,482,487 people (a 3.1% decrease with respect to 2011), with an important concentration in the coastal region, where the population density is estimated to be about three times higher than the density in hinterland zones [33].

The case study area includes the coastal zone of Greece, spanning through 12 out of the 13 regions (Nuts 2), with the addition of the independent Monastic community of Mount Athos. The South Aegean region (R9 on the map) contains the largest share of the coastal zone as a percentage of its total area (44.7%), followed by the North Aegean (31.6%, R5 on the map), Ionian Islands (37.5%, R13 on the map), Athos (30.9%, R3 on the map) and the Attica region (22.6%, R10 on the map), Central Greece (8.8%, R7 on the map), West Greece (8.3%, R8 on the map), and the Peloponnese (8.1%, R12 on the map); the smallest percentages are in Thessaly (3.5%, R6 on the map), Epirus (3.5%, R4 on the map), East Macedonia and Thrace (4.6%, R1 on the map), and Central Macedonia (5.1%, R2 on the map).

Population varies considerably among the different regions, with Attica and the Central Macedonia region presenting the highest population densities, as they contain the two major metropolitan centers of Greece (Athens and Thessaloniki, respectively). Only the South Aegean region and Crete present population growth in the 2011–2021 period, with all the other regions characterized by population decline. Major coastal cities include Thessaloniki, Piraeus (forming part of the Urban Agglomeration of Athens), Patras, Volos, Kalamata, and Kavala in mainland Greece, with the addition of important island cities like Herakleion, Chania, and Rethymno in Crete; Rodos, Kos, and Ermoupoli in the

South Aegean region; Kerkyra in the Ionian region; and Chios and Mytilene in the North Aegean region.

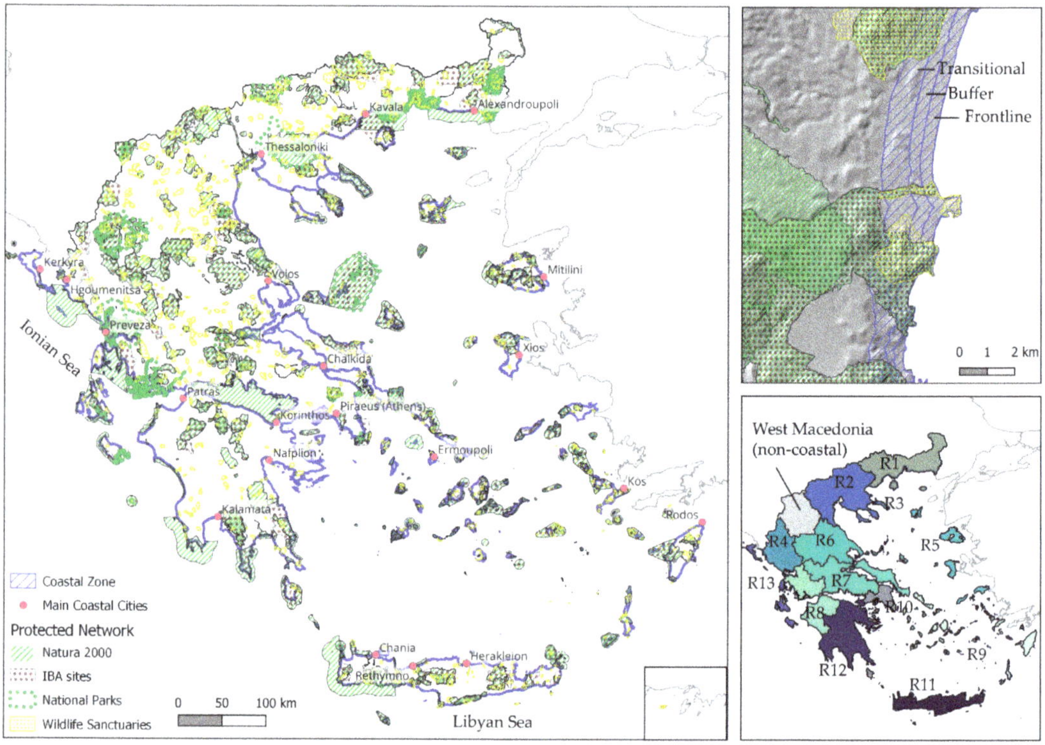

Figure 4. Map of the case study area including the regional administrative division (R1–R13 as defined in text), highways, and the network of protected areas.

The case study area includes a dense network of protected areas: 29.3% of the total coastal zone is part of the Natura 2000 Network (not counting the marine part of Natura 2000 areas), 25.3% is part of the Important Bird Areas' (IBA) network, 8.0% belongs to Wildlife Sanctuaries, and 11.4% in National Parks. As these areas usually overlap, a combined analysis showed that about 40% of the total coastal zone of Greece lies within protected areas. The largest share of the protected coastal zone was identified to be in Athos (100%), Epirus (59% of regional coastal zone), the East Macedonia and Thrace region (58.6%), and the South Aegean region (53.5%), while in the Ionian Islands, Attica, Central Greece, and the Peloponnese, less than one third of the coastal zone is protected. The coastal geomorphology is considerably diverse, including mainland and island parts, delta rivers, low-elevation land, steep slopes, mountain cliffs, and gorges, in close proximity to the seashore. According to the methodology adopted, high-slope and/or elevated areas close to the coast (up to 300 m) were included in the analysis as they form an integral part of the coastline zone; such areas are crucial for biodiversity, and are also usually perceived by real estate and market forces as candidate land for second housing and tourism development due to the views they offer towards the sea.

Scenarios reveal that Greece will be largely affected by climate change, with most of the consequences characterized as negative [89]. The mean temperature is expected to increase by over 2 °C and 2.5 °C for scenarios RCP 2.6 and 4.5, respectively, and by over 3 °C under scenario RCP 8.5. However, the increase in high temperatures is expected to be lower in the coastal and insular zone than in the hinterland flatlands and semimountainous or mountainous areas [89]. More periods of drought during the spring and summer are expected in certain parts like Crete, the South Peloponnese, and the South Aegean [89], with an increase in days of extreme heatwaves and related fire risk [90]. This is a crucial parameter, as drought is already severely affecting certain parts of these specific regions, with increasing urbanization and mass tourism posing particular stress on local water resources, exceeding the carrying capacity of places. Moreover, wildfires in coastal parts of Greece have already led to deforestation and the loss of natural land and vegetation, causing large-scale ecological disasters, characteristic examples being the recent forest fires in islands like Chios (in 2016), Samos (in 2021), Evvoia (in 2021), and Rodos (in 2023). Coastal settlement formations that have been developed illegally are particularly vulnerable to natural disasters, as proved by the recent wildfire in the settlement of Mati in Attica (2018) with 104 lives lost, partly attributed to problematic urban planning and a lack of evacuation options [91].

Decreased rainfall will mostly affect Crete, the Southeast Peloponnese, Central Greece, and the coastal part of Thessaly [89], while according to Representative Concentration Pathway (RCP) scenarios related to climate change, more specifically RCP 2.6 and 4.5 scenarios, precipitation could increase in certain coastal parts of Northern Greece (Thessaloniki, Chalkidiki, Thrace, and the North Aegean Islands), with a similar estimation for the coastal zone of West Greece, Epirus, and the Ionian Islands. Soil moisture will decrease mostly in the southern and western part of Greece, while under all scenarios, season length is expected to increase in the hinterland's mountainous areas, while it will remain relatively stable in the insular and coastal territories. A major problem is the increasing frequency and severity of "flash-floods" and Mediterranean cyclones leading to disasters in low-land areas, like in the case of the "Ianos" cyclone hitting the Ionian Islands, the Peloponnese, and West Greece in 2021, and the recent "Daniel" storm in September 2023, with unprecedented volumes of water destroying settlements and infrastructure in Thessaly, including coastal and hinterland areas.

Moreover, a large part of the Greek coastline presents high vulnerability to sea level rise [92], which is estimated to be close to 4 mm/year in the Aegean for the 1992–2014 period, and with a worst-case scenario of a 102 cm rise between now and 2100 [93]. Erosion data [94] show that about 30% of the Greek coastline was already affected by erosion in 2004. By co-estimating parameters like areas of high ecological value and population living within the area of influence of coastal erosion, high exposure was reported for the Attica region and moderate exposure was reported for West Greece, Central Macedonia, and Thrace [94]. Local geological and geomorphological factors, including vertical ground motions, also affect vulnerability to sea level rise and coastal erosion, rendering projections for the Greek territory more complex. However, even in the more optimistic scenario, a large part of coastal wetlands will be affected due to marine flooding and water salinization, and a considerable loss of beach area can be expected, with impacts on coastal infrastructure.

As already noted in Section 1, the time span covered in this work coincides to a large extent with the economic crisis period of 2009–2018, related to the public debt crisis of the Greek state. Austerity measures imposed by the IMF, the European Union, and the European Central Band as part of the bail-out program [95,96] plunged Greek society into general poverty, especially for the low and middle socioeconomic classes, who experienced a heavy loss of income and unprecedented unemployment [97]. Major effects of the crisis included the collapse of the construction sector, the devaluation of land properties, and a dramatic cut-down of public expenses, including expenses in the social welfare sector, public sector, and relevant infrastructure [98]. Moreover, this period was marked by radical changes regarding the general socioeconomic and political structure, which also affected the spatial planning framework and related policies [99,100].

While metropolitan areas in Greece were severely affected, exhibiting increasing spatial segregation, unemployment, and social exclusion [101], tourism locations and particularly islands were more resistant [102]. Domestic tourism was particularly hit by the crisis, yet international tourist arrivals in Greece increased from 15 million in 2010 to

31.3 million in 2019 [103]. This fact has been linked to the spread of Airbnb rentals in urban and coastal territories. Airbnb had grown rapidly in Greece since 2012, reaching up to about 126,000 registrations in 2018 [104,105], with over half of them currently located in the islands (a crude estimation based on Airdna platform data, https://www.airdna.co/ accessed on 24 September 2023). While the official tourism sector presented a relatively slow growth rate, with a 13% increase in bed capacity in the 2010–2020 period [104], Airbnb came to fill the "gap" in terms of supply and demand, mostly utilizing and upgrading existing building stock, but also driving new construction in island and coastal territories. Therefore, recent evidence shows that despite the spatial and developmental impacts of the economic recession period, followed by the COVID-19 pandemic that has largely affected incoming flows of visitors during 2019 and 2020, dynamic processes affecting coastal and insular areas in Greece are still in place.

3. Results

3.1. Impervious Land Expansion at the Regional Scale

To analyze imperviousness changes in the 2006–2018 period, the methodology described in Section 2 was used, based on the *NC* (new cover) and *NCR* (new cover rate) indexes. Based on the obtained results, impervious land expansion (new cover) in the coastal zone at a nationwide scale is 5228 ha, corresponding to a 5.8% increase in the existing impervious surface compared to 2006 (i.e., 89,570 ha). Analysis based on the three sequential coastline zones showed that 41% of new impervious cover was located in the frontline zone, 23.6% in the buffer zone, and 34.4% in the transitional zone. The most dynamic regions in terms of the *NC* index in the frontline zone were the two heavily urbanized "metropolitan" regions (Attica and Central Macedonia) plus the island regions of Crete, Ionian Islands, and the South Aegean, with the addition of the Peloponnese. Table 1 presents the obtained index scores for each region.

Table 1. Results per region and coastal subzone (frontline, buffer, and transitional).

	Region	*NC*			*NCR*			% Impervious Land		
		Z1 *	Z2 *	Z3 *	Z1	Z2	Z3	Z1	Z2	Z3
R1	East Macedonia and Thrace	0.5	0.2	0.1	4.1	3.7	2.7	11.8	6.6	5.0
R2	Central Macedonia	0.6	0.3	0.3	3.5	2.4	2.9	19.4	13.8	9.1
R3	Athos (Mon. community)	0.2	0.0	0.0	12.7	20.9	10.3	1.5	0.1	0.3
R4	Epirus	0.5	0.3	0.4	5.3	4.0	7.2	9.1	8.2	5.6
R5	North Aegean	0.3	0.2	0.1	6.1	6.8	7.8	6.0	2.7	1.9
R6	Thessaly	0.6	0.3	0.2	5.0	5.8	4.4	11.5	5.4	3.6
R7	Central Greece	0.6	0.3	0.3	6.1	6.9	11.2	10.9	4.9	3.2
R8	West Greece	0.6	0.6	0.5	4.9	7.1	9.4	12.8	9.2	6.3
R9	South Aegean	0.6	0.3	0.3	10.4	9.7	10.9	6.9	4.0	3.1
R10	Attica	1.2	0.6	0.5	4.0	2.5	3.1	30.9	24.3	17.1
R11	Crete	1.0	0.6	0.4	6.7	5.8	6.6	16.0	10.3	6.4
R12	Peloponnese	1.0	0.6	0.3	9.4	11.4	9.6	11.2	6.1	3.8
R13	Ionian Islands	0.7	0.3	0.2	5.1	4.1	5.4	15.2	7.9	4.8
GR	Greece (mean)	0.7	0.4	0.3	5.9	5.4	6.2	12.5	7.7	5.3

* Z1: frontline, Z2: buffer, and Z3: transitional.

The *NCR* index in the frontline zone was considerably dynamic in the South Aegean (10.4) and the Peloponnese (9.4), followed by Crete (6.7), North Aegean (6.1), and Central Greece (6.1), while the *NCR* was considerably lower in the two "metropolitan" regions (Attica and Central Macedonia), where new impervious cover was only a small percentage of already developed land (3.3% and 3%, respectively); this fact is most probably attributed to the economic crisis that particularly affected large urban centers. The *NCR* was particularly high in the South Aegean and the Peloponnese, not only in the frontline zone but also in the buffer and the transitional zones; this fact is related to the considerable expansion of development towards the hinterland, while in Central Greece, high *NCR* values were reported in the transitional zone. The monastic community of Athos presented high *NCR* scores (12.2); however, this is attributed to the very small imperviousness level it presents (with only 0.43% of total land developed until 2018) and cannot be classified as truly "dynamic". As shown by the data presented in Table 1, 30.9% of the frontline zone in Attica was covered by impervious areas in 2018, with this percentage reaching 19.4% in Central Macedonia, 16% in Crete, and 15.2% in the Ionian Islands. With the exception of Athos, in all regions, a gradual decrease in imperviousness was observed moving from the frontline to the transitional zone, with this decrease being more rapid in Attica, Epirus, and Central Macedonia.

A combined analysis of both indexes (Figure 5a) reveals that the Peloponnese and West Greece were the regions with simultaneously high *NC* and *NCR* values (over the national mean). The South Aegean and Central Greece regions presented high *NCR* values (over the national mean) and *NC* values close to the national mean, while Crete presented high *NC* values and *NCR* values close to the national mean. On the other hand, Thessaly and East Macedonia and Thrace were the regions presenting the lowest rates of impervious land expansion, in terms of both *NC* and *NCR*, in the coastal zone. When *NC* is plotted against impervious area per region (2006), a linear relationship is observed (Figure 5b); therefore, new cover is correlated to the overall level of development, with Attica, Central Macedonia, and the Peloponnese more notably departing from this linear trend.

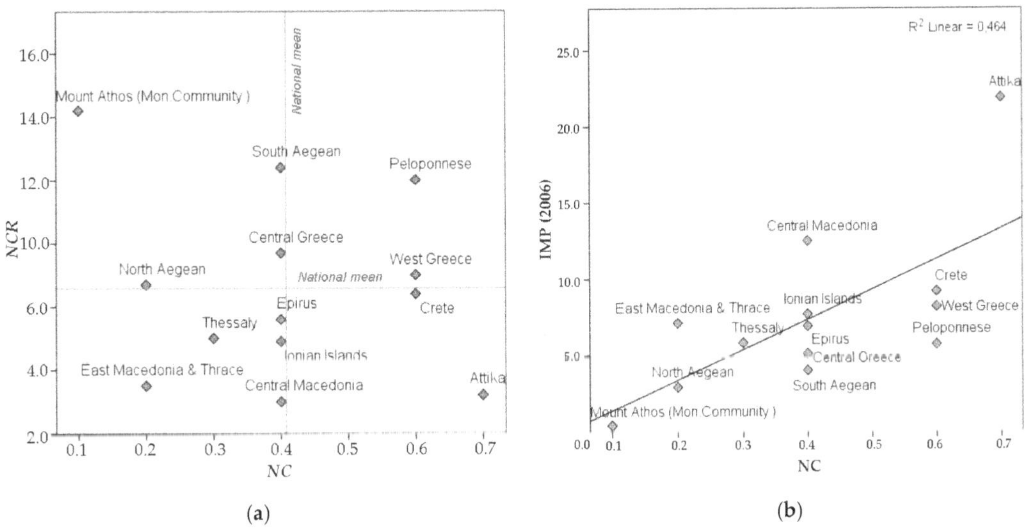

Figure 5. (**a**) Mean *NC* and *NCR* per region plotted against each other. (**b**) Mean *NC* plotted against % imperviousness coverage of year 2006.

In the period of 2006–2018, new highways influencing coastal areas included the A8 highway (Olympia Odos) crossing through the North Peloponnese (and part of the West Greece region), the enlargement of the A1 highway around the Maliakos Gulf, the A5 highway (Ionia Odos) connecting the western part of Greece from south to north, the A9 highway in Northern Crete, and finally, the small coastal part of the A2 highway (Egnatia Odos) and the Thessaloniki–Moudania highway in Central Macedonia. However, analysis showed that highway construction only marginally affected the results. By controlling for a buffer of 50 m on both sides of the national highway network, it was identified that new cover related to highway construction was only 2.4% of new impervious cover during the 2006–2018 period, mostly affecting the Central Macedonia region (16% of new cover was highway-related), West Greece (3.3%), Epirus (3.2%), and Crete (2.6%).

3.2. Impervious Land Expansion at the Local Scale

Regional level data show that Greek coastal territories still posess a large amount of undeveloped land. However, overall statistics might obscure pressures exerted in certain areas, as impervious land is not distributed evenly. As certain coastal zones are still affected to a considerable degree, local-scale analysis is critical in order to identify coastline sectors where impervious land expansion is accelerating. Consequently, results are also analyzed based on a division of coastline sectors according to local administrative boundaries, where impervious land changes and relevant *NC* and *NCR* indexes are estimated and compared. Local/municipal entity boundaries (i.e., the first-level division of municipalities in Greece) are used, as they usually correspond to the geographical level of local spatial planning as specified by general urban plans.

According to the sector analysis, the *NC* index during the 2006–2018 period (Figure 6) revealed a marginal increase in impervious land expansion in most coastal territories of Greece, with several notable exceptions. First of all, *NC* was higher than 1 (corresponding to over 1% of coastal-land affected by new cover in the 2006–2018 period) in 81 out of the 480 local/municipal communities containing coastal areas. These include the following: (a) Coastal urban areas in Athens, Thessaloniki, Chalkida, and Volos, with urban land expansion attributed to industrial/commercial development and relative infrastructure, including port facilities. (b) Highway-related development, mostly affecting municipalities along the highway axes mentioned above. (c) Tourism development, which was the dominant type of coastal land use conversion during the 2006–2018 period.

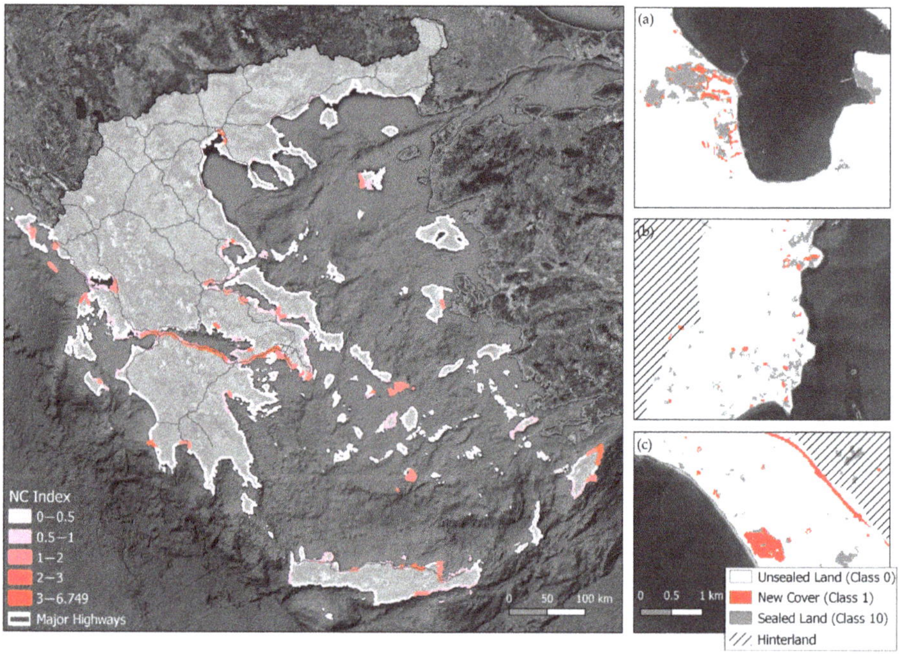

Figure 6. *NC* index and characteristic locations of (**a**) industrial activity (Almyros), (**b**) scattered tourism development (Paros), and (**c**) organized tourism development and road construction (Costa Navarino resort).

More specifically, tourism-related artificial land expansion was recorded in island parts like Thasos (eastern part), Limnos (western part), Skiathos, Paxoi, Zakynthos (western part), Kerkyra (western part), Lefkada (northern part), Rodos (northern and southeastern part), Mykonos, Paros, Kos, Syros, and Santorini. In Crete, important impervious land expansion was recorded in certain parts of the northern coast like Kolympari (Chania), Chersonisos-Malia (Heraklion), and Elounta (Lasithi). All these are reputable mass tourism destinations where, despite the economic crisis, new construction continues. Tourism-related impervious land expansion was also reported in the Pylos area (attributed to the construction of the Costa Navarino resort), and along the South Attica coast towards Cape Sounio. Figure 7 presents the *NCR* distribution, and a characterization of coastal sectors based on combined criteria: (a) "Intense coastal development", i.e., areas with impervious land in 2018 exceeding the national mean ($\geq$7.8%), (b) "Coastal development & expansion", i.e., areas with existing development (impervious land exceeding half of the national mean, $\geq$3.9%) and the simultaneous condition of *NC* and *NCR* exceeding the national mean ($NC \geq 0.43$ and $NCR \geq 5.8$), (c) "Expansion in less developed areas", i.e., areas with *NC* and *NCR* exceeding half of the national mean, and impervious land in 2018 less than half the national mean ($\leq$3.9%).

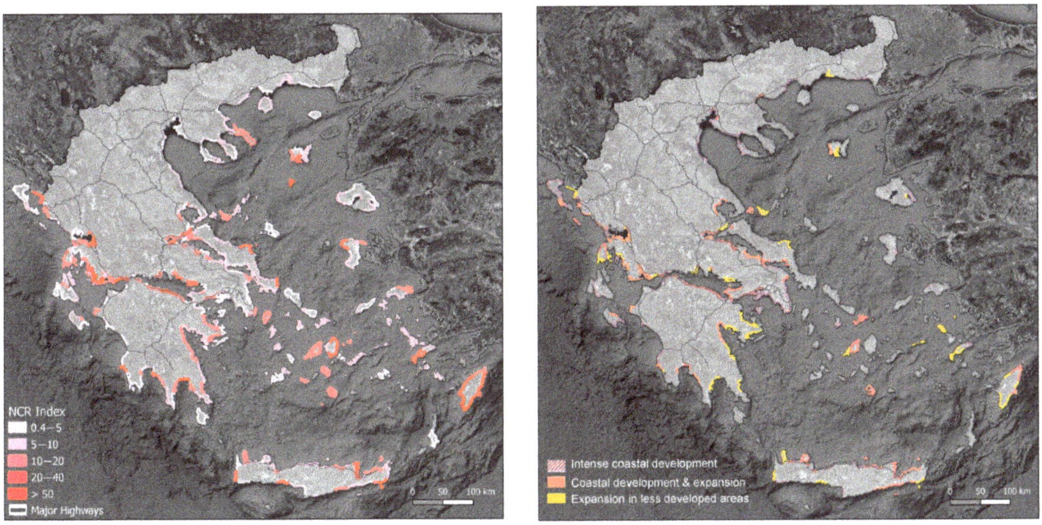

(**a**) (**b**)

Figure 7. (**a**) *NCR* index; (**b**) characterization of coastal sectors based on combined criteria.

An inspection of the corresponding maps showed that high rates of impervious land expansion (*NCR* index) were recorded in most of the islands, and also in the South Peloponnese, along the Corinthian Bay, the Maliakos Gulf, and on the Aetolia-Acarnania coast (those last two areas were partly affected by road construction). As displayed in Figure 7b, areas with high levels of existing coastal development with continuing expansion were identified (a) in Aegean islands with mass tourism (Mykonos, Paros, Santorini, Skiathos, and the northern part of Rodos and Lindos); (b) Ionian islands like Lefkada and Paxoi and the central part of Kerkyra, with the addition of several parts of the Epirus coast (close to the city of Preveza and in Parga-Syvota); (c) parts of Crete including the periurban area of Chania, Agios Nikolaos–Elounda, and Ierapetra; (d) the urbanized corridor along the Corinthian Bay, as well as touristic urban zones in Nafplion and Calamata in the Peloponnese. On the other hand, the results confirm that the northern part of the Greek coast, including the highly developed touristic zones of Pieria, Chalkidiki, and Kavala, presents developmental stability.

3.3. Impervious Land Expansion within Protected Areas

One of the most important issues regarding climate change adaptation is related to impervious land expansion in less developed areas, coinciding in many cases with areas of the protected natural network (or in close proximity to it). It is important to note that even small percentages of imperviousness within protected areas can considerably affect the ecosystem due to the multiple types of landscape fragmentation provoked by low-density scattered development. The results show that impervious land expanded by 798 ha (or by 8.6%) in the period of 2006–2018 within coastal protected areas (Table 2). The protected areas mostly affected by impervious land expansion during the 2006–2018 period were identified in the South Aegean (NCR_{pr} = 11.8), West Greece (NCR_{pr} = 14.0), and Crete (NCR_{pr} = 13.6). Additionally, the results show that in Attica, the Ionian Islands, Epirus, and West Greece, protected land is significantly affected by imperviousness, with 5.78%, 4.65%, 3.38%, and 3.13% of protected coastal land affected by 2018, respectively (with 2.14% being the mean value at the nationwide level). These imperviousness figures could include old rural settlements, suburban sprawl, illegal second housing formations, tourism development, and relative infrastructure, together creating a hybrid spatial pattern that can be linked to considerable ecosystem degradation.

Table 2. Results regarding protected coastal network.

	Region	Protected Area (km^2)	% of Coastal Zone	Impervious Expansion (ha)	NC_{pr} *	NCR_{pr} *	% of Protected Land Affected
R1	East Macedonia and Thrace	379	58.6	44.1	0.12	5.8	2.14
R2	Central Macedonia	270	28.3	0.0	0.12	0.1	2.21
R3	Athos (Mon. community)	102	99.8	5.2	0.05	12.2	0.48
R4	Epirus	193	59.9	34.6	0.18	5.6	3.41
R5	North Aegean	562	47.0	64.8	0.12	9.3	1.36
R6	Thessaly	213	38.6	30.8	0.14	7.4	2.10
R7	Central Greece	386	28.2	80.8	0.21	8.8	2.58
R8	West Greece	399	42.9	153.2	0.39	14.0	3.16
R9	South Aegean	1246	53.5	150.6	0.12	11.8	1.14
R10	Attica	185	21.6	46.8	0.25	4.6	5.78
R11	Crete	364	34.8	69.0	0.19	13.6	1.58
R12	Peloponnese	387	31.0	54.0	0.14	7.8	1.92
R13	Ionian Islands	247	29.1	69.1	0.28	6.4	4.65
GR	Greece (mean)	4933	39.8	802.9	0.17	8.6	2.14

* *NC* and *NCR* indexes adjusted for areas within the protected network.

This issue is becoming more and more important in the islands of Greece, where tourism-related development encroached on Natura 2000 and wildlife sanctuary areas in the period of 2006–2018 under a highly dispersed spatial pattern scheme, radically altering existing land cover, while rendering new infrastructure particularly vulnerable to CC impacts. The results show that development can also be intense on land in close proximity (up to 1–2 km) to the protected network. This is important as tourism facilities in many cases chose to be located exactly at the border of protected areas; nevertheless, they still considerably affect the ecosystem by altering water resources, biodiversity, microclimatic conditions, etc. This is, for example, the case for the Agios Prokopios area on Naxos Island (Figure 8), where scattered development surrounds protected wetlands within disaster-prone (flood risk) low-elevation land close to the coast.

Figure 8. Tourism-related expansion in Agios Prokopios (Naxos Island), affecting Natura 2000 wetlands and the wildlife sanctuary.

4. Discussion

The research results confirm the hypothesis of a "moratorium" on coastal urban sprawl in Greece during the period of 2006–2018, with low rates of impervious land expansion. At an overall level, a 5.8% increase in impervious land was reported, with less than 0.5% of previously undeveloped areas in the coastal zone affected by new construction. The highest *NC* index values were reported for the frontline zone, corresponding to the areas with immediate accessibility to the sea, while there was a gradual decrease in imperviousness in the buffer and transitional zones.

At the regional level, the South Aegean, Crete, and the Peloponnese were the more "dynamic" regions; downscaling results to a local level revealed continuing trends of impervious land expansion in specific coastal sectors, especially those affected by tourism, while other areas were affected by industrial/commercial sprawl as well as highway construction. The results also show that in the 2006–2018 period, protected natural areas (comprising up to 40% of land in the coastal zone) continued to be affected by impervious land expansion. This is evaluated as an alarming fact: natural areas like wetlands, sand dunes, forests, and scrub vegetation are abundant along the Greek coastline and play a crucial role in climate change adaptation, acting as barriers to coastal erosion, preventing floods, and acting as carbon sinks, while preserving biodiversity. Impervious land encroachment could considerably degrade their ecosystemic role and render coastal areas particularly vulnerable to climate change impacts.

Therefore, the results obtained in this work could inform national- and regional-level policies regarding CC adaptation in many ways: First of all, they highlight areas of intense development at the coastal front, where specific actions should be prioritized to handle increasing environmental pressures and natural hazards. A characteristic example is the urbanized corridor along Corinthian Bay, an area where impervious land expansion still continued in the 2006–2018 period and which was identified as an area heavily affected by coastal erosion, putting buildings and infrastructure in proximity to the sea in danger. Secondly, the results identified natural areas where new development is encroaching, like on highly touristic islands, in the absence of a strict framework handling climate-related vulnerabilities. An example is the Lindos Municipality in Rodos, identified by the results as a sector of coastal impervious expansion and coinciding with the area of a destructive forest wildfire which led to the large-scale evacuation of hotel resorts in July 2023. Similar locations where impervious land expansion is simultaneously increasing vulnerabilities to natural disasters (i.e., to flood and fire risk) and decreasing the ability of natural coastal ecosystems to mitigate the effects of climate change were identified in many parts along the case study area; therefore, providing specific planning solutions for these areas could be an important expansion of this work.

It is important to note that the overall slowdown of impervious land expansion implied by the results should be considered as a temporal effect related to the general socioeconomic and political context of the examined period and the sociospatial manifestations of the Greek crisis. Among the major impacts of the economic recession has been the sudden interruption of the real estate boom that had peaked in Greece in the previous period, with construction activity gradually collapsing after 2008. This was accompanied by the devaluation of land properties; the inability of households to pay off home loans taken up during the previous period due to the dramatic cut-down of wages, pensions, and public expenses; and the bankruptcy of construction companies in the private sector.

The economic crisis period has also largely impacted spatial planning issues in Greece. One of the first political actions of the economic crisis period was the legalization of millions of illegal construction projects [106–108], mainly aiming to provide income for the state budget. Illegal houses in Greece include structures along the shore, within forest land, and/or natural areas. These are usually areas of spontaneous development, presenting increasing vulnerability to natural disasters. Moreover, the studied period coincides with a significant reform of the Greek Spatial Planning Framework [109], existing since the 2000s, with the insertion of a new set of spatial planning tools largely imposed by the IMF and EU partners [76] and adopted by successive Greek governments to accelerate projects, making procedures more quick, flexible, and favorable to large-scale investments [110,111]. These tools include Special Spatial Plans for Strategic Investments (entitled ESXASE in Greek, Law 3894/2010) [112], forming part of the newly defined Special Urban Plans (Law 4269/2014, Law 4447/2016, and Law 4759/2020) [113–115]. Spatial planning reform also involves legislative instruments to capitalize on public land property [79] (Special Spatial Development Plans for Public Property, entitled ESXADA in Greek, Law 3986/2011) [116], with many of them targeted at high-value coastal territories. These new sets of spatial planning tools, legislated during the crisis period, are currently being activated by the mass tourism sector to obtain permission for a considerable number of future projects within privileged locations on the coastal front (e.g., recent ESXADA projects in Afantou in Rodos, Paliouri, and Ag. Ioannis in Chalkidiki, and Kassiopi in Kerkyra; ESXASE projects in Kavousi, Elounda, Cavo Sidero, and Kissamos Bay in Crete, etc.).

At the same time, the restructuring of the real estate market has led to profit opportunities for foreign speculators and investors buying off cheap land and property during the economic crisis period [117]; this process is also driven by the golden visa policies initiated in 2013 in Greece, and a growing global demand for high-quality vacation housing on the Mediterranean coast [21]. Moreover, the spread of Airbnb in coastal and island territories, coupled with steadily increasing international tourism arrivals, has been one of the most important new facts during the crisis period; this trend is still rapidly accelerating. In Greece, this can be further fueled by new legislation regarding complex tourism resorts and mixed small-scale tourism resorts that allows the incorporation of private housing formations under favorable building regulations. Such private vacation houses within hotel zones can be rented, sold, or exploited by their owners in a flexible way, while taking advantage of being located in privileged natural locations or even within Natura 2000 areas at a minimum distance from the shoreline that has been exceptionally set to 30 m (from 50 m, which is normally permitted for tourism facilities).

In light of these remarks, we can deduce that as Greece enters the post-crisis period (or more correctly, the post-memorandum period, as the effects of the crisis are still very important for Greek society), a "rebound" of coastal urban sprawl might be expected. Recent data show that construction activity is finally picking up after a downturn for over

10 years, while after considerable fluctuations during the COVID-19 pandemic (2020–2021), construction activity has been accelerating since 2022 (Figure 9). Especially in the case of mass tourism destinations like the South Aegean, this increase has been impressive, reaching precrisis figures. This "rebound", combined with increasing demand for coastal space and a lack of strict land use planning regulations prioritizing nature conservation, is evaluated as an alarming issue, adding another challenge for climate adaptation strategies that still remain in their infancy in Greece, with an effective mechanism to coordinate the interconnections between spatial plans and climate change adaptation policies still missing [118].

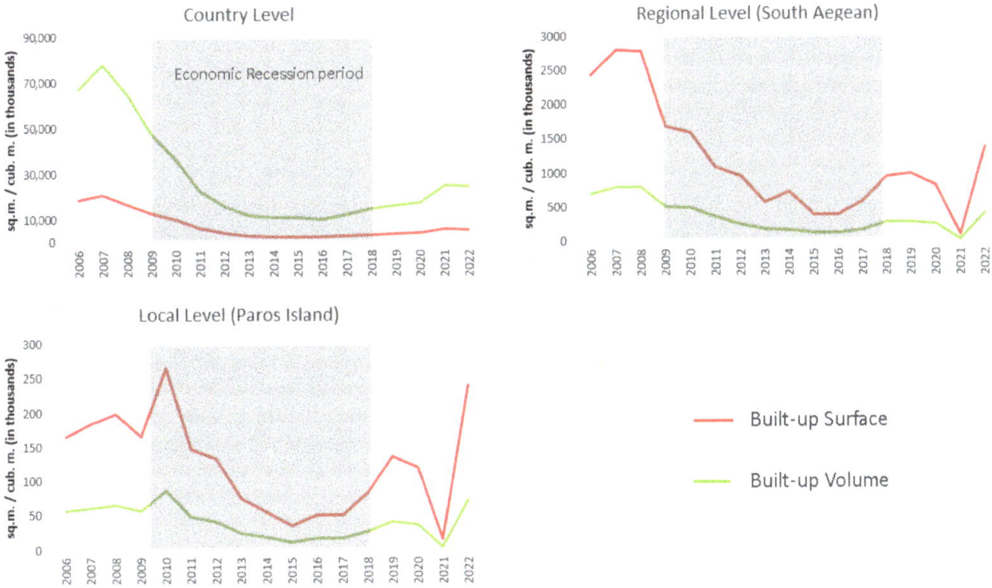

Figure 9. The collapse of construction activity in Greece during the economic recession period and the recent "rebound" in highly touristic regions like South Aegean and places like Paros Island (own elaboration of data from the Hellenic Statistical Authority).

5. Conclusions

In the Mediterranean, the rapid acceleration of urban sprawl is fueled by periurbanization processes and mass tourism development, particularly concentrated in the coastal zone. Continuing imperviousness expansion could lead to landscape fragmentation, increase the urban heat island effect, reduce natural land and vegetation, accelerate coastal erosion, and lead to the overexploitation of land and natural resources.

Regarding the effect of such processes on coastal territories, our work implies that imperviousness is a better control parameter than built-up footprints as it also accounts for road construction, parking lots, pools, and other related structures that affect ecosystems. Reducing imperviousness can be considered as a "green" adaptation measure, directly affecting how a territory will function in the case of a flashflood, a forest fire, coastal erosion, and other impacts of climate change; therefore, strategies prioritizing biodiversity preservation in coastal zones are needed. Towards this direction, it is of major importance to utilize high-resolution spatial data to assess the impact development has already had on coastal territories, to evaluate climate change threats, and to provide tools for monitoring land cover changes.

Herein, geospatial methods were applied to examine the spatial outcome of the recession period in Greece in terms of imperviousness land expansion, and to assess the way it has affected development in coastal zones. The results provide a full-scale estimation of recent impervious land expansion in Greece, which is able to inform and support national- and regional-level spatial planning decisions and strategies.

The results reveal that, at an overall level, Greece still possesses a large percentage of undeveloped land. This can be a valuable asset in order to combat climate change at its frontline, i.e., the coastal zone. However, as the country emerges from the crisis period, a notable "rebound" in the construction sector can be expected, which, coupled with the continuing dynamic of the mass tourism sector, Airbnb expansion, and global real estate interest in investing in coastal locations, is imposing particular stress on coastal and insular territories, undermining recent national and regional efforts towards climate resilience. This is in line with the persistent view of tourism as a "heavy industry" in Greece, related to an unsustainable model of coastal tourism development, increasing social inequalities and environmental vulnerabilities. In the absence of a coherent national-level spatial planning framework guiding development in coastal areas, as well as enforcing directions for nature preservation, it can be expected that an unsustainable business-as-usual scenario could take place. This can be alarming, and could also mark a possible future expansion of this work, placing particular focus on protected areas under stress, with the execution of in-depth case study analyses to specifically evaluate climate change impacts and probable solution and adaptation measures.

In conclusion, it is our strong belief that spatial planning and land use policies remain an irreplaceable tool in preventing undesirable land use/land cover changes and securing a climate-resilient future. The management of urban sprawl should be related to reduced rates of land take, with this issue being particularly relevant to coastal zone management, where coastal retreat policies with strict nature preservation should be enforced. Currently, as important steps towards climate change adaptation policies are made on the European and international scales, scientific inputs such as those provided by the present research could be particularly relevant for properly handling spatial planning decisions, through a climate resilience approach, to secure a sustainable pathway for the Mediterranean.

Funding: This research received no external funding.

Data Availability Statement: The data used to support the findings of this study can be made available by the corresponding author upon request.

Conflicts of Interest: The authors declare no conflict of interest.

References

1. Lavalle, C.; Rocha Gomes, C.; Baranzelli, C.; Batista, E.; Silva, F. *Coastal Zones Policy Alternatives on European Coastal Zones 2000–2050*; EUR 24792 EN; Publications Office of the European Union: Luxembourg, 2011. Available online: https://citeseerx.ist.psu.edu/document?repid=rep1&type=pdf&doi=e7a543af907c22030afca985f3e55b8baffc8f95 (accessed on 24 September 2023).
2. MedECC. *Climate and Environmental Change in the Mediterranean Basin–Current Situation and Risks for the Future. First Mediterranean Assessment Report*; Cramer, W., Guiot, J., Marini, K., Eds.; Union for the Mediterranean, Plan Bleu, UNEP/MAP: Marseille, France, 2020; ISBN 978-2-9577416-0-1. [CrossRef]
3. Vimal, R.; Geniaux, G.; Pluvinet, P.; Napoleone, C.; Lepart, J. Detecting threatened biodiversity by urbanization at regional and local scales using an urban sprawl simulation approach: Application on the French Mediterranean region. *Landsc. Urban Plan.* **2012**, *104*, 343–355. [CrossRef]
4. European Environmental Agency (EEA). *The Changing Faces of Europe's Coastal Areas*; EEA Report No 6/2006; Office for Official Publications of the European Communities: Luxembourg, 2006; ISBN 92-9167-842-2.
5. Satta, A.; Puddu, M.; Venturini, S.; Giupponi, C. Assessment of coastal risks to climate change related impacts at the regional scale: The case of the Mediterranean region. *Int. J. Disaster Risk Reduct.* **2017**, *24*, 284–296. [CrossRef]
6. Morote, A.-F.; Hernández, M. Urban sprawl and its effects on water demand: A case study of Alicante, Spain. *Land Use Policy* **2016**, *50*, 352–362. [CrossRef]
7. European Commission. Biodiversity Strategy for 2030. 22 December 2021. Available online: https://environment.ec.europa.eu/strategy/biodiversity-strategy-2030_en (accessed on 24 September 2023).

8. IPCC. *Climate Change and Land: An IPCC Special Report on Climate Change, Desertification, Land Degradation, Sustainable Land Management, Food Security, and Greenhouse Gas Fluxes in Terrestrial Ecosystems*; Shukla, P.R., Skea, J., Buendia, E.C., Masson-Delmotte, V., Pörtner, H.-O., Roberts, D.C., Zhai, P., Slade, R., Connors, S., van Diemen, R., et al., Eds.; Cambridge University Press: Cambridge, UK; New York, NY, USA, 2019. [CrossRef]
9. Smiraglia, D.; Cavalli, A.; Giuliani, C.; Assennato, F. The Increasing Coastal Urbanization in the Mediterranean Environment: The State of the Art in Italy. *Land* **2023**, *12*, 1017. [CrossRef]
10. Official Journal L 034, Protocol on Integrated Coastal Zone Management in the Mediterranean 04/02/2009 P. 0019–0028. Available online: https://eur-lex.europa.eu/legal-content/EN/TXT/?uri=CELEX:22009A0204(01) (accessed on 24 September 2023).
11. Egidi, G.; Cividino, S.; Vinci, S.; Sateriano, A.; Salvia, R. Towards local forms of sprawl: A brief reflection on Mediterranean urbanization. *Sustainability* **2020**, *12*, 582. [CrossRef]
12. Lagarias, A.; Sayas, I. Urban sprawl in the Mediterranean: Evidence from coastal medium-sized cities. *Reg. Sci. Inq.* **2018**, *December 2018*, 15–32.
13. Soto, M.T.R.; Clavé, S.A. Second homes and urban landscape patterns in Mediterranean coastal tourism destinations. *Land Use Policy* **2017**, *68*, 117–132. [CrossRef]
14. Mejjad, N.; Rossi, A.; Pavel, A.B. The coastal tourism industry in the Mediterranean: A critical review of the socio-economic and environmental pressures & impacts. *Tour. Manag. Perspect.* **2022**, *44*, 101007. [CrossRef]
15. Kizos, T.; Tsilimigkas, G.; Karampela, S. What Drives Built-Up Area Expansion on Islands? Using Soil Sealing Indicators to Estimate Built-Up Area Patterns on Aegean Islands, Greece. *Tijdschr. Voor Econ. En Soc. Geogr.* **2017**, *108*, 836–853. [CrossRef]
16. Simpson, M.C.; Gössling, S.; Scott, D.; Hall, C.M.; Gladin, E. *Climate Change Adaptation and Mitigation in the Tourism Sector: Frameworks, Tools and Practices*; UNEP, University of Oxford, UNWTO, WMO: Paris, France, 2008.
17. UNEP/MAP—United Nations Environment Programme/Mediterranean Action Plan. State of play of tourism in the Mediterranean: A Roadmap for a Greener, Inclusive & Resilient Tourism in the Mediterranean. Sustainable Tourism Community-Interreg Med. 2022. Available online: https://planbleu.org/wp-content/uploads/2022/11/EN_VF_stateoftourism_PLANBLEU.pdf (accessed on 24 September 2023).
18. Lagarias, A.; Stratigea, A. Coastalization patterns in the Mediterranean: A spatiotemporal analysis of coastal urban sprawl in tourism destination areas. *Geo. J.* **2023**, *88*, 2529–2552. [CrossRef]
19. Romano, B.; Zullo, F. The urban transformation of Italy's Adriatic coastal strip: Fifty years of unsustainability. *Land Use Policy* **2014**, *38*, 26–36. [CrossRef]
20. Liziard, S. Littoralisation de l'Arc latin: Analyse spatio-temporelle de la répartition de la population à une échelle fine. *Espace Popul. Soc.* **2013**, *1*, 21–40. [CrossRef]
21. Membrado, J.C.; Huete, R.; Mantecón, A. Urban sprawl and northern European residential tourism in the Spanish Mediterranean coast. *Tour. Rev.* **2016**, *10*. [CrossRef]
22. Colaninno, N.; Cerda, J.F.; Roca, J. Spatial patterns of land use: Morphology and demography, in a dynamic evaluation of urban sprawl phenomena along the Spanish Mediterranean coast. In Proceedings of the European Regional Science Association Congress-51st European Congress of the Regional Science Association International, Barcelona, Spain, 30 August–2 September 2011; pp. 1–18.
23. Malerba, A., Castro Noblejas, H.; Sortino Barrionuevo, J.F.; Mérida Rodríguez, M. Urban Transformation of the Coastline from a Landscape Perspective. Analysis of Cases on the Costa del Sol (Spain). In *INTERNATIONAL SYMPOSIUM: New Metropolitan Perspectives*; Springer International Publishing: Cham, Switzerland, 2022; pp. 1671–1682.
24. Hof, A.; Blázquez-Salom, M. The Linkages between Real Estate Tourism and Urban Sprawl in Majorca (Balearic Islands, Spain). *Land* **2013**, *2*, 252–277. [CrossRef]
25. Wolff, C.; Nikoletopoulos, T.; Hinkel, J.; Vafeidis, A. Future urban development exacerbates coastal exposure in the Mediterranean. *Sci. Rep.* **2020**, *10*, 14420. [CrossRef] [PubMed]
26. Vardopoulos, I.; Ioannides, S.; Georgiou, M.; Voukkali, I.; Salvati, L.; Doukas, Y.E. Shaping Sustainable Cities: A Long-Term GIS-Emanated Spatial Analysis of Settlement Growth and Planning in a Coastal Mediterranean European City. *Sustainability* **2023**, *15*, 11202. [CrossRef]
27. Braičić, Z.; Lončar, J. Economic-geographic analysis of differentiated development in croatian coastal region. *J. Geogr./Rev. Za Geogr.* **2015**, *10*, 7–24.
28. Kranjčević, J.; Hajdinjak, S. Tourism urbanization in Croatia. The cases of Poreč in Istria and Makarska in Dalmatia. *Comp. Southeast Eur. Stud.* **2019**, *67*, 393–420. [CrossRef]
29. Sarantakou, E.; Terkenli, T. Non-Institutionalized Forms of Tourism Accommodation and Overtourism Impacts on the Landscape: The Case of Santorini, Greece. *Tour. Plan. Dev.* **2019**, *16*, 411–433. [CrossRef]
30. Panousi, S.; Petrakos, G. Overtourism and tourism carrying capacity—A regional perspective for Greece. In *Culture and Tourism in a Smart, Globalized, and Sustainable World*; Katsoni, V., Ciná van Zyl, C., Eds.; Springer: Cham, Switzerland, 2021; pp. 215–229. [CrossRef]
31. Lagarias, A.; Stratigea, A.; Theodora, Y. Overtourism as an emerging threat for sustainable island communities–Exploring indicative examples from the South Aegean Region, Greece. In Proceedings of the International Conference on Computational Science and Its Applications, Athens, Greece, 3–6 July 2023; Springer Nature: Cham, Switzerland, 2023; pp. 404–421. [CrossRef]

32. Vandarakis, D.; Malliouri, D.; Petrakis, S.; Kapsimalis, V.; Moraitis, V.; Hatiris, G.A.; Panagiotopoulos, I. Carrying Capacity and Assessment of the Tourism Sector in the South Aegean Region. Greece. *Water* **2016**, *15*, 2616. [CrossRef]
33. Lagarias, A.; Zacharakis, I.; Stratigea, A. Assessing the coastal vs hinterland divide by use of multitemporal data: Case study in Corinthia, Greece. *Eur. J. Geogr.* **2022**, *13*, 001–026. [CrossRef]
34. Wang, J.; Chen, J.; Wen, Y.; Fan, W.; Liu, Q.; Tarolli, P. Monitoring the coastal wetlands dynamics in Northeast Italy from 1984 to 2016. *Ecol. Indic.* **2021**, *129*, 107906. [CrossRef]
35. Theodora, Y.; Stratigea, A. Climate Change and Strategic Adaptation Planning in Mediterranean Insular Territories: Gathering Methodological Insights from Greek Experiences. In Proceedings of the Computational Science and Its Applications–ICCSA 2021: 21st International Conference, Cagliari, Italy, 13–16 September 2021; Proceedings, Part X 21, 2021. Springer International Publishing: Cham, Switzerland, 2021. [CrossRef]
36. Minetos, D.; Polyzos, S.; Sdrolias, L. Features and spatial analysis of illegal housing in Greece. *Trans. Line* **2017**, *1*, 86–107.
37. Yoyas, N. Illegal constructions in the post-memorandum Greece: Changes and constants in a chronic phenomenon. *Reg. Sci. Inq.* **2018**, *10*, 153–161.
38. Kousis, M. Marine and Coastal Issues in Local Environmental Conflict: Greece, Spain, and Portugal1. In *Contesting the Foreshore Tourism, Society, and Politics on the Coast*; Boissevain, J., Selwyn, T., Eds.; Amsterdam University Press: Amsterdam, The Netherlands, 2004.
39. Evelpidou, N.; Petropoulos, A.; Karkani, A.; Saitis, G. Evidence of Coastal Changes in the West Coast of Naxos Island, Cyclades, Greece. *J. Mar. Sci. Eng.* **2021**, *9*, 1427. [CrossRef]
40. García-Ayllón, S. La Manga case study: Consequences from short-term urban planning in a tourism mass destiny of the Spanish Mediterranean coast. *Cities* **2015**, *43*, 141–151. [CrossRef]
41. Rodríguez-Santalla, I.; Navarro, N. Main Threats in Mediterranean Coastal Wetlands. The Ebro Delta Case. *J. Mar. Sci. Eng.* **2021**, *9*, 1190. [CrossRef]
42. Atik, M.; Altan, T.; Artar, M. Land Use Changes in Relation to Coastal Tourism Developments in Turkish Mediterranean. *Pol. J. Environ. Stud.* **2010**, *19*, 21–33.
43. Briguglio, L.; Avellino, M. Overtourism, Environmental Degradation and Governance in Small Islands with Special Reference to Malta. In *Shaping the Future of Small Islands*; Palgrave Macmillan: Singapore, 2021; pp. 301–322. [CrossRef]
44. European Environmental Agency (EEA). *Urban Adaptation to Climate Change in Europe, Report No 2/2012*; European Environmental Agency (EEA): Copenhagen, Denmark, 2012.
45. Baker, E.; Sevaldsen, P.; Kurvits, T. *The State of the Mediterranean Marine and Coastal Environment*; Grid-Arendal: Tyholmen, Norway, 2013. Available online: https://policycommons.net/artifacts/2390445/the-state-of-the-mediterranean-marine-and-coastal-environment/3411769/ (accessed on 24 September 2023).
46. Leka, A.; Lagarias, A.; Panagiotopoulou, M.; Stratigea, A. Development of a Tourism Carrying Capacity Index (TCCI) for sustainable management of coastal areas in Mediterranean islands–Case study Naxos, Greece. *Ocean Coast. Manag.* **2022**, *216*, 105978. [CrossRef]
47. European Environmental Agency (EEA). *Land and Soil in Europe. Why We Need to Use These Vital and Finite Resources Sustainability*; European Environmental Agency (EEA): Copenhagen, Denmark, 2019. [CrossRef]
48. Communication from the Commission to the European Parliament, the European Council, the Council, the European Economic and Social Committee and the Committee of the Regions. EU Soil Strategy for 2030-Reaping the Benefits of Healthy Soils for People, Food, Nature and Climate. COM(2021)699. Available online: https://eur-lex.europa.eu/legal-content/EN/TXT/?uri=CELEX:52021DC0699 (accessed on 24 September 2023).
49. Communication from the Commission to the European Parliament, the European Council, the Council, the European Economic and Social Committee and the Committee of the Regions. Proposal for a Regulation of the European Parliament and of the Council on Nature Restoration; COM(2022)304. Available online: https://eur-lex.europa.eu/legal-content/EN/TXT/?uri=CELEX:52022PC0304 (accessed on 24 September 2023).
50. UNFCCC; CBD; IISD; GIZ; UNEP; SwedBio. *Promoting Synergies between Climate Change Adaptation and Biodiversity through the National Adaptation Plan (NAP) and National Biodiversity Strategies and Action Plan (NBSAP) Processes*; United Nations Climate Change Secretariat: Bonn, Germany, 2022.
51. Bauer, M.E.; Loffelholz, B.C.; Wilson, B. Estimating and mapping impervious surface area by regression analysis of Landsat imagery. In *Remote Sensing of Impervious Surfaces*; CRC Press: Boca Raton, FL, USA, 2007; pp. 31–48. ISBN 9780429138768.
52. El Garouani, A.; Mulla, D.J.; El Garouani, S.; Knight, J. Analysis of urban growth and sprawl from remote sensing data: Case of Fez, Morocco. *Int. J. Sustain. Built Environ.* **2017**, *6*, 160–169. [CrossRef]
53. Jat, M.K.; Garg, P.K.; Khare, D. Monitoring and modelling of urban sprawl using remote sensing and GIS techniques. *Int. J. Appl. Earth Obs. Geoinf.* **2008**, *10*, 26–43. [CrossRef]
54. Kuang, W. Mapping global impervious surface area and green space within urban environments. *Sci. China Earth Sci.* **2019**, *62*, 1591–1606. [CrossRef]
55. Torbick, N.; Corbiere, M. Mapping urban sprawl and impervious surfaces in the northeast United States for the past four decades. *GISci. Remote Sens.* **2015**, *52*, 746–764. [CrossRef]
56. Zhou, Y.; Wang, Y.Q. An Assessment of Impervious Surface Areas in Rhode Island. *Northeast Nat.* **2007**, *14*, 643–650. [CrossRef]
57. Xian, G.; Klaver, L.Y.; Hossain, N. Measuring urban sprawl and extent through multitemporal. *US Geol. Surv.* **2006**, *1726*, 57.

58. Ghazaryan, G.; Rienow, A.; Oldenburg, C.; Thonfeld, F.; Trampnau, B.; Sticksel, S.; Jürgens, C. Monitoring of urban sprawl and densification processes in western Germany in the light of SDG indicator 11.3. 1 based on an automated retrospective classification approach. *Remote Sens.* **2021**, *13*, 1694. [CrossRef]
59. Li, J.; He, L.; Dai, J.; Li, X.; Fu, Q. Urban Sprawl Study Based on Impervious Surface Percent. *Remote Sens. Technol. Appl.* **2008**, *23*, 424–427. [CrossRef]
60. Salvati, L. The spatial pattern of soil sealing along the urban-rural gradient in a Mediterranean region. *J. Environ. Plan. Manag.* **2014**, *57*, 848–861. [CrossRef]
61. Salvati, L. Soil sealing, population structure and the socioeconomic context: A local-scale assessment. *Geo. J.* **2016**, *81*, 77–88. [CrossRef]
62. Fox, D.; Youssaf, Z.; Adnès, C.; Delestre, O. Relating imperviousness to building growth and developed area in order to model the impact of peri-urbanization on runoff in a Mediterranean catchment (1964–2014). *J. Land Use Sci.* **2019**, *14*, 210–224. [CrossRef]
63. Tombolini, I.; Zambon, I.; Ippolito, A.; Grigoriadis, S.; Serra, P.; Salvati, L. Revisiting "Southern" Sprawl: Urban Growth, Socio-Spatial Structure and the Influence of Local Economic Contexts. *Economies* **2015**, *3*, 237–259. [CrossRef]
64. Lagarias, A.; Prastacos, P. Comparing the urban form of South European cities using fractal dimensions. *Environ. Plan. B Urban Anal. City Sci.* **2020**, *47*, 1149–1166. [CrossRef]
65. Salvati, L.; Carlucci, M. The way towards land consumption: Soil sealing and polycentric development in Barcelona. *Urban Stud.* **2016**, *53*, 418–440. [CrossRef]
66. Salvati, L.; Quatrini, V.; Barbati, A.; Tomao, A.; Mavrakis, A.; Serra, P.; Sabbi, A.; Merlini, P.; Corona, P. Soil occupation efficiency and landscape conservation in four Mediterranean urban regions. *Urban For. Urban Green.* **2016**, *20*, 419–427. [CrossRef]
67. Gramaglia, C.; Salles, C.; Rio, M.; Tournoud, M.G. Reducing the imperviousness of urban soils: A way of improving the quality of runoff that is struggling to impose itself in the Mediterranean area (No. IAHS2022-432). In Proceedings of the Copernicus Meetings, Montpellier, France, 29 May–3 June 2022.
68. Ferreira, C.S.; Walsh, R.P.; Steenhuis, T.S.; Ferreira, A.J. Effect of peri-urban development and lithology on streamflow in a Mediterranean catchment. *Land Degrad. Dev.* **2018**, *29*, 1141–1153. [CrossRef]
69. Mehr, A.D.; Akdegirmen, O. Estimation of urban imperviousness and its impacts on flashfloods in Gazipaşa, Turkey. *Knowl.-Based Eng. Sci.* **2021**, *2*, 9–17. [CrossRef]
70. Morabito, M.; Crisci, A.; Guerri, G.; Messeri, A.; Congedo, L.; Munafò, M. Surface urban heat islands in Italian metropolitan cities: Tree cover and impervious surface influences. *Sci. Total Environ.* **2021**, *751*, 142334. [CrossRef]
71. Andreu, V.; Gimeno-García, E.; Pascual, J.; Vazquez-Roig, P.A.; Picó, Y. Presence of pharmaceuticals and heavy metals in the waters of a Mediterranean coastal wetland: Potential interactions and the influence of the environment. *Sci. Total Environ.* **2016**, *540*, 278–286. [CrossRef]
72. Kati, V.; Kassara, C.; Vrontisi, Z.; Moustakas, A. The biodiversity-wind energy-land use nexus in a global biodiversity hotspot. *Sci. Total Environ.* **2021**, *768*, 144471. [CrossRef]
73. Communication from the Commission to the European Parliament, the Council, the European Economic and Social Committee and the Committee of the Regions—Roadmap to a Resource Efficient Europe, COM(2011) 571 final, 2011. Available online: https://eur-lex.europa.eu/legal-content/EN/TXT/?uri=CELEX%3A52011DC0571 (accessed on 24 September 2023).
74. Communication from the Commission to the European Parliament, the Council, the European Economic and Social Committee and the Committee of the Regions—Green Infrastructure (GI): Enhancing Europe's Natural Capital, COM(2013) 249 final, 2013. Available online: https://eur-lex.europa.eu/legal-content/EN/TXT/?uri=celex%3A52013DC0249 (accessed on 24 September 2023).
75. González-García, A.; Palomo, I.; Arboledas, M.; González, J.A.; Múgica, M.; Mata, R.; Montes, C. Protected areas as a double edge sword: An analysis of factors driving urbanisation in their surroundings. *Glob. Environ. Chang.* **2022**, *74*, 102522. [CrossRef]
76. SIOSE, National Technical Team, High Resolution Land Cover/Land Use Information System in Spain (HR SIOSE) Technical Document, Dirección Gral. del Instituto Geográfico Nacional. Subdirección Gral. de Cartografía y Observación del Territorio. Available online: http://www.siose.es/SIOSEtheme-theme/documentos/pdf/D0_220615_HR_SIOSE_Technical_Document_v1_EN.pdf\ (accessed on 24 September 2023).
77. Gounaridis, D.; Chorianopoulos, I.; Koukoulas, S. Exploring prospective urban growth trends under different economic outlooks and land-use planning scenarios: The case of Athens. *Appl. Geogr.* **2018**, *90*, 134–144. [CrossRef]
78. Tsilimigkas, G.; Gourgiotis, A.; Derdemezi, E.T. Spatial planning incompetence to discourage urban sprawl on Greek Islands. Evidence from Paros, Greece. *J. Coast. Conserv.* **2022**, *26*, 11. [CrossRef]
79. *Technical Specifications of Local Urban Plans*; 3545/B/03.08.2021; Official Government Gazette: Athens, Greece, 2021.
80. *Technical Specifications of Regional Spatial Frameworks of Law 4447/2016 (A'241)*; 2818/B/17.07.2021; Official Government Gazette: Athens, Greece, 2021.
81. European Environment Agency (EEA). High Resolution Layers. European Union, Copernicus Land Monitoring Service, 2021. Available online: https://land.copernicus.eu/en/products (accessed on 24 September 2023).
82. Kleinewillinghöfer, L.; Olofsson, P.; Pebesma, E.; Meyer, H.; Buck, O.; Haub, C.; Eiselt, B. Unbiased Area Estimation Using Copernicus High Resolution Layers and Reference Data. *Remote Sens.* **2022**, *14*, 4903. [CrossRef]

83. Langanke, T. Copernicus Land Monitoring Service–High Resolution Layer Imperviousness: Product Specifications Document, Publisher: Copernicus team at EEA (European Environment Agency). 2016. Available online: https://land.copernicus.eu/en/technical-library (accessed on 24 September 2023).
84. Agardy, T.; Alder, J. Coastal systems. In *Ecosystems and Human Well-Being: Current State and Trends*; Hassan, R., Scholes, R., Ash, N., Eds.; Island Press: Washington, DC, USA, 2005; Volume 1, pp. 513–549.
85. McGranahan, G.; Balk, D.; Anderson, B. The rising tide: Assessing the risks of climate change and human settlements in low elevation coastal zones. *Environ. Urban.* **2007**, *19*, 17–37. [CrossRef]
86. Lagarias, A.; Stratigea, A. High-resolution spatial data analysis for monitoring urban sprawl in coastal zones: A case study in Crete Island. In Proceedings of the Computational Science and Its Applications–ICCSA 2021: 21st International Conference, Cagliari, Italy, 13–16 September 2021; Proceedings, Part X 21. Springer International Publishing: Cham, Switzerland, 2021; pp. 75–90. [CrossRef]
87. BirdLife International. *Important Bird and Biodiversity Area (IBA) Digital Boundaries: March 2023 Version*; BirdLife International: Cambridge, UK; Available online: http://datazone.birdlife.org/site/requestgis (accessed on 24 September 2023).
88. Hellenic Statistical Authority, Population-Housing Census. 2021. Available online: https://www.statistics.gr/en/2021-census-pop-hous (accessed on 24 September 2023).
89. Kartalis, K.; Kokkosis, X.; Filippopoulos, K.; Polydoros, A.; Lappa, K.; Mavrakou, T. Integrating Climate Change in the Transformation of Developmental Model in Greece, Online 2021. Available online: https://www.dianeosis.org/wp-content/uploads/2021/10/Climate_Change_2021.pdf (accessed on 24 September 2023). (In Greek).
90. European Environmental Agency. *Climate Change, Impacts and Vulnerability in Europe*; EEA Report 2016, No 1/20172016; European Environmental Agency (EEA): Copenhagen, Denmark, 2016.
91. Dandoulaki, M.; Lazoglou, M.; Pangas, N.; Serraos, K. Disaster Risk Management and Spatial Planning: Evidence from the Fire-Stricken Area of Mati, Greece. *Sustainability* **2023**, *15*, 9776. [CrossRef]
92. Ministry of Environment and Energy. National Strategy for Climate Change in Greece. 2016. Available online: https://www.depa.gr/wp-content/uploads/2020/02/06.04.2016-espka-teliko_.pdf (accessed on 24 September 2023). (In Greek)
93. Marcos, M.; Jorda, G.; Cozannet, G. Sea level rise and its impacts on the Mediterranean. In *Jean-Paul Moatti et Stéphane Thiébault (dir.) The Mediterranean Region under Climate Change, a Scientific Update*; IRD Éditions: Marseille, France, 2016; pp. 265–275. [CrossRef]
94. Eurosion, 2004, Living with Coastal Erosion in Europe: Sediment and Space for Sustainability PART I-Major Findings and Policy Recommendations of the EUROSION Project (2004). Available online: http://www.eurosion.org/ (accessed on 24 September 2023).
95. Zambarloukou, S. Greece After the Crisis: Still a south European welfare model? *Eur. Soc.* **2015**, *17*, 653–673. [CrossRef]
96. Tulumello, S.; Cotella, G.; Othengrafen, F. Spatial planning and territorial governance in Southern Europe between economic crisis and austerity policies. *Int. Plan. Stud.* **2020**, *25*, 72–87. [CrossRef]
97. Hardouvelis, G.; Gkionis, I. A decade long economic crisis: Cyprus versus Greece. *Cyprus Econ. Policy Rev.* **2016**, *10*, 3–40.
98. Mitrakos, T. Inequality, Poverty and Social Welfare in Greece: Distributional Effects of Austerity. Released by Bank of Greece 2014, Working Paper 174. Available online: https://ssrn.com/abstract=4184595 (accessed on 24 September 2023). [CrossRef]
99. Gemenetzi, G. Restructuring Local-Level Spatial Planning in Greece Amid the Recession and Recovery Period: Trends and Challenges. *Plan. Pract. Res.* **2022**, *38*, 564–582. [CrossRef]
100. Vitopoulou, A.; Yiannakou, A. Public land policy and urban planning in Greece: Diachronic continuities and abrupt reversals in a context of crisis. *Eur. Urban Reg. Stud.* **2020**, *27*, 259–275. [CrossRef]
101. Corella, G.; Othengraphen, F.; Papaioannou, A.; Tulumello, S. Socio-political and socio-spatial implications of the economic crisis and austerity politics in Southern European cities, spatial impacts of the economic crisis. In *Cities in Crisis: Socio-Spatial Impacts of the Economic Crisis in Southern European Cities*; Knieling, J., Othengrafen, F., Eds.; Routledge: London, UK, 2015.
102. Psycharis, Y.; Kallioras, D.; Pantazis, P. Economic crisis and regional resilience: Detecting the 'geographical footprint'of economic crisis in Greece. *Reg. Sci. Policy Pract.* **2014**, *6*, 121–141. [CrossRef]
103. Institute of Greek Tourism Confederation. Available online: https://insete.gr/ (accessed on 24 September 2023).
104. Boutsioukis, G.; Fasianos, A.; Petrohilos-Andrianos, Y. The spatial distribution of short-term rental listings in Greece: A regional graphic. *Reg. Stud.* **2019**, *6*, 455–459. [CrossRef]
105. Athanassiou, E.; Cholezas, I.; Lychnaras, V.; Skintzi, G. *Greek Economic Outlook 2018/37*; Centre of Planning and Economic Research (KEPE): Athens, Greece, 2018.
106. *Law 4014/2011-Environmental Licensing of Works and Activities, Regulation of Illegal Constructions in Respect to Environmental Balance and Other Arrangements of the Ministry of Environment*; Issue A 209/21.9.2011; OGG (Official Government Gazette): Athens, Greece, 2011. (In Greek)
107. *Law 4178/2013-Confronting of Illegal Construction: Environmental Balance and Other Arrangements*; Issue A 174/08.08.2013; OGG (Official Government Gazette): Athens, Greece, 2013. (In Greek)
108. *Law 4495/2017-Control and Protection of Built Environment and Other Arrangements*; Issue A'167/3-11-2017; OGG (Official Government Gazette): Athens, Greece, 2017. (In Greek)
109. Asprogerakas, E.; Melissas, D. Reflections on the hierarchy of the spatial planning system in Greece (1999–2020). *Int. Plan. Stud.* **2023**. [CrossRef]

110. Papageorgiou, M. Spatial planning in transition in Greece: A critical overview. *Eur. Plan. Stud.* **2017**, *25*, 1818–1833. [CrossRef]
111. Vezyriannidou, S.; Portokalidis, K. Special Spatial Plans VS Local Spatial Plans. Towards a new vision of planning system at the local level in Greece, during the period of economic crisis. *RELAND Int. J. Real Estate Land Plan.* **2018**, *1*, 402–408.
112. *Law 3894/2010-Acceleration and Transparency for the Implementation of Strategic Investments*; Issue A 204/2.12.2010; OGG (Official Government Gazette): Athens, Greece, 2010. (In Greek)
113. *Law 4269/2014-Regional and Urban Planning Reform–Sustainable Development*; Issue A 142/28.6.2014; OGG (Official Government Gazette): Athens, Greece, 2014. (In Greek)
114. *Law 4447/2016-Spatial Planning, Sustainable Development and Other Provisions*; Issue A 241/23.12.2016; OGG (Official Government Gazette): Athens, Greece, 2016. (In Greek)
115. *Law 4759/2020-Modernization of Regional and Urban Planning Legislation, and Other Arrangements*; Issue A 245/09.12.2020; OGG (Official Government Gazette): Athens, Greece, 2020. (In Greek)
116. *Law 3986/2011–Urgent Measures for the Implementation of the Medium-Term Fiscal Strategy 2012–2015*; Issue A 152/1-7-2011; OGG (Official Government Gazette): Athens, Greece, 2011. (In Greek)
117. Hadjimichalis, C. Crisis and land dispossession in Greece as part of the global 'land fever'. *City* **2014**, *18*, 502–508. [CrossRef]
118. Lazoglou, M.; Serraos, K. Climate change adaptation through spatial planning: The case study of the region of Western Macedonia. *IOP Publ. Conf. Ser. Earth Environ. Sci.* **2021**, *899*, 012021. [CrossRef]

Disclaimer/Publisher's Note: The statements, opinions and data contained in all publications are solely those of the individual author(s) and contributor(s) and not of MDPI and/or the editor(s). MDPI and/or the editor(s) disclaim responsibility for any injury to people or property resulting from any ideas, methods, instructions or products referred to in the content.

land MDPI

Article

Optimization of Microclimate Conditions Considering Urban Morphology and Trees Using ENVI-Met: A Case Study of Cairo City

Ahmed Yasser Abdelmejeed [1,2,*] and Dietwald Gruehn [1]

[1] Research Group—Landscape Ecology and Landscape Planning (LLP), Department of Spatial Planning, TU Dortmund University, 44227 Dortmund, Germany; dietwald.gruehn@tu-dortmund.de
[2] Faculty of Urban and Regional Planning, Cairo University, Cairo 11562, Egypt
* Correspondence: ahmed.abdelmejeed@tu-dortmund.de

Abstract: This research aims to optimize the use of trees to enhance microclimate conditions, which has become necessary because of climate change and its impacts, especially for cities suffering from extreme heat stress, such as Cairo. It considers elements of urban morphology, such as the aspect ratio and orientation of canyons, which play an important role in changing microclimate conditions. It also considers both sides of each canyon because the urban shading is based on the orientation and the aspect ratio, which can provide good shade on one side of the canyon but leave the other side exposed to direct and indirect radiation, to ensure a complete assessment of how the use of trees can be optimized. As Cairo city is very large and has a variety of urban morphologies, a total of 144 theoretical cases have been tested for Cairo city using ENVI-met to cover the majority of the urban cases within the city (Stage 1). Then, the same tree scenarios used in the theoretical study are applied to an existing urban area in downtown Cairo with many urban morphology varieties to validate the results of the theoretical study (Stage 2). After testing all cases in both stages, it became very clear that the addition of trees cannot be the same for the different aspect ratios, orientations, and sides of the different canyons. For example, eastern roads should have more trees than other orientations for all aspect ratios, but the required number of trees is greater for the northern side than the southern side, as the southern side is partially shaded for a few hours of the day by buildings in moderate and deep canyons. Northern streets require a very limited number of trees, even in shallow canyons, on both sides. The correlation between the number of trees on each side for the different orientations and aspect ratios shows a strong negative relationship, but the correlation values change between the different sides and orientations. The results of applying trees to an existing urban area show almost the same results as the theoretical study's results, with very slight differences occurring because of the irregularity of the existing study area. This proves that when adding trees, not only the aspect ratio and orientation but also the side of each canyon should be considered to ensure that pedestrians, in all cases, have better microclimate conditions and that the use of trees is optimized.

Keywords: urban heat island (UHI); outdoor thermal comfort; urban trees; urban shading; street canyon aspect ratio and orientation; greenery effect; ENVI-met greenery simulation; urban cooling strategies

1. Introduction

1.1. Heat Stress and UHI Appearance and Causes

Cities grow over many decades with different urban fabrics and morphologies, and they continue to evolve in various shapes to meet their populations' growing needs. Because of rapid expansion and increases in population density, cities absorb more heat and suffer from heat stress and the effects of urban heat islands (UHIs), which disturb human health [1]. Cities' centers have been observed to be hotter than their suburbs. This is due to urban areas absorbing and trapping longwave radiation, which supports UHIs [2]. In one study, the air and surface temperatures in city centers were higher than those in suburban areas by 5.0 to 5.5 C° [3]. In another study conducted in downtown Cairo, an increase of 0.5 to 3.5 C° in the surface temperature was observed and reached 10 C° as the maximum difference in comparison to the suburbs [4,5]. The urban climate is an important issue pertaining to local and global climates, and it is influenced by several urban design factors, such as urban morphology and density, properties of urban surfaces, and different types of vegetation cover [6]. The process of trapping longwave radiation mainly controls UHIs. When buildings are taller and streets are narrower, urban canyons absorb less longwave radiation; however, they trap the absorbed heat [2]. Distance from the city center, surface albedo, aspect ratio, and vegetation density are major predictors of the UHI response. In one study, it was found that every 500 m increase in the distance from the city center reduced the interurban heat island by 0.13 C°. Increasing the surface albedo by 0.01 decreased the UHI by 0.18 C°, whereas increasing the vegetation density ratio by 0.10 yielded a 0.17 C° reduction in the UHI. A 10% increase in the aspect ratio increased the UHI by 0.17 C° [7].

1.2. Urban Morphology Relating to Urban Shading and PET Parameters

Urban morphology, sky view factor (SVF), and shading are the major factors that have a significant role in enhancing microclimate conditions and reducing UHI effects [2,8]. The shadow-cast effect produced by buildings helps reduce pedestrian radiant load and, consequently, improves thermal comfort, especially in high-density cities, although ventilation is reduced [9,10]. Shallow canyons are susceptible to worse thermal conditions than their deeper counterparts with similar aspect ratio values [9]. Asymmetrical streets are better than low, symmetrical streets at enhancing wind flow and blocking solar radiation [11]. Increasing the SVF in the selection of an urban configuration reduces UHI intensity [12]. Deep urban canyons can reduce the amount of direct solar radiation during the daytime. Therefore, the level of thermal comfort in an open space (i.e., high SVF) is lower than that in a shaded space (i.e., low SVF) [13]. The results of the analysis prove that thermal comfort is mainly affected by exposure to solar radiation [14]. In conclusion, shading from direct radiation is more important than the increase in absorbed radiation due to urban reflectance.

The physiological equivalent temperature (PET) meteorological parameters (air temperature, wind speed, radiant temperature, and humidity) [15] can be controlled and enhanced using urban morphology. The mean radiant temperature (TMRT) is a key meteorological parameter governing human energy balance and is used to evaluate the thermal comfort of humans [16]. The target is to keep the TMRT below 45 C° [8]. Air temperature and specific humidity have emerged as the least effective, suggesting that urban configurations can alter their values only to a limited extent [10,17]. The outdoor thermal comfort level significantly depends on the speed and direction of the urban wind flow [18]. Wind speed has been widely reported to have an influence on urban heating, and there is a strong negative correlation between wind speed and air temperature [19]. All of these PET meteorological parameters, along with urban shading and SVF, can be optimized using urban morphology and urban geometry elements, such as by adjusting the street canyon aspect ratio and orientation, in addition to using different densities of vegetation [10,16,20,21].

1.2.1. Aspect Ratio Effect

The aspect ratio (AR), or a canyon's height-to-width ratio (H/W), is an important parameter that is usually used to investigate the influence of urban geometry on an outdoor environment, especially temperature and building energy demand [17,22]. The aspect ratio is the dominant factor for daily net solar radiation gains on road and wall surfaces. The effects of shadows on surrounding buildings are also important factors for the radiation environment in urban street canyons [23], as the enhancement of shade due to increased H/W ratios is capable of producing significant reductions in the PET [16,24]. There is a strong relationship between UHIs and the aspect ratio during the night, as the effect of the street canyon on UHI intensity is significant. According to one study conducted in the city of Basel, Switzerland, the intensity of the maximum nighttime UHI has a linear relationship with the SVF, which is controlled by the aspect ratio [25]. In addition, the lowest daytime mean radiant temperatures result from the high aspect ratios of streets. Air temperatures decrease slightly with an increase in aspect ratios, but the radiation fluxes expressed by the mean radiant temperature are, by far, more decisive [26].

In Osaka, the daily net solar radiation gains are large for roads in which the aspect ratio is greater than approximately 1.5 (H/W). Roads in which the aspect ratio is between 1.0 and 1.5 (H/W) are also in the target range for effective urban heat island mitigation measures, and particular attention is needed for the north sides of east–west roads and the centers of north–south roads [23]. In Malaysia, for the six asymmetrical aspect ratios of Putrajaya Boulevard, an aspect ratio of 2–0.8, which reduces the temperature of surfaces by 10 to 14 C° and the air by 4.7 C°, is recommended for enhancing the boulevard's microclimates and mitigating tropical heat islands. In the northeast to southwest direction, aspect ratios of 0.8–2 reduce the morning microclimate and night heat islands, yet the negative effects during the day are greater than the positive effects during the nighttime [11]. Along Wall Street, New York City, the outcomes of winter and summer analyses show high values of daytime air temperatures along the widest street canyon (aspect ratio = 0.33) [27]. In the center of Camagüey, Cuba, aspect ratios higher than one are advisable, as they contribute to improving the thermal conditions of courtyards in the summer. When the aspect ratio of a courtyard is H/W = 0.5, no variations in the TMRT are obtained when using different orientations because most of the courtyards have surfaces that are exposed to direct solar radiation during the critical period of the day (11:00 h and 14:00 h), and particular subzones of the courtyard that are adjacent to the surrounding facades are more comfortable than the central subzone, increasing the aspect ratio from 0.5 to 3 and reducing the TMRT by 15.7 C° [8]. An increase of 0.5 in the aspect ratio's values can decrease the maximum mean radiant temperature by 2.90 C° on average in the early morning and late afternoon and, consequently, decrease the PET [17]. Regarding the impact of the AR on UHIs, streets featuring a lower aspect ratio have a high frequency of heat stress in the daytime but low PET in the nighttime [10]. Comparing both east–west- and north–south-oriented streets against surface temperature measurements in Tokyo, it was found that the shading effect of a tall building in north–south street canyons had less of an impact on solar gains than that in east-west streets. Tall buildings and narrow canyons reduce the SVF and increase the amount of shaded area on the surface, resulting in lower temperatures in canyons during the daytime but higher temperatures at night [24].

1.2.2. Street Orientation Effect

Street orientation is considered to play an influential role in altering the microclimate in urban areas, and it influences the exposure of canyon surfaces to direct solar radiation. A north–south (N–S) street orientation will be fully exposed to solar radiation at midday but mostly shaded in the early morning and late afternoon. This is contrary to an east–west (E–W) street orientation, which is fully exposed in the early morning and late afternoon [17,22,28,29]. North–south-oriented streets are cooler than those with an east–west orientation, and the comfort level in these areas increases along with their H/W ratio [20], because east–west-oriented canyons are exposed to sunlight throughout the day regardless of their H/W ratio, whereas north–south-oriented canyons are only exposed to sunlight during certain times of the day [20]. One study conducted on urban heat island mitigation measures found that the top priorities are the north side of east–west roads and the center of north–south roads [23]. The NW–SE orientation shows a slightly lower PET level than N–S and E–W orientations [30]. It has been found that an orientation angle between 30° and 60° with wind direction and a canyon aspect ratio of 2.5 can reduce the PET value by 5 to 9 C° throughout most of the study area during midafternoon on a summer day [18].

In a study conducted in Sydney [17], it was concluded that streets situated on the north–south axis offer a superior level of thermal comfort than east–west-oriented streets. The PET values presented a comfortable range of 12.33% during the daytime; streets on the NE–SW axis provided the highest level of thermal comfort, on average at 24.95%, and the worst option was evaluated as the NW–SE orientation. As the duration of solar exposure and the average mean radiant temperature (TMRT) increase, mainly because the wind velocity decreases, outdoor users face a lack of thermal satisfaction.

1.2.3. Combined Effect of Aspect Ratio and Street Orientation

The orientation and canyon aspect ratio have a profound influence on the urban microclimate that directly impacts street-level thermal comfort, as PETs at the street level strongly depend on the aspect ratio and street orientation [14,18,28–30]. Street geometry and orientation influence the amount of solar radiation received by street surfaces, as well as the airflow in urban canyons [6]. For E–W orientations, streets with an H/W greater than two should be fully shadowed only during the hottest and coolest months of the year [20]. Streets on the E–W axis present the worst conditions for all H/W ratios (up to 3.0). An increase in the H/W ratio on an E–W street does not improve PET levels [14]. It is difficult to mitigate the heat stress along an E–W-oriented street. The walls provide only a limited amount of shading, even for proportions with an H/W ratio of 4:1. In comparison, an N–S orientation combined with a high aspect ratio, equal to or greater than an H/W ratio of 2:1, provides a much better thermal environment with lower PET maxima and shorter periods of high stress [14,20]. Thermal stress can be reduced in a street canyon with a northwest–southeast orientation combined with an aspect ratio of at least 1.5, and these street configurations can reduce heat stress, increase the frequency of comfortable thermal conditions, and enable solar access throughout the year in the midlatitudes [10]. An orientation angle of 30–60° in the wind direction and a canyon aspect ratio of 2.5 can reduce PETs by 5–9 C° during the midafternoon on a summer day [20].

1.2.4. Urban Trees

The effectiveness of trees in enhancing daytime thermal comfort decreases as urban density increases and vice versa at night [9,22]. Urban trees can reduce the effects of the surrounding building mass and help create a low-SVF environment that is cooler during both daytime and nighttime [13]. It has been demonstrated that urban morphology and urban vegetation shading affect solar radiation storage during the day in the summer, and urban shading significantly contributes to UHI mitigation [13]. Significant temperature differences between vegetated and non-vegetated areas have been observed, which can be explained by both the shading and evapotranspiration effects of trees [31]. As a rule, on summer days, outdoor activities in unshaded areas are not recommended between 10:00 h and 15:00 h. Therefore, the provision of shade, using canopies and vegetation, is necessary if outdoor activities are to occur during this time of day [8]. In open-set high-rise urban areas, the presence of trees could produce a relevant reduction in thermal stress at the pedestrian level [30]. Trees can be considered as a solution to improve the thermal condition of streets, especially in streets designed along nonoptimal orientations with low-rise buildings [17]. In a previous study, an area that was at an angle of 30° from the north with an aspect ratio of 1.0 was found to not require any plantings, as a continuous shaded zone was created by the buildings along both streets in the parallax and perpendicular directions. However, decreasing the aspect ratio creates a need for shade-providing trees for the streets in the perpendicular direction, as the distance between the buildings increases [18]. A study [14] demonstrated that trees have a much more considerable effect on E–W streets. The reduction in PET values is very significant, especially for the side of the street facing south. Increasing 10% of the urban vegetation can reduce Ta and MRT throughout the entire day and nighttime by up to 0.8 C° [16]. PET has been found to be approximately 10 C° lower under trees than in green areas (38 C°) and at least 25 C° lower than in enclosed areas (48 C°) [10]. Increasing the density of vegetation in shallow urban canyons (H/W = 0.5) increased PET enhancement by more than 4% compared with low-density vegetation in the same urban canyon [17].

1.3. Cairo City: The Case Study

1.3.1. Cairo City's Climate

Cairo is in a subtropical climatic region with a dry climate. During summer (June to August), it is hot and dry with a maximum mean temperature of 28 C° [5]. Cairo city receives an enormous amount of solar radiation, which causes the city to experience massive heat stress.

1.3.2. City Urban Morphology

The Greater Cairo Metropolitan Area boasts the largest urban area in Africa and ranks as the 11th largest city in the world [4]. A study found that Cairo's land cover was 233.78 km^2 in 1973, growing to 557.87 km^2 in 2006, which means that it has more than doubled in size, and the rate of urbanization is 9.8 km^2 per year [32]. Figure 1 shows its rapid urban growth over 22 years on both agricultural and desert land [33]. A total area of 187.32 km^2 of agricultural lands has been lost because of this urban expansion [32]. The New Urban Community Authority considers 25% of Cairo to be informal, but some research considers 50% or 66.6% of Cairo to be informal [34]. This formal and informal rapid growth has increased the city's density and size, in addition to creating different urban morphologies and fabrics (i.e., different aspect ratios, from very shallow to shallow, moderate, deep, and very deep) as shown in Figure 2, which react in different ways to the meteorological parameters, creating different microclimate conditions inside the different urban canyons.

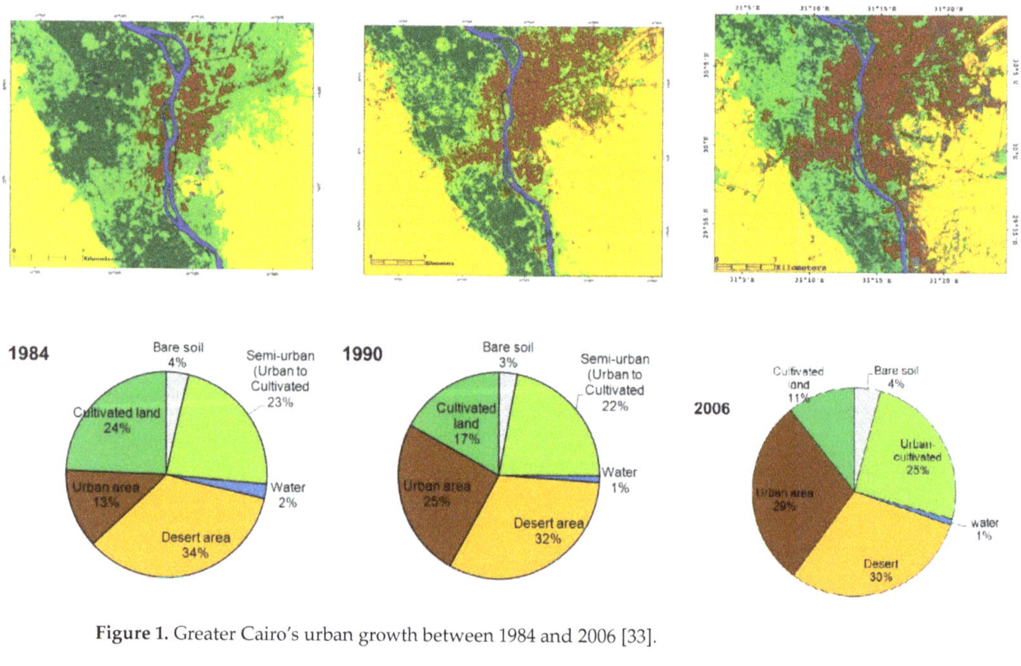

Figure 1. Greater Cairo's urban growth between 1984 and 2006 [33].

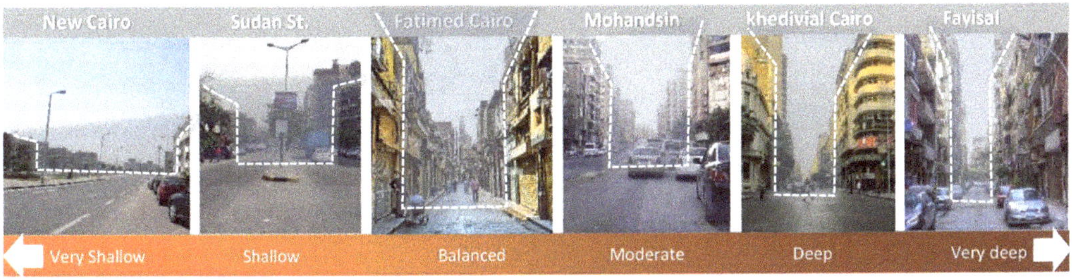

Figure 2. Examples of different urban canyon aspect ratios in Cairo city.

1.4. Research Gap and Target

This research aimed to optimize the integration of many different urban morphologies (with varying aspect ratios and orientations) and urban trees (low and high densities) to understand the correlation between them and maximize the enhancement of the microclimate conditions for the case study (i.e., Cairo city). The objectives of this research were not only to fill the gaps in the understanding of different urban canyons and how they respond to harsh microclimate conditions but also how to integrate various urban canyons with different densities of urban trees, as well as how these trees would perform inside different urban canyons. In addition, as previous studies mostly focused on the enhancement of the whole canyon, in this study, the main focus is on the sides of different canyons, as they are quite important because canyon sides are the places where people walk, sit, and stand, while the rest of the canyon is mainly for vehicles. The findings of this research should help urban designers and landscape architects choose an aspect ratio, street orientation, and tree density from an urban climate point of view while developing urban projects in Greater Cairo.

2. Materials and Methods

To achieve the research target, a study analyzing and testing the urban morphology's characteristics with and without different tree densities was conducted. The method of testing and analyzing the relationship was performed in two stages, as shown in Figure 3. Stage one was the creation of a theoretical model representing the different common urban canyons in Cairo city along different orientations. This model was tested first without trees and then with different tree densities. Stage two involved testing the theoretical model's outcomes when applied to an existing case study in downtown Cairo with similar aspect ratios and orientations. For a better understanding of the relationship and to go further in depth regarding the details of the urban canyons, this study conducted both stages on both sides of the urban canyon; this will help to understand how both sides of an urban canyon (i.e., where people are walking) react to climate conditions and how changing the aspect ratio and orientation, as well as adding different tree densities, will impact the thermal comfort and the UHI effects at the pedestrian level on each side. Studying both sides of a street provides more accurate and detailed results because in some canyons, one side is shaded by buildings but the other is totally exposed to direct sun radiation [35], which affects the control and optimization of the number of added trees; this is in line with water efficiency approaches, which is quite important in Egypt's case, as it suffers from water scarcity [36].

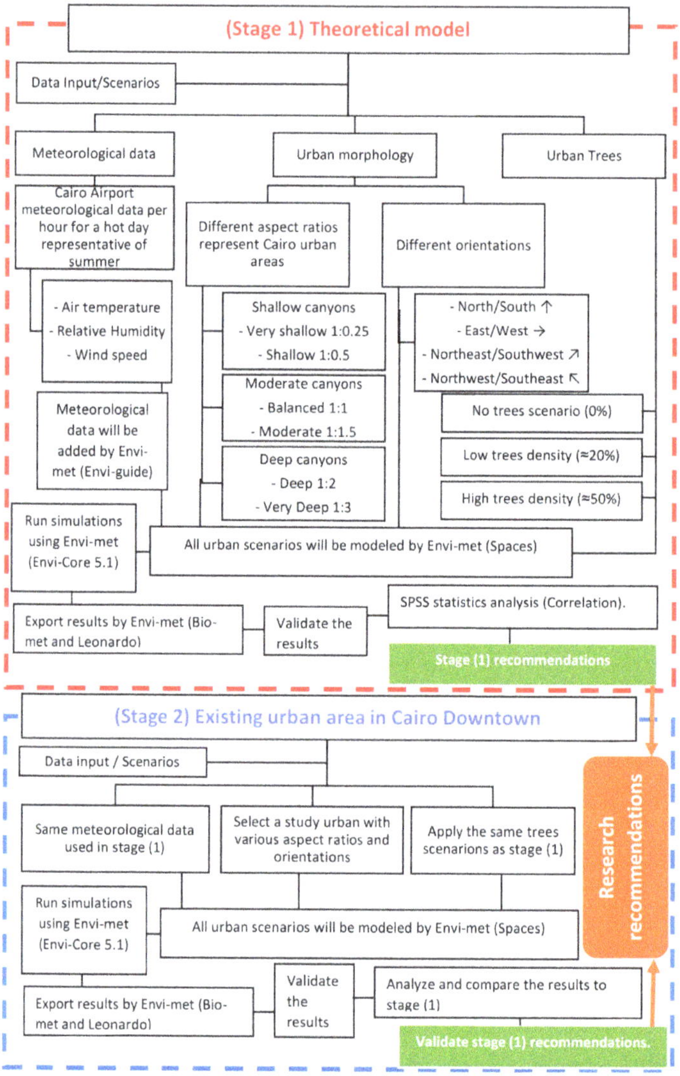

Figure 3. Research methodology.

2.1. Stage (1): Theoretical Model

In Stage One, a theoretical model was developed representing the different aspect ratios and orientations that are very common in Cairo city (Step a). Then, tree scenarios were applied to the base case (Step b) to understand the effects of the trees after comparing the results of both steps.

2.1.1. Stage (1): Step (a) Theoretical Model (Base Case)

The range of aspect ratios that were evaluated varied between very shallow and very deep, and six aspect ratios were developed and evaluated (3:1, 2:1, 1:1.5, 1:1, 0.5:1, and 0.25:1), which covers the majority of the various urban canyons in Cairo city. The six aspect ratios developed were oriented to four different orientations every 45 degrees, and the different orientations represented the main orientations (i.e., north–south, east–west, northeast–southwest, and northwest–southeast); that is, the total number of cases in the theoretical model base case was 24. Figure 4 shows the base case of the theoretical model, and Figure 5 shows the shading analysis for the different aspect ratios and orientations. As the aim was to understand how each urban canyon reacts and performs, each side of an urban canyon was analyzed and compared to fully comprehend each urban canyon using receptors (i.e., measuring points), as shown in Figure 4A; hence, the total number of cases in the base case was 48.

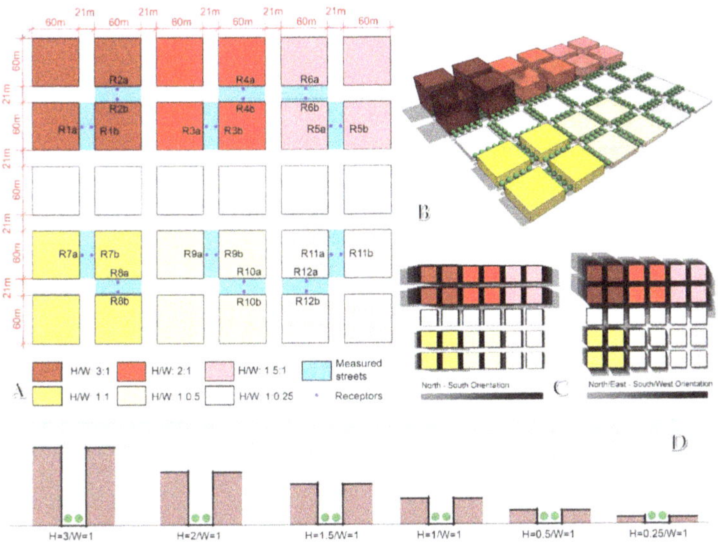

Figure 4. The theoretical model's urban geometry: (**A**) theoretical model's plan showing the measured canyons and receptors' locations; (**B**) 3D view of the theoretical model with trees; (**C**) full day shading analysis for both orientations; (**D**) AR cross-sections.

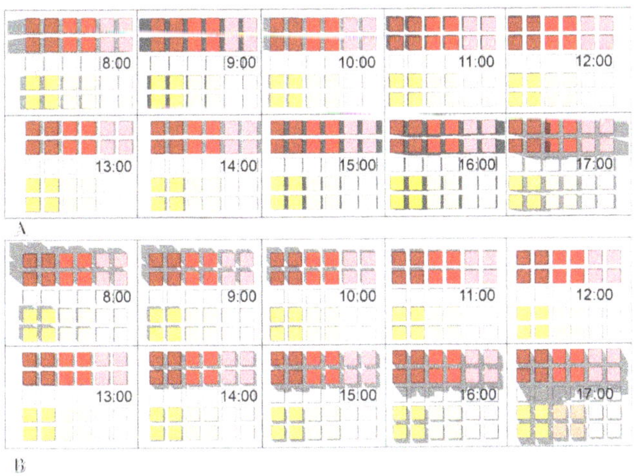

Figure 5. Shading distribution for each AR per hour from morning to sunset: (**A**) north orientation; (**B**) northeast orientation. Created using SketchUp after aligning the model to the original location of Cairo city.

2.1.2. Stage (1): Step (b) Tree Scenarios

The tree scenarios were developed and added to the base case to compare the tree results to the base case results without trees to understand the effect of the trees on every street on each side. Two tree scenarios were developed that represented two tree densities (low tree density ≈ 20%; high tree density ≈ 50%), as shown in Figure 6. The total number of cases after adding the tree scenarios, in addition to the base case scenarios, was 144 (base cases = 48; tree scenario cases = 96), as shown in Table 1, which is sufficient for comparing various urban cases with different aspect ratios and orientations, as well as for the integration of different tree scenarios. The tree scenarios were applied to all urban canyons with different aspect ratios and orientations, and both street canyon sides were measured for all the tree scenarios.

Figure 6. Tree density scenarios: 0%; low density (20%); high density (50%).

Table 1. All scenarios for the theoretical model.

Urban Canyon	Canyon Side (a or b)	Aspect Ratio	Orientation 1	Orientation 2	Tree Scenario 1	Tree Scenario 2	Tree Scenario 3	Total Number of Cases
R1	R1a	3 to 1	North–South	NE–SW	0%	20%	50%	6
	R1b	3 to 1	North–South	NE–SW	0%	20%	50%	6
R2	R2a	3 to 1	East–West	NW–SE	0%	20%	50%	6
	R2b	3 to 1	East–West	NW–SE	0%	20%	50%	6
R3	R3a	1 to 2	North–South	NE–SW	0%	20%	50%	6
	R3b	1 to 2	North–South	NE–SW	0%	20%	50%	6
R4	R4a	1 to 2	East–West	NW–SE	0%	20%	50%	6
	R4b	1 to 2	East–West	NW–SE	0%	20%	50%	6
R5	R5a	1.5 to 1	North–South	NE–SW	0%	20%	50%	6
	R5b	1.5 to 1	North–South	NE–SW	0%	20%	50%	6
R6	R6a	1.5 to 1	East–West	NW–SE	0%	20%	50%	6
	R6b	1.5 to 1	East–West	NW–SE	0%	20%	50%	6
R7	R7a	1 to 1	North–South	NE–SW	0%	20%	50%	6
	R7b	1 to 1	North–South	NE–SW	0%	20%	50%	6
R8	R8a	1 to 1	East–West	NW–SE	0%	20%	50%	6
	R8b	1 to 1	East–West	NW–SE	0%	20%	50%	6
R9	R9a	0.5 to 1	North–South	NE–SW	0%	20%	50%	6
	R9b	0.5 to 1	North–South	NE–SW	0%	20%	50%	6
R10	R10a	0.5 to 1	East–West	NW–SE	0%	20%	50%	6
	R10b	0.5 to 1	East–West	NW–SE	0%	20%	50%	6
R11	R11a	0.25 to 1	North–South	NE–SW	0%	20%	50%	6
	R11b	0.25 to 1	North–South	NE–SW	0%	20%	50%	6
R12	R12a	0.25 to 1	East–West	NW–SE	0%	20%	50%	6
	R12b	0.25 to 1	East–West	NW–SE	0%	20%	50%	6
			Total number of cases					**144**

2.1.3. Result Measurement

The results for both sides of a canyon in each scenario were compared to better comprehend the performance and the relationship between different urban canyons and trees. This comparison provides extensive information to aid urban planners and landscape architects in making decisions during the development of urban areas in Cairo city, particularly after implementing the findings in an existing urban area (Stage 2) in downtown Cairo.

2.2. Stage (2): Existing Case Study

The purpose of this section is to assess an existing urban area and compare its findings with those of the theoretical model. This will help validate the results of the theoretical model, which is highly symmetrical and uniform; however, in reality, urban areas, especially in old city zones, are not that uniform. Thus, comparing the results of both theoretical and existing case studies will provide insight into the tolerance and accuracy of the findings. This evidence can then be used to apply the research recommendations.

2.2.1. Study Area Location and Urban Characteristics

The selected study area should be located in the center of Cairo city so that it will be under the influence of the UHI and represent the urban density of the city downtown [4,5]. The selected study area, located in Khedival Cairo, in the city's downtown area, has varying street widths due to its hierarchical road systems, resulting in different aspect ratios and orientations. These urban varieties make the case study suitable for studying and representing the majority of urban cases in the theoretical study. Figure 7 illustrates the selected study area's location and urban characteristics.

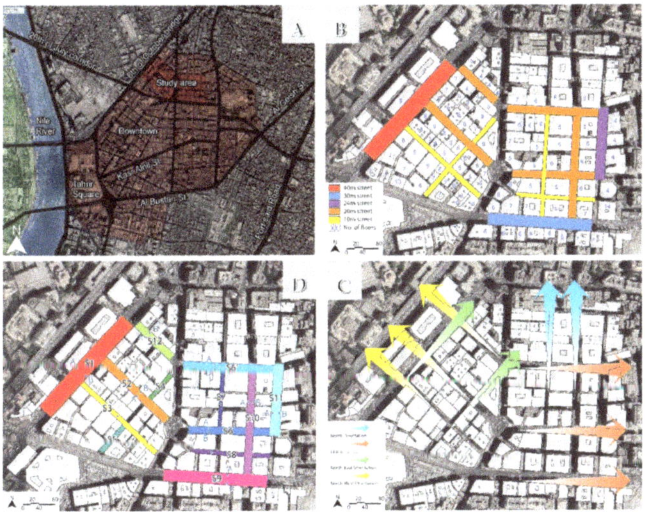

Figure 7. Existing urban case study: (**A**) location of the study area; (**B**) different street widths and building heights; (**C**) different street orientations; (**D**) streets selected for this study and location of the receptors for each street on both canyon sides.

As depicted in Figure 7B and Table 2, the study area contains numerous urban varieties that offer many different cases for comparison with the theoretical model. This study encompassed streets with different widths from 10 m for local pedestrian streets to 40 m for major roads within the study area (Rameses St., S1), with different building heights per street, leading to many aspect ratios ranging from 0.5:1 to 2:1, as shown in Table 2. In addition, the study area's streets have four urban orientations, as shown in Figure 7C and Table 2, providing urban canyons with various orientations (N–S, E–W, NE–SW, and NW–SE), covering all orientations in the theoretical model. This large variety within the study area aided in representing and validating the theoretical cases. Table 3 displays the number of theoretical study cases that were covered in the study area, with 12 cases from the theoretical model being covered, representing approximately 75% of the total number of cases after excluding those with very deep and very shallow aspect ratios, which are uncommon in Cairo city. This demonstrates that the study area represented the urban varieties found in the theoretical model well. Furthermore, it was well suited to this research and significantly contributed to the achievement of this study's objectives. Figure 7D and Table 2 provide information on the streets selected for this study.

Table 2. Selected urban canyons in the study area and their aspect ratios and orientations.

Abbreviation in Figure 7D	Street Name	Avg. Width (m)	Avg. Height (m)	Aspect Ratio (H/W)	Orientation	No. of Cases
S1	Rameses St.	40	20	0.5 to 1	NE ↗	6
S2	Sayed Anbar St.	20	21	1.05 to 1	NW ↖	6
S3	Souq Al-Tawfikiya St.	12	17	1.42 to 1	NW ↖	3
S4	Al-Boursa Al-Kadyima St.	12	24	2 to 1	NE ↗	3
S5	Al-Boursa Al-Kadyima St.	12	18	1.5 to 1	NE ↗	3
S6	Mohamed Bek Al-Alfy St.	19	18	0.95 to 1	E →	6
S6′	Waked St.	20	19	0.95 to 1	E →	6
S7	Zakriya Ahmed St.	9	16	1.8 to 1	N ↑	3
S8	Saraya Al-Azbakiya St.	10	19	1.9 to 1	E →	3
S9	26 July St.	29	18	0.6 to 1	E →	6
S10	Emad Al-Din St.	19	17	0.9 to 1	N ↑	6
S11	Bostan Al-Dekkah St.	24	14	0.6 to 1	N ↑	6
S12	Suliman Al-Halabi St.	20	14	0.7 to 1	NW ↖	6
		Total number of cases				**63**

Table 3. Cases of the theoretical model covered in the study area (highlighted in green).

N	Street	NE	Street	E	Street	NW	Street
1 to 3		1 to 3		1 to 3		1 to 3	
1 to 2	**S7**	1 to 2	**S4**	1 to 2	**S8**	1 to 2	
1 to 1.5		1 to 1.5	**S5**	1 to 1.5		1 to 1.5	**S3**
1 to 1	**S10**	1 to 1		1 to 1	**S6 and S6′**	1 to 1	**S2**
1 to 0.5	**S11**	1 to 0.5	**S1**	1 to 0.5	**S9**	1 to 0.5	**S12**
1 to 0.25		1 to 0.25		1 to 0.25		1 to 0.25	

2.2.2. Study Area Tree Scenarios

The tree scenarios that were applied to the study area were the same as those in the theoretical model. In total, three different tree scenarios were applied: no trees at 0%, a low density of trees at 20%, and a high density of trees at 50%, as shown in Figure 8. Applying the same tree scenarios as in the theoretical model allowed for a comparison of the impact of the trees in different canyons between the theoretical model and the case study. Because of the presence of both wide and narrow streets in the study area, certain streets (S3, S4, S5, S7, and S8) only had a single row of trees located in the middle of the street. This is because the width of these streets is 10 m, making it impossible to accommodate two rows of trees. Conversely, in streets that are exceptionally wide, such as S1, three rows of trees were planted to achieve 50% tree coverage.

Figure 8. Tree scenarios for this study: (**A**) 0%; (**B**) 20%; (**C**) 50%.

2.3. Data Input, Model Setup, and Measuring Points

The software used to run the simulations for the different cases in the theoretical model and the study area was ENVI-met V5.1, 2022. All required information for the model was provided, representing existing urban configurations (material, soil) of common materials in Cairo, as well as meteorological data.

2.3.1. Model Setup and Geometry

The model's location was 30.02 latitude and 31.22 longitude. These coordinates were obtained from the ENVI-met V5.1 2022 database by selecting the location of Greater Cairo from the ENVI-met application "Spaces". As shown in Table 4, for both the theoretical and case study models, the study area's model was created using the following model geometry: materials, soil, and trees.

Table 4. Model setup and geometry for both the theoretical model and the case study.

Modeling Information	Theoretical Model	Case Study Model
Area size	X = 165, Y = 140, Z = 30	X = 212, Y = 151, Z = 22
Grid resolution	X = 3, Y = 3, Z = 3	X = 3, Y = 3, Z = 3
Orientation	Model (1) = 0, Model (2) = −45	0
Split lower grid box into 5 sub cells	Yes	Yes
Telescoping applied	Telescoping factor of 20%, starting at a 63 m height	Not applied, as the maximum building height is not very tall
Maximum model height *	198 m	66 m
Nesting grids **	5 Grids, sandy soil	5 Grids, sandy soil
DEM	Not applied, as the site is flat	Not applied, as the site is flat
Soil	Asphalt for roads, concrete for sidewalks, and sand under buildings	Asphalt for roads, concrete for sidewalks, and sand under buildings
Buildings materials	Default wall—moderate insulation	Default wall—moderate insulation
Tree model and size	*Latin name: Acer Platanoides* *** Height = 15 m; crown width = 7 m	*Latin name: Acer Platanoides* *** Height = 15 m; crown width = 7 m

* The model's height is more than double the height of the tallest building, as recommended by the software [37]. ** Nesting grids were added, in addition to 5 cells from the boundary sides of the model being kept empty, as recommended by the software developers [37]. *** *Acer Platanoides* was selected from the ENVI-met database, as it met the required criteria of having a large canopy, high LAD, and good canopy height, matching the recommendations in [38,39].

2.3.2. Simulation Configuration

The simulation configuration was conducted for a representative summer day (i.e., hottest day of July: 28 July 2022). The air temperature, relative humidity, and wind speed were added to ENVI-met for each hour of the simulated day, and climate data were imported from the Cairo Airport weather station [40,41]. The starting time of the simulation was 1:00 a.m., and the total duration of the simulation was 24 h. ENVI-met simple forcing was used for the metrological data. Output data were extracted every hour and converted using the ENVI-met "Bio-met" to calculate the PET values at each receptor for each hour in all scenarios. The different PET values were the main factors in the comparison. They were compared among all scenarios to assess the effect of each scenario. In addition, other main parameters were measured, such as wind speed (WS) and total mean radiant temperature (TMRT). As Figure 9 illustrates, the PET and its parameters were measured at one given point located at the center of each street's side and centralized between trees to avoid the direct shade of the tree canopies. The receptors were located on both sides of each selected street, exactly in the middle, and for the tree scenarios, they were shifted to be exactly in the middle between trees, helping to ensure that the results would not be affected by the direct shade of the trees and that the results that are compared represent the indirect impact of the tree scenarios.

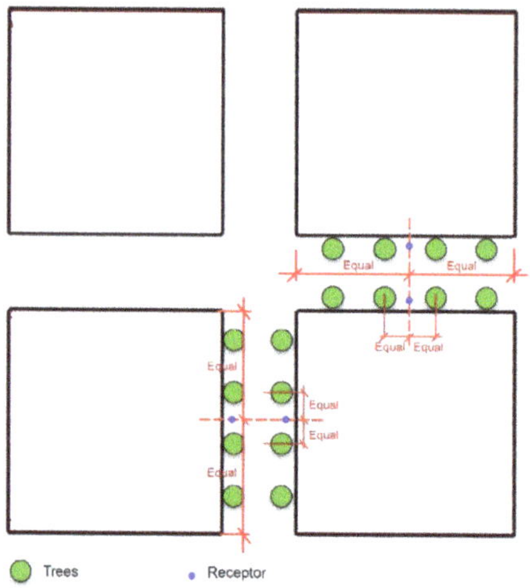

Figure 9. Locations of the receptors (i.e., measuring points) in the middle of each canyon and centralized between trees.

2.4. Validation of the Results

After comparing the meteorological data measured in Cairo city on the 28th of July [40,41] with the ENVI-met outputs for the potential air temperature and relative humidity for both the theoretical models and the case study model at an empty point located in the center of each model, it was found that there was a good match between the measured data and the output data, as shown in Figure 10a,c. Also, the root mean square error (RMSE) and the index of agreement (d) were calculated for the measured and simulated air temperature and relative humidity, and as shown in Figure 10b,d, the RMSE ranged between 1.04 and 0.662 for the air temperature and between 9.06 and 5.5 for the relative humidity. The index of agreement ranged between 0.977 and 0.993 for air temperature and between 0.903 and 0.974 for relative humidity. This means that all models accurately represented the weather conditions in Cairo city, and the results for the different scenarios and different urban cases are reliable and represent how these canyons would perform in real-life situations.

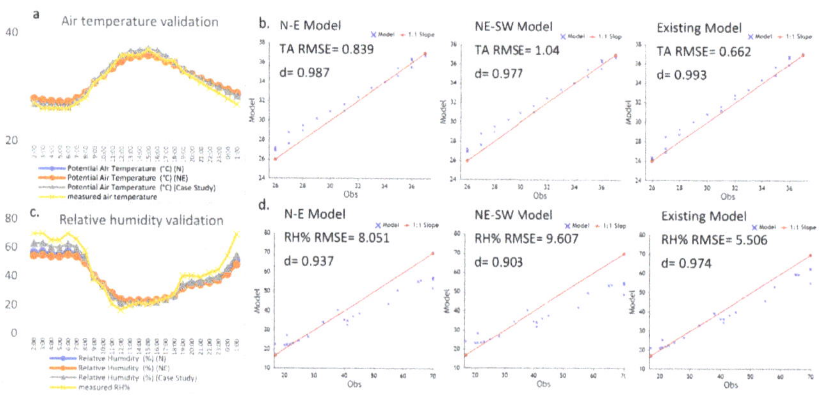

Figure 10. Results of the validation: (**a**,**c**) comparison between the measured and simulated air temperature and relative humidity; (**b**) TA RMSE and index of agreement for each model; (**d**) RH% RMSE and index of agreement for each model.

3. Results

The results are divided into three parts. The first part concerns the theoretical model's results (Stage 1), which present the results of the different aspect ratios with different orientations with and without applying the tree scenarios. The second part presents the results of the existing study area (Stage 2) and how the different canyons perform with and without trees. The final part is a comparison between both parts to gain a better understanding of the results and validate the results of the theoretical model. The ΔPET, ΔTMRT, and ΔWind speed were the major parameters in the comparison, as recommended in many studies, because they have the greatest impact on thermal comfort and they change significantly between different types of canyons, both with and without trees [16–18].

3.1. Theoretical Model Results

The results are presented in two stages, with the first stage explaining how the different canyons performed by comparing the PET values of each side of the different street canyons with various aspect ratios and orientations, and the second stage comparing the trees' impact on each side of the street canyons. The purpose of the first stage is to understand how changing the aspect ratio, orientation, and side of the canyon affect thermal comfort and changing PET values before adding trees, as well as to understand which canyons do not need trees and which require the addition of trees.

3.1.1. Stage One: Results for the 0% Tree Scenario

By comparing the PET values for each side of all the urban canyons, as shown in Figure 11 for the northern and eastern canyons and Figure 12 for the northwestern and northeastern canyons, it was found that the PET values varied between the different aspect ratios, orientations, and canyon sides. The worst PET values were measured on the (a) sides of the eastern streets, as shown in Figure 11c,d, where the PET values reached more than 50 C° for all aspect ratios over many hours during the daytime, in addition to reaching the highest PET value of 58 C° (R2a) at 15:00. On the (b) sides of the eastern canyons, the PET values were slightly lower than those on the (a) sides; however, with different aspect ratios, both sides were under extreme heat stress, and the effect of the different aspect ratios in reducing the PET values was very limited, by around 2–3 C°, and only in moderate to deep urban canyons on the (b) side (i.e., R4b, R6b, and R8b). For other orientations, the aspect ratio played an important role in reducing PET values by providing good urban shading, which is increased in moderate and deep canyons. As shown in Figure 11a,b, in the northern canyons, on the (a) street sides, an increase in the aspect ratio reduced the PET by approximately 5 C° in R1a for 4 h, in R3a and R5a for 7 h, in R7a for 6 h, and in R9a for 3 h compared with R11a, which is the shallowest urban canyon. On the (b) side, the impact of the aspect ratio was less than that on the (a) side, as the decrease in PET was lower, both in terms of the value and reduction in the number of hours. As shown in Figure 12a–d, the performance of the different aspect ratios on both sides of the canyon is very good, and the urban shading is very effective, especially for the northwest orientation's (a and b) sides, as the PET in all cases did not exceed 50 C° and decreased by 5 C° or more with an increase in the aspect ratio. Moreover, the number of hours with a reduced PET increased in R9a,b for 4 h, in R7a,b for 6 h, in R5a,b for 7 h, in R3a,b for 8 h, and in R1a,b for 3 h compared with R11a,b, which is the shallowest canyon. Similar PET enhancements occurred along the northeast orientation on both canyon sides with an increase in the aspect ratio, especially during the afternoon hours, when PET values decreased on average by 5 C° in R2a,b for 8 h, in R4a,b for 9 h, in R6a,b for 7 h, in R8a,b for 7 h, and in R10a,b for 4 h and 2 h compared with the shallowest canyon (R12a,b). In conclusion, the effect of the urban aspect ratio appears clearly in the northern, northeast, and northwest orientations, especially on side (a) for the northern streets and side (b) for the northeast and northwest street canyons.

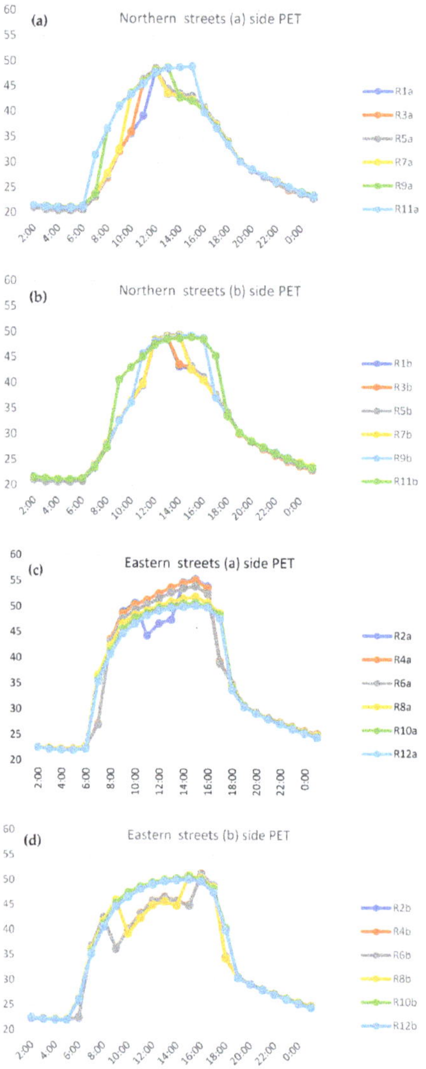

Figure 11. PET values for all aspect ratios of the northern and eastern canyons on both sides at Z = 1.5 m.

To better understand the differences in the PETs on each canyon side for the different orientations and aspect ratios, a detailed study of the main PET parameters (total mean radiant temperature (TMRT) and wind speed) [17] was applied to each side of the different street canyons.

As shown in Figure 13, the TMRT values and the charts' shapes are quite similar to those of the PET charts, which signifies that the PET reduction was mainly driven by the decrease in the TMRT [10]. The highest TMRT values with a limited TMRT reduction via a change in the aspect ratio were measured for the eastern canyons. The reduction on the (a) sides was limited in R2 to 10 C° for 3 h only, and it was slightly better on side (b) for R4, R6, and R8, with the same reduction range of 7 h as in R4 and R6, and for 5 h in R8.

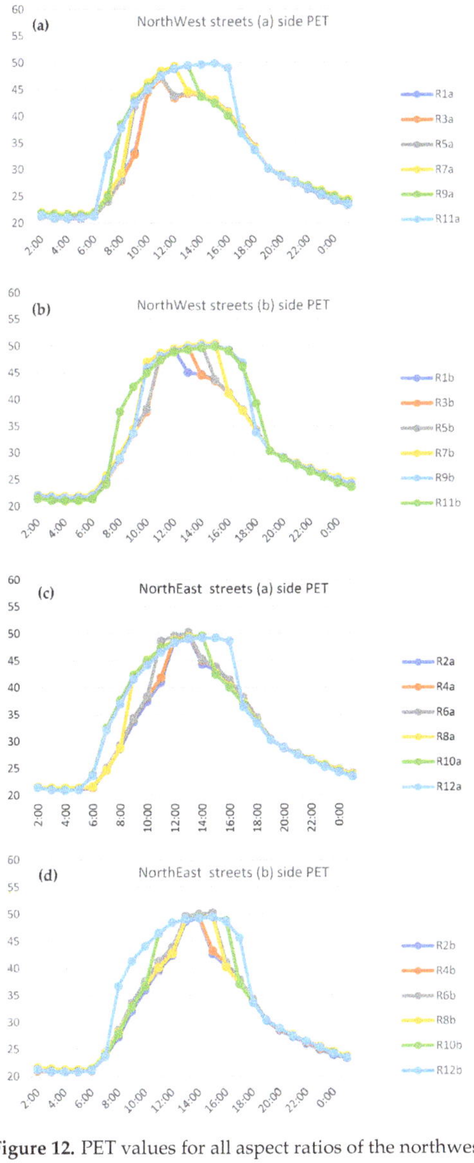

Figure 12. PET values for all aspect ratios of the northwest and northeast canyons on both sides at Z = 1.5 m.

The reduction in the TMRT values was significant in the northern canyons on side (b), northwest canyons on side (b), and northeast canyons on side (b) during the afternoon hours. The reduction exceeded 10 C° for many hours (more than 5 h) in the moderate and deep canyons, and the TMRT reduction decreased gradually as the canyons became shallower. In the northern canyons on side (a), northwest canyons on side (a), and northeast canyons on side (a), the main TMRT reduction took place during the morning to noon hours. The TMRT values decreased by more than 10 C° for many hours, up to 5 h for deep canyons, and the reduction and number of reduction hours gradually decreased with a decrease in the aspect ratio from deep to shallow.

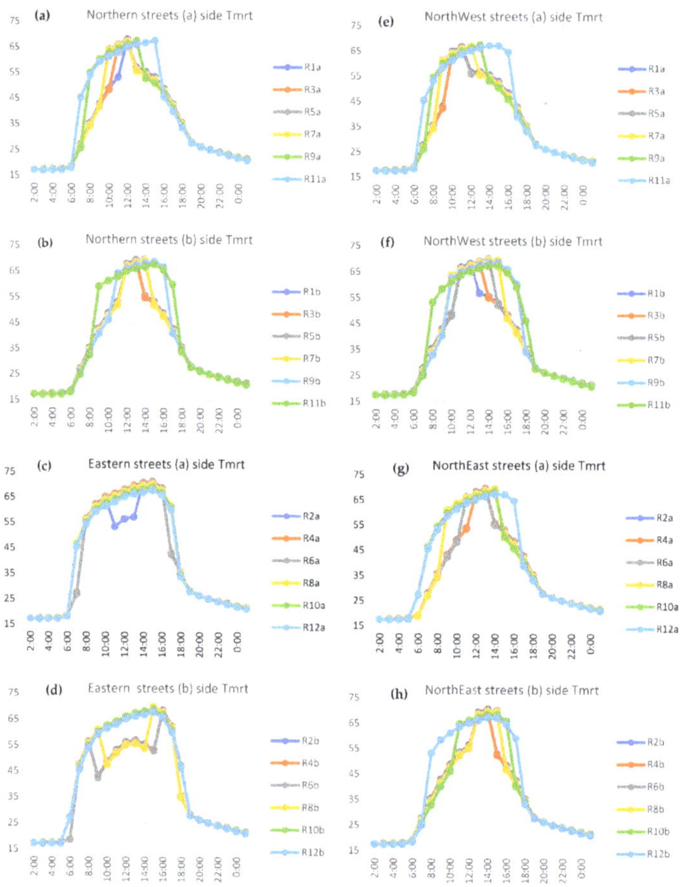

Figure 13. TMRT values for all aspect ratios of the northern, eastern, northeast, and northwest canyons on both sides at Z = 1.5 m.

Studying the wind speed values is crucial for gaining a better understanding of the performance of each canyon orientation, as it helps in the investigation and comprehension of the reasons behind the varying PET values and aids in further research on different orientations, aspect ratios, and canyon sides. As shown in Figure 14, the wind speed changed significantly among the different orientations and changed slightly between the different aspect ratios and canyon sides of the same orientation. A significant change in the wind speed primarily occurred between the northern and eastern streets; as shown in Figure 14a,b, the drop in the wind speed was significant, as the values decreased from a range of 3.5 to 2 m/s in the northern streets to a range of 0.6 to 0.2 m/s in the eastern streets for the different aspect ratios and street sides. This large drop clarifies an additional reason for the high PET values in the eastern roads in general. In the northwest and northeast roads, on side (a), the wind speed for all aspect ratios were within the range of 1–2 m/s, which implies that the wind speed for this orientation is within a middle range, which did not significantly differ with a change in any of their orientations (NW or NE). However, on the (b) side of the northeast and northwest orientations, the difference in the range of wind speed increased from 1 to 2.8 m/s, and the highest value was measured for the northeast orientations.

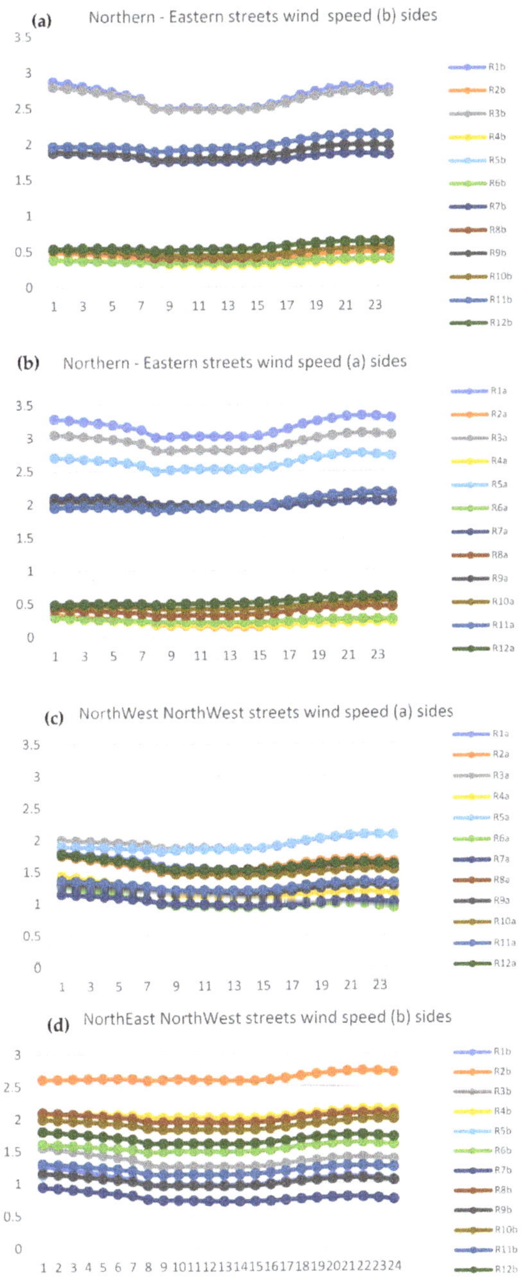

Figure 14. Wind speed values for all aspect ratios of all orientations on both sides at Z = 1.5 m.

By measuring the TMRT and wind speed for the different aspect ratios and orientations on both canyon sides, the reason behind the difference in the PET values is elaborated: the urban shading based on the sun's path and angle in relation to the aspect ratio is not significant along the eastern roads on both sides; however, it is very significant in the northern, northwest, and northeast on side (b) during the afternoon hours. It has a good impact on the northern, northeast, and northwest canyons on side (a) during the morning to noon hours. Also, the wind speed is very low or almost non-noticeable on the eastern roads. However, it is strong on northern roads and moderate in the northeast and northwest canyons.

Figure 15 shows the ΔPET between each side of each street canyon with different aspect ratios and orientations, and it clarifies the significant change that occurred for the eastern roads between side (a) and side (b), as the maximum PET difference reached 14 C° in the deep and moderate canyons. This significant difference within the same canyon is because of the effect of the aspect ratio and orientation, which reduces the PET on the (b) side much more than that of the (a) side for most of the daytime. The differences in the northern, northeast, and northwest canyons vary between the morning and afternoon hours; thus, the aspect ratio plays an important role that varies during the day because of the sun's path, which changes its angle during the daytime. The main enhancement in the morning hours was measured in the northeast roads, and the main enhancement in the afternoon hours was measured in the northern canyons. The northwest canyons showed a balanced enhancement between the morning and afternoon hours. Shallow canyons, such as R11 and R12, in all orientations showed a very minor change between sides. This signifies that the effect of the aspect ratio was not considered in the shallow canyons; however, in the moderate and deep canyons, the difference between the canyon sides was significant in the N, NE, and NW and varied based on the orientations.

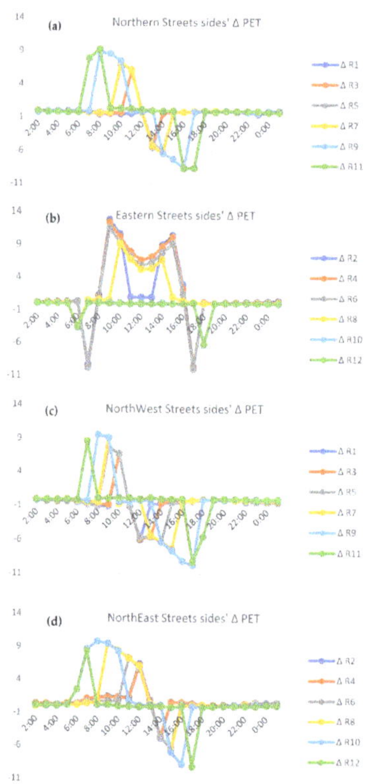

Figure 15. ΔPET for both sides of each canyon (a side–b side) for the different aspect ratios and orientations at Z = 1.5 m.

This concludes how the different sides of the aspect ratios and orientations perform without adding trees. This provides a clear understanding and some insight into the expected performance of trees when they are added to different sides with various aspect ratios and orientations.

3.1.2. Stage Two: Results of the Tree Scenarios

In this stage, the effect of adding different tree percentages was measured by comparing the PET values in each tree scenario for each side of the road for different aspect ratios and orientations. As clarified, the two tree densities (20% and 50%) were tested on different sides of various street canyons to understand the relationship between trees and different sides of various street canyons.

As shown in Figure 16, tree performance varied based on aspect ratio, orientation, and street side. Trees reduced PET values significantly in two cases. The first case involved all aspect ratios of the eastern canyons, and the second case involved shallow urban canyons in all other orientations. In Figure 16, there are two shapes of charts: M-like shape and A-like shape. The M-like shape represents a significant performance of the trees in charts (b, d, f, h, i, j, k, and l). These charts depict the PET reduction reaching a maximum, which remained for most of the daytime, except during the noon hours. This led to an M-like shape for PET reduction, which occurred for most of the time. In these canyons, the ΔPET reached up to 17 C° as the maximum and an average of 12–15 C°, and this average reduction lasted, on average, for 8–10 h, which is almost the entire daytime period and a significant PET reduction. However, in the northern orientation in the moderate and deep canyons, all of the charts showed an A-like shape, which represents a limited enhancement of the PET compared to other cases because the ΔPET reached 13 C° as the maximum for only 3 h in very few cases, and the average ΔPET reached between 4 and 7 C° for most of the daytime. The results in Figure 16 also show that increasing the tree percentage from 20% to 50% in shallow canyons is very promising for both sides of shallow canyons and on side (a) in moderate and deep eastern canyons. However, increasing the tree density in deep and moderate northern canyons did not lead to any significant enhancements.

In Figure 17, the northwest and northeast canyons showed equal tree performance, with a significant enhancement in the shallow canyons for both orientations on both sides, as shown in canyons R9, R10, R11, and R12. For the moderate and deep aspect ratios, the tree performance of the canyons in both orientations was almost the same and very similar to the performance of the trees in the moderate and deep canyons of the northern orientation. In addition, the maximum ΔPET reduction reached an average of 12–14 C°, and the average reduction reached 4–7 C°. The performance of the trees on both sides of the canyons was almost the same, with a slight enhancement on side (b). Increasing the density of the trees from 20% to 50% played an important role in reducing the PET values in both orientations in the shallow canyons, as the ΔPET between the 20% and 50% tree scenarios reached 10 C° in some hours in the shallow canyons. In the moderate and deep canyons, increasing the tree percentage led to minor PET reductions, with an average of 2–3 C° only between the 20% and 50% tree scenarios.

Figure 18 shows the ΔPET for the 20% and 50% tree scenarios compared to the 0% tree scenario for all aspect ratios and orientations at 12:00 and 15:00. At noon (12:00), the 20% tree performance in all orientations was almost the same; also, the 50% tree performance in all orientations was almost the same. This is due to the sun's location at this hour, which is almost perpendicular to the urban canyons, and the role of the aspect ratio almost vanishes. The 50% tree enhancement of the PET values is significant compared with the enhancement produced by the 20% tree scenario at that hour due to the provision of more shading to the urban canyons, and the only difference was in the deepest canyon (H:W = 3:1) because a slight appearance of urban shading occurred for all orientations, reducing the significance of the high-density tree scenario's enhancement slightly.

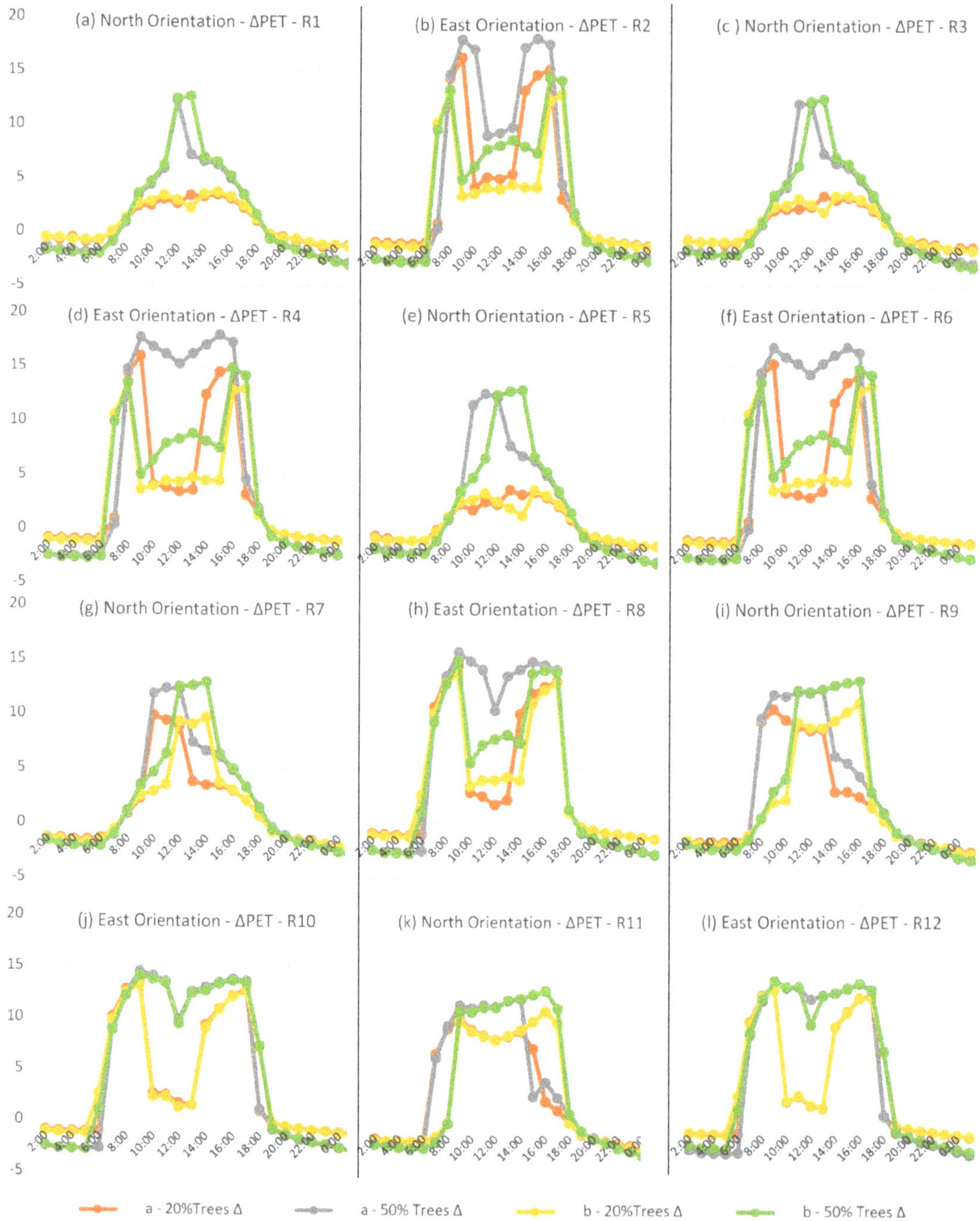

Figure 16. ΔPET with the 20% and 50% tree scenarios in comparison with the 0% tree scenario for northern and eastern canyons on both sides (a and b) at Z = 1.5 m.

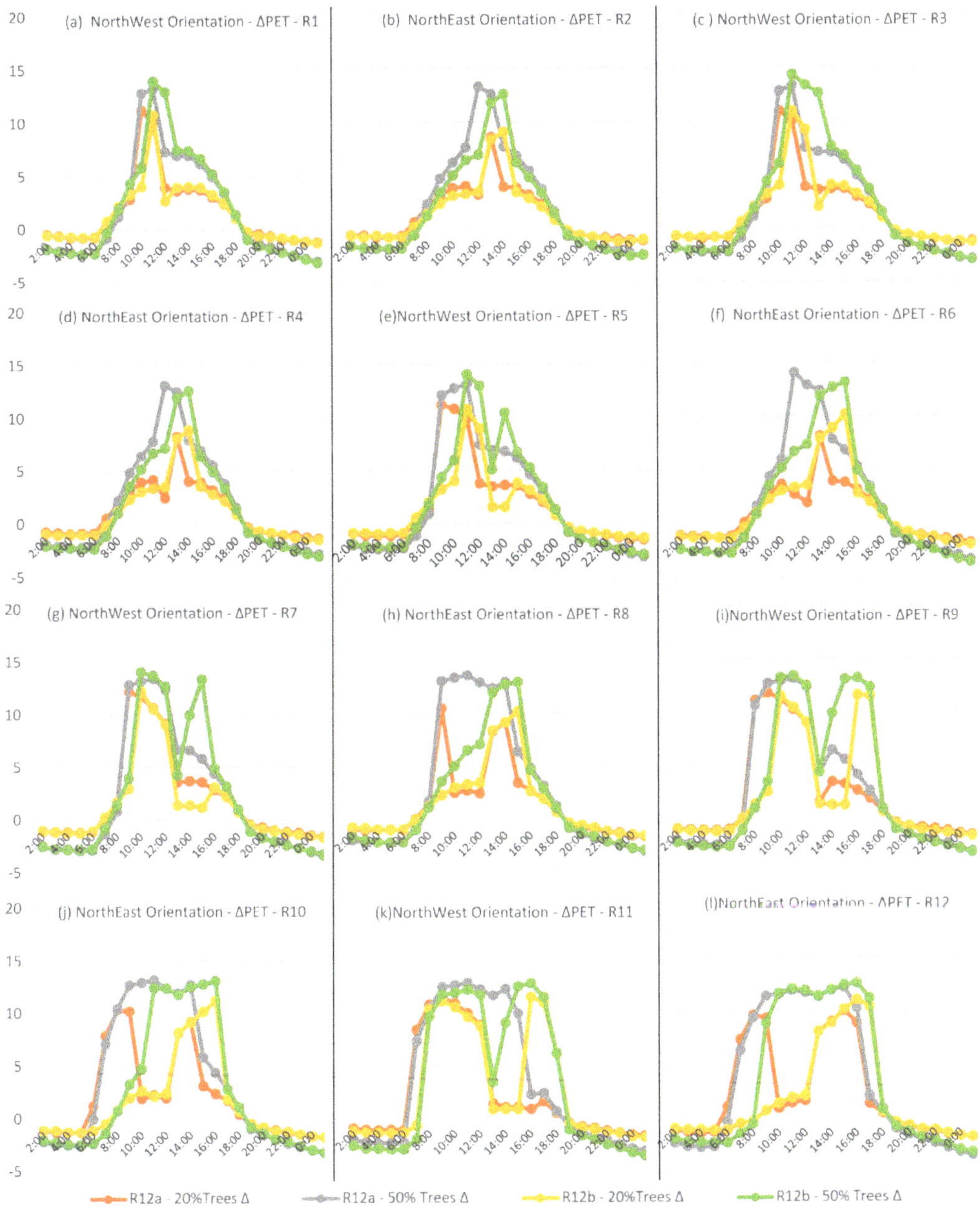

Figure 17. ΔPET with the 20% and 50% tree scenarios in comparison with the 0% tree scenario for the northeast and northwest canyons on both sides (a and b) at Z = 1.5 m.

Figure 18. ΔPET at Z = 1.5 m for the 20% and 50% tree scenarios compared to the 0% tree scenario for all aspect ratios and orientations at 12:00 (**a**,**c**,**e**,**g**) and 15:00 (**b**,**d**,**f**,**h**).

In the same figure, in the afternoon (15:00), when tree shading is mixed with urban canyon shading, the actual importance of the trees appears in some canyons and disappears in other canyons. In the shallow canyons, the dark green category, which represents a reduction of 12 to 14 C°, appears clearly in most of the shallow canyons for the 50% tree scenario, and this changed gradually from dark green to orange, which represents a reduction of 6 to 8 C°, in the moderate canyons. Then, it gradually changed to a red color, which represents a reduction of 2 to 4 C°, in the deep canyons, especially in the north, northeast, and northwest orientations, with the eastern orientation showing a greater reduction. It was better than the other orientations, even in the deep canyons. Also, the 20% tree scenario showed the same gradual enhancements with lower values, as it ranged between light green (10 to 12 C°) and red (2 to 4 C°), and the effect is presented on the map in a dotted form, covering an area that is not fully continuous, as shown for the 50% scenario.

Figures 19 and 20 show a detailed comparison of both sides of the various street canyons with different orientations and aspect ratios that were applied in two ways. The first method compared the average ΔPET during the peak hours in the daytime (from 11:00 to 16:00), as presented in Figure 19. The second method compared how many hours the PET was reduced by 8 C° or more. This reduction helps change the level of thermal comfort by almost two thermal zones and should be considered a significant enhancement [30]. Both comparisons (Figures 19 and 20) show the significant impact of applying high tree densities on shallow streets and in eastern canyons on both sides. In addition to explaining the gradual performance of the trees with various aspect ratios (moderate to deep), in the other orientations, the trees' performance decreased with an increase in the aspect ratio (H/W). It also demonstrates that for deep and moderate canyons with northern, northeast, and northwest orientations, increasing the tree density from 20% to 50% does not lead to significant PET reductions and remains almost the same in some cases.

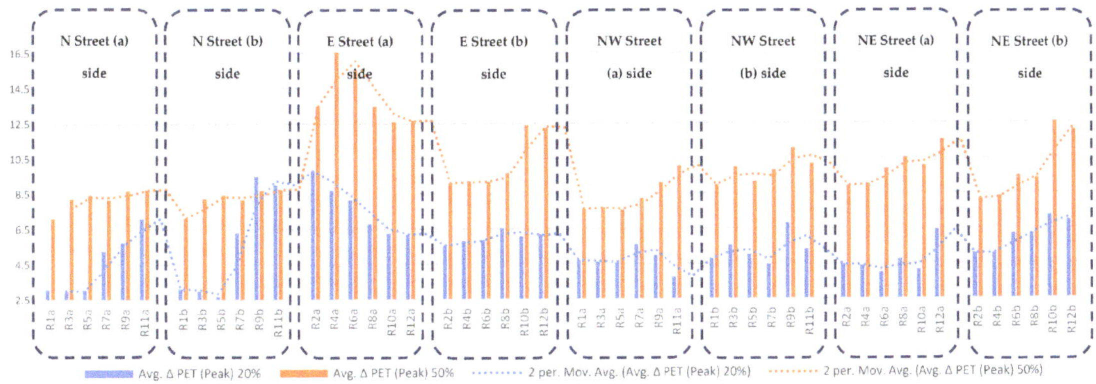

Figure 19. Comparing the average ΔPET during the peak daytime period (from 11:00 to 16:00) on both sides for all aspect ratios and orientations at Z = 1.5 m.

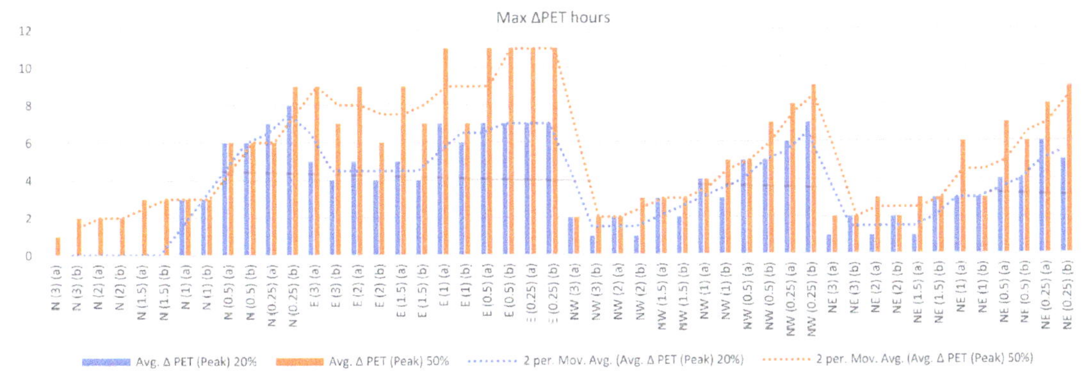

Figure 20. Comparing the number of hours that the PET was reduced by 8 C° or more on both sides for all aspect ratios and orientations at Z = 1.5 m.

For a better understanding of the relationship between the aspect ratio and tree scenario performance for different orientations on both canyon sides, an additional statistical analysis was performed. Different aspect ratios were analyzed against the reduced hours with different tree densities for each street orientation on each side using SPSS analysis to calculate the correlation, mean, standard deviation, and regression. Before calculating the correlation between both mentioned variables, as shown in Table 5, all input data for the aspect ratio and avg. ΔPET (peak hours) for all orientations on both sides were normally distributed based on the Kolmogorov–Smirnov (KS) test, except the ΔPET values for the 50% tree scenario for a north orientation on the "b" side and the 50% tree scenario for an east orientation on the "b" side therefore Pearson correlation was applied for all cases, except for the mentioned cases in which Spearman correlation was applied. Table 5 proves that there was a very strong negative correlation between an increase in the aspect ratio and the ΔPET for peak hours with different tree percentages for all orientations on both sides. However, the correlation value was not the same for all orientations and sides. For example, the correlation for eastern roads was lower than that for the other orientations at −0.849, −0.849, −0.878, and −0.802, which reached −0.971 to −0.900 in most cases. Also, the NE on side (b) and NW on side (a) showed a less significant correlation, and this result is very similar to that of the eastern orientation, which is in line with the measured PET reduction. This means that the trees' performance was less dependent on the urban canyon shading because the urban shading is very minor in this orientation, and the other elements, such as the wind speed, do not help; therefore, the trees' presence is very important in these canyons. In addition, the mean values presented in Table 5 represent the huge difference between the eastern cases and the rest of the orientations; this is also in line with the PET reduction and correlation results. The correlation results could show more significant changes if the results of the shallow aspect ratios (0.5 and 0.25) were excluded because the results of both ratios are the same for the eastern orientation, whereby the impact of the aspect ratio disappears. Figure 21 shows linear scatter plots and regressions of the aspect ratio and ΔPET of the number of 8 C° or greater reduction hours for both tree scenarios (20% and 50%). This clearly shows that there was a negative linear relationship between the aspect ratio and the ΔPET for both tree densities; however, this varied based on the street's aspect ratio, orientation, and canyon side, in addition to the percentage of trees changing the overall values in some cases. The 50% tree scenario showed a stronger relationship than the 20% tree scenario in all orientations. The northern, northwest, and northwest showed higher R^2 values between the ΔPET and aspect ratio, with values $R^2 \geq 0.8$ for the 50% tree scenario; however, in the eastern canyons, on both sides, the values were much lower, reaching $R^2 \leq 0.72$ for side (a) and $R^2 \leq 0.77$ for side (b).

Table 5. Statistical analysis of the correlation between the aspect ratio and the 20% and 50% tree scenarios' avg. ΔPET (peak hours) for all orientations on both sides.

SPSS Statistics Analysis	N(a) Avg. ΔPET (Peak) 20% Trees	N(a) Avg. ΔPET (Peak) 50% Trees	N(b) Avg. ΔPET (Peak) 20% Trees	N(b) Avg. ΔPET (Peak) 50% Trees	E(a) Avg. ΔPET (Peak) 20% Trees	E(a) Avg. ΔPET (Peak) 50% Trees	E(b) Avg. ΔPET (Peak) 20% Trees	E(b) Avg. ΔPET (Peak) 50% Trees	NW(a) Avg. ΔPET (Peak) 20% Trees	NW(a) Avg. ΔPET (Peak) 50% Trees	NW(b) Avg. ΔPET (Peak) 20% Trees	NW(b) Avg. ΔPET (Peak) 50% Trees	NE(a) Avg. ΔPET (Peak) 20% Trees	NE(a) Avg. ΔPET (Peak) 50% Trees	NE(b) Avg. ΔPET (Peak) 20% Trees	NE(b) Avg. ΔPET (Peak) 50% Trees
Mean	2.667	3.500	2.833	4.167	6.000	10.000	5.333	8.167	3.667	4.000	3.167	4.833	2.667	4.833	3.167	4.000
Std. Deviation	3.204	2.074	3.488	2.787	1.095	1.095	1.506	2.229	1.633	2.280	2.401	2.714	2.066	2.483	1.169	2.898
(K-S) Test, c	0.105	0.200 e	0.121	0.041	0.056	0.056	0.069	0.012	0.200 e	0.200 e	0.200 e	0.200 e	0.125	0.197	0.200 e	0.094
Correlation AR	−0.871 *	−0.932 **	−0.863 *	−0.971 **	−0.849 *	−0.849 *	−0.878 *	−0.802	−0.929 **	−0.858 *	−0.887 *	−0.911 *	−0.853 *	−0.936 **	−0.900 *	−0.794
Sig. (2-tailed)	0.024	0.007	0.027	0.001	0.033	0.033	0.022	0.055	0.007	0.029	0.019	0.012	0.031	0.006	0.014	0.059

* Correlation is significant at the 0.05 level (2-tailed).
** Correlation is significant at the 0.01 level (2-tailed).
(highlighted) Kolmogorov–Smirnov test: data are not normally distributed.
Spearman correlation.
Pearson correlation.
c Lilliefors significance correction.
e This is the lower bound of the true significance.

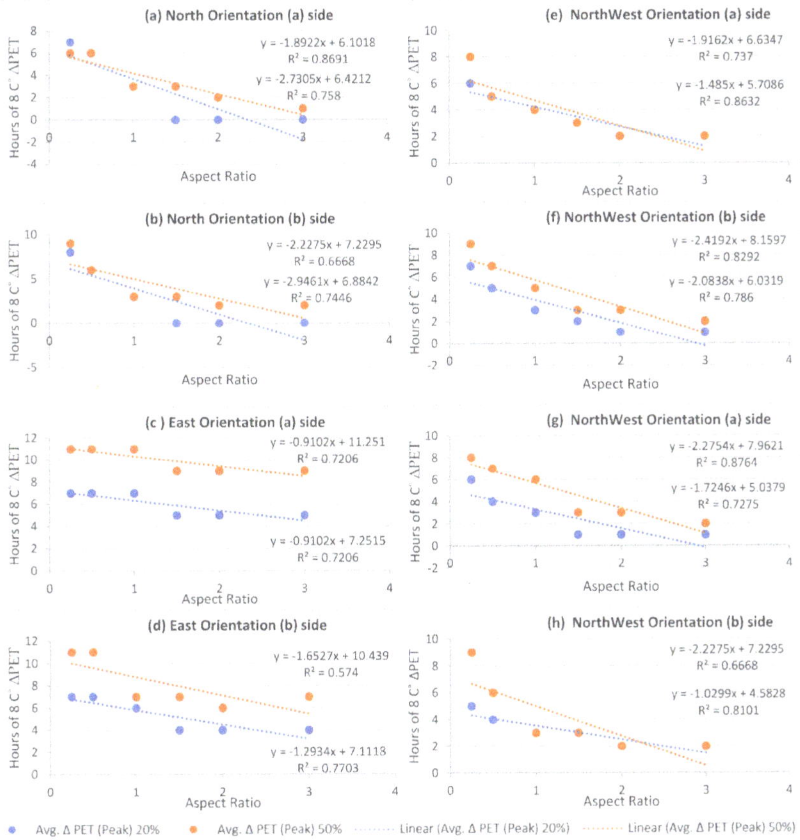

Figure 21. The correlation between the aspect ratio and ΔPET of each orientation for both sides (a and b) at Z = 1.5 m.

3.1.3. Conclusion of the Trees' Performance in the Theoretical Model

On the basis of the analysis of all the results, it has been proven that the performance of the trees varied depending on the aspect ratio, orientation, and side of the urban canyon. The eastern roads are the most in need of a high density of trees. This is because the canyon experiences heat stress throughout the day, and the urban shading on both sides is insufficient to improve the microclimate conditions for the varying aspect ratios. Additionally, the wind speed is extremely low, exacerbating the already poor thermal comfort along this orientation. Northern streets are more shaded by buildings; therefore, increasing the aspect ratio for these orientations is very efficient. Adding trees to moderate to deep canyons will not help enhance the thermal comfort significantly. In addition, the good wind speed helps to enhance thermal comfort when combined with the shade provided by buildings. The northeast and northwest orientations are very similar to the northern canyon, with good canyon shading and moderate wind speed, and the role of trees is not as significant as in the eastern orientation.

3.2. Case Study Results

In order to validate the results of the theoretical study, all the findings were compared to the outcomes by implementing the same tree scenarios in the case study area located in downtown Cairo. As shown in Figure 22, after applying different tree densities to different canyon aspect ratios and orientations on both sides, the results are quite similar to the theoretical model's results for the same tree scenarios. The trees' impact is very significant in shallow canyons and eastern canyons, as charts (f and i) in Figure 22 for the eastern canyons show significant PET reductions on both sides, in which the PET is reduced by approximately 18 C° as a maximum and reaching an average of a 13 to 14 C° reduction for most of the daytime hours. Significant results were mainly found for the 50% tree scenario with an average of 4 to 5 C° more than in the 20% tree scenario. PET reduction was limited in the other orientations (northern, northeast, and northwest) and hardly exceeded 10 C° PET reductions in very few hours in some canyons. The average PET reduction reached from 4 to 7 C°, which is very similar to the theoretical model's results. A slight change was measured in S1 (northeast) and S12 (northwest) because the PET reduction was slightly better than the theoretical model's results but still much less than the enhancement of the eastern canyons.

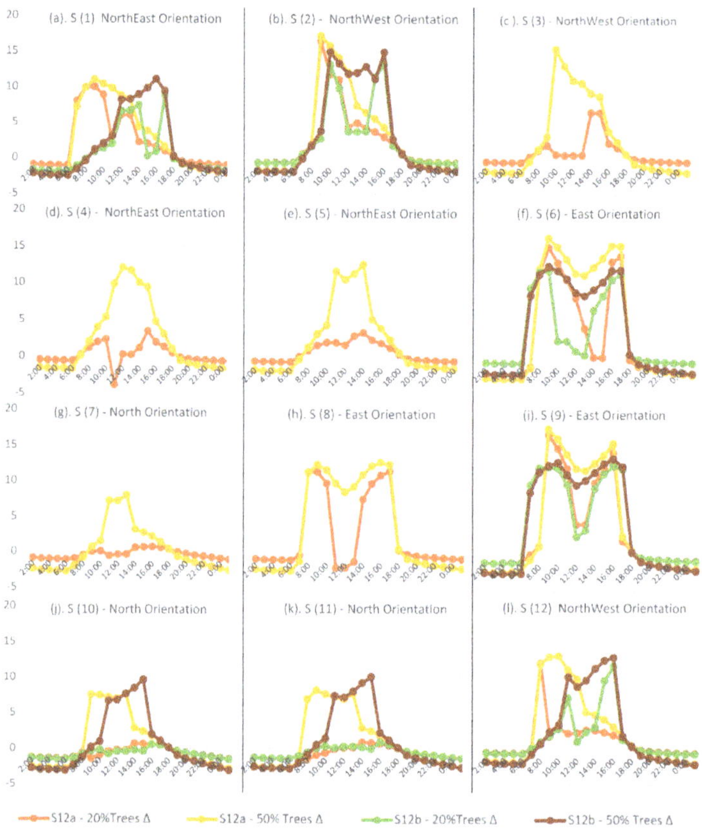

Figure 22. ΔPET with the 20% and 50% tree scenarios compared with the 0% tree scenario for all canyons on both sides (a and b) at Z = 1.5 m.

Figure 23 demonstrates the ΔPET for the same canyon orientation and aspect ratio (S6 and S6′) on both sides (a and b), as shown in Table 2. Although the street widths and buildings heights are irregular, both canyons with the same aspect ratio showed almost the same results, with slight differences occurring most probably because of the irregular building heights in each canyon. The ΔPET difference between both canyons reached 1 to 3 C° in only a very few hours in both canyons on both sides, and this clarifies that as long as the aspect ratio is the same between different canyons, the performance will be very similar regardless of any slight change in the height of the buildings and width of the street, as long as the canyon aspect ratio is still the same.

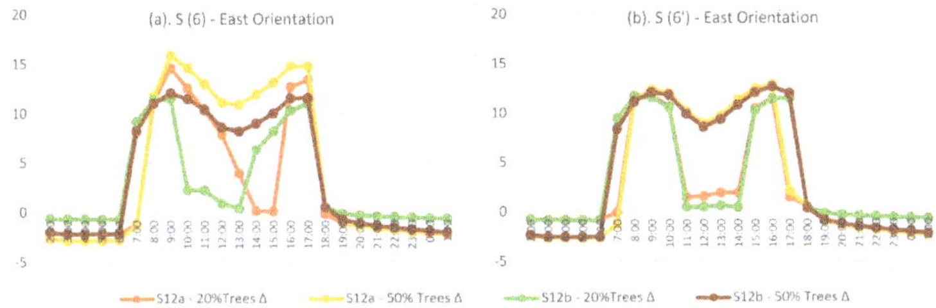

Figure 23. ΔPET for the 20% and 50% tree scenarios in comparison with the 0% tree scenario for S6, and S6′ on both sides (a and b) at Z = 1.5 m.

Figure 24 illustrates the impact of the trees on the entire study area for the 20% and 50% tree scenarios at 12:00 and 15:00. This shows that increasing the tree density is crucial for the eastern canyons, and the blue color (12 to 14 C° ΔPET) represents the four eastern canyons (S6, S6′, S8, and S9). The blue color in Figure 24d for the 50% tree scenario at 15:00in S1 on the (b) side, shows that the trees' impact was significant because this canyon is very shallow. The performance of the 20% and 50% tree scenarios varied with the different aspect ratios and orientations at 15:00, when the tree performance mixed with the various effects of urban shading based on the aspect ratio and orientation. The maps also show the equal impact of the trees at 12:00 for both tree scenarios when the sun is almost perpendicular, and the effect of the different aspect ratios and orientations disappeared. These maps/results are quite similar to the theoretical study's maps/results, and slight differences in a very few cases occurred due to irregular aspect ratios, as the buildings' heights in the existing study area are not exactly the same as those in the theoretical model, which is 100% regular in terms of the building heights and street widths.

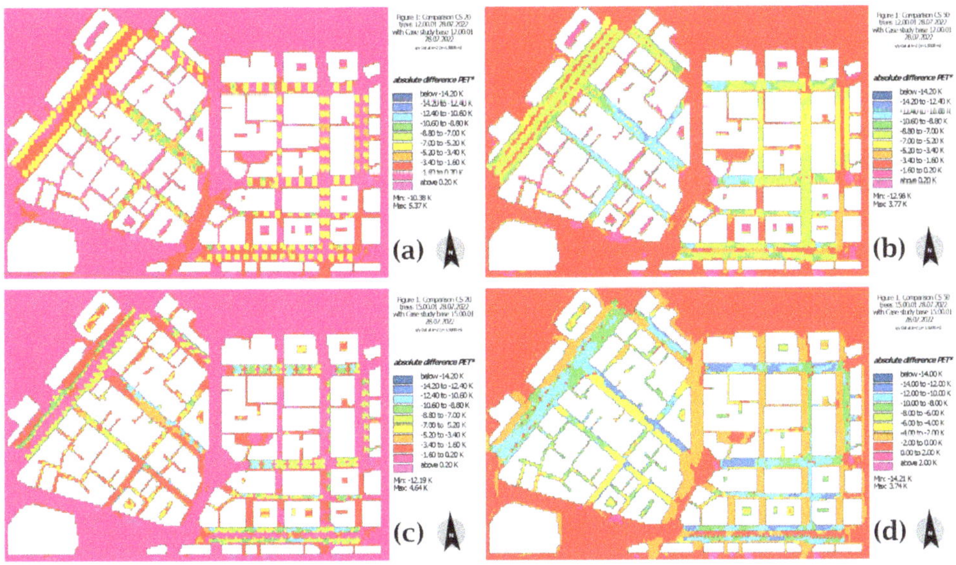

Figure 24. ΔPET at Z = 1.5 m for the 20% and 50% tree scenarios compared to the 0% tree scenario for all aspect ratios and orientations at 12:00 (**a**,**b**) and at 15:00 (**c**,**d**).

By reviewing Figures 25 and 26, which show a comparison of the average ΔPET for the peak time (from 11:00 to 16:00) on both sides for the various aspect ratios and orientations in Figure 25 and a detailed comparison between the total hours of PET reduction of 7 h or more on both sides for the various aspect ratios and orientations in Figure 25, both studies prove that the impact of the trees varied significantly for all shallow canyons in addition to the eastern canyons on both sides. The overall linear shape of each chart in Figures 25 and 26 is very similar to the same charts of the theoretical model analysis, as shown in Figures 19 and 20. This validates the results and findings of the theoretical model; however, in the case study, there are slight differences compared to the theoretical model. This is a normal occurrence that happens because of the irregularity of the existing study area in terms of the aspect ratios and various heights of buildings inside each urban canyon; the existing study area cannot have the same regularity as the theoretical model. Despite this, the overall results match the theoretical model and the existing study area.

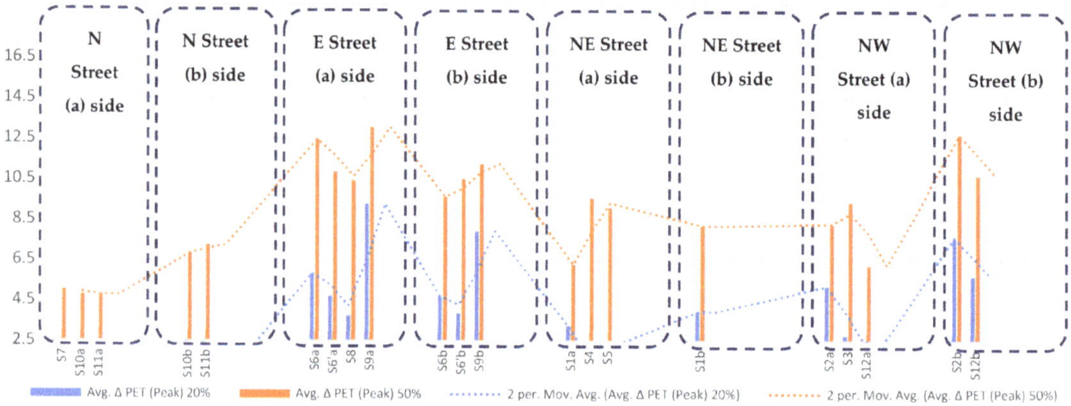

Figure 25. Comparing the average ΔPET for the peak daytime hours (from 11:00 to 16:00) on both sides for all aspect ratios and orientations at Z = 1.5 m.

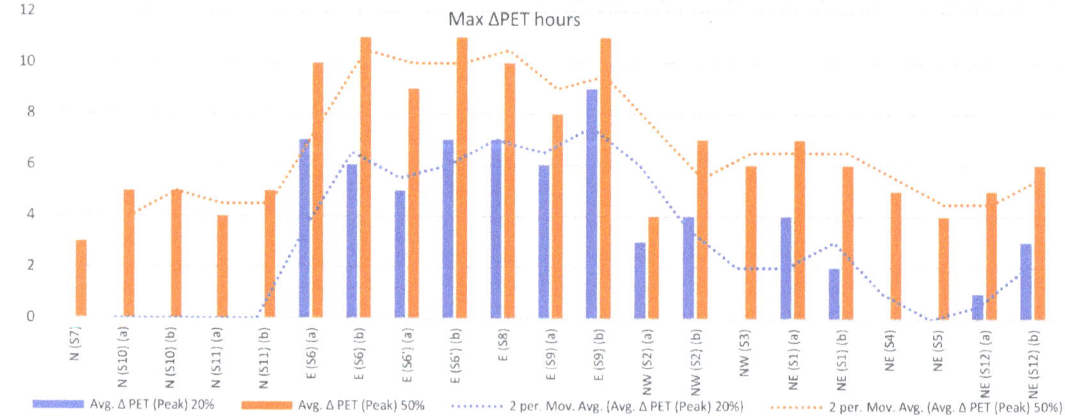

Figure 26. Comparing the number of hours that the PET was reduced by 7 C° or more on both sides for all aspect ratios and orientations at Z = 1.5 m.

4. Discussion

After reviewing the theoretical and case study results, it is obvious that trees are important and add value to different urban canyons. This importance varies depending on the orientation, aspect ratio, and side of the canyon.

The eastern canyon requires a high density of trees, and this is applicable to both sides, particularly the northern side. Moreover, this finding is consistent with those of previous studies [8,14,20]. The northern, northeast, and northwest canyons showed better performance than the eastern canyons. The addition of trees to the Northern canyons improved it slightly, but it was not as significant as the effect of adding trees to the eastern canyons. This aligns with previous findings from other studies [20,30]. For some canyons in the northern, northeastern, and northwestern areas, having tall buildings (moderate to deep canyons) helps keep the area cooler. This is because the buildings block the hot sun from shining directly on the streets. These canyons also have strong winds, which help make them feel cooler. Adding trees to these canyons (moderate and deep canyons in these orientations) might not make much of a difference in how comfortable it feels, and it might even block some of the wind. However, it can be helpful to add trees to one side of the canyons, which is side (b). The trees on side (b) provide shade in the afternoon and make the canyons feel cooler without blocking the wind. Shallow canyons in all orientations and all aspect ratios on both sides of the canyons do not provide enough shading at all. In addition, placing high densities of trees in these canyons is mandatory to enhance thermal comfort conditions, as the canyon effect completely disappears, especially on the northern side of the eastern canyon (side a). Applying a low density of trees on one side of the road (side b) could be useful in deep canyons with northern, northwestern, and northeastern orientations. On the other hand, high density is necessary on both sides of the eastern and shallow canyons. When comparing the results of the theoretical study and the existing case study, it becomes apparent that the unity and regularity of urban canyons will yield slightly different outcomes compared to the irregular canyons found in existing urban areas. However, the main results are similar but not exactly the same due to changes in the urban conditions at the measurement point. Any alteration to the aspect ratio, orientation, or other urban elements will have an impact, albeit not a significant one. It is not feasible to obtain exact numbers from a uniform and regular theoretical model. Slight variations in the results were observed when altering the street widths and building heights while maintaining the same aspect ratio and orientation, which aligns with [42].

5. Conclusions

This study applied an ENVI-met simulation to explore the effect of aspect ratios, orientations, and trees on the microclimate of different sides of urban canyons in Cairo city. Firstly, a total of 144 theoretical cases were simulated to reveal the microclimate's relationship with urban morphologies and tree densities. Then, a real neighborhood in Cairo was simulated to verify the results obtained from the theoretical cases. The results showed a significant tree performance; this performance became more significant based on the aspect ratio (PET reduction reached 12 C° using a high density of trees in shallow canyons), the canyon orientation (PET reduction reached 14 C° in the eastern orientation compared to the northern orientation of the same deep canyon), and the side of the canyon (PET reduction at the northern side of the eastern canyon reached 6 to 7 C° more than the other side of the same canyon by adding a high density of trees). On the basis of the results of this study, the addition of trees to urban canyons should be based on urban morphology characteristics, such as aspect ratio and orientation. Additionally, careful consideration should be given to the side on which trees are planted, as any alteration to these three elements of urban geometry necessitates a wholly different approach. This guidance is of paramount importance, particularly for cities grappling with significant challenges like heat stress and water scarcity, as is the case in Cairo. The findings of this study are broad and cover the majority of urban cases in Cairo city. They should be considered by stakeholders and decision makers in Cairo's authorities and municipalities, as this study could be part of design guidelines and city management while developing new communities or upgrading existing areas because the study provides a detailed approach on how applying trees can be optimized to enhance microclimate conditions for pedestrians. The research methodology and approach could be also applied to different urban cases in Cairo city which are not covered in this research, especially cases in which the urban canyon is not fully defined, to reach the same target of optimizing the use of trees.

Author Contributions: Conceptualization, A.Y.A.; Methodology, A.Y.A.; Writing—original draft, A.Y.A.; Writing—review & editing, D.G.; Visualization, A.Y.A.; Supervision, D.G. All authors have read and agreed to the published version of the manuscript.

Funding: The research is not receiving any fund only the APC is funded by the TU Dortmund University.

Data Availability Statement: The data that support the findings of this study (All Envi-met files) are available upon request.

Conflicts of Interest: The authors declare no conflict of interest.

Abbreviations

UHI	Urban Heat Island
PET	Physiological Equivalent Temperature
TMRT	Total Mean Radiant Temperature
WS	Wind Speed
RH%	Relative Humidity
AR	Aspect Ratio
RMSE	Root Mean Square Error
d	Index of Agreement

References

1. Yin, C.; Yuan, M.; Lu, Y.; Huang, Y.; Liu, Y. Effects of urban form on the urban heat island effect based on spatial regression model. *Sci. Total Environ.* **2018**, *634*, 696–704. [CrossRef]
2. Theeuwes, N.E.; Steeneveld, G.J.; Ronda, R.J.; Heusinkveld, B.G.; van Hove, L.W.A.; Holtslag, A.A.M. Seasonal dependence of the urban heat island on the street canyon aspect ratio. *Q. J. R. Meteorol. Soc.* **2014**, *140*, 2197–2210. [CrossRef]
3. Dimoudi, A.; Kantzioura, A.; Zoras, S.; Pallas, C.; Kosmopoulos, P. Investigation of urban microclimate parameters in an urban center. *Energy Build.* **2013**, *64*, 1–9. [CrossRef]
4. Taheri Shahraiyni, H.; Sodoudi, S.; El-Zafarany, A.; Abou El Seoud, T.; Ashraf, H.; Krone, K. A Comprehensive Statistical Study on Daytime Surface Urban Heat Island during Summer in Urban Areas, Case Study: Cairo and Its New Towns. *Remote Sens.* **2016**, *8*, 643. [CrossRef]
5. Abutaleb, K.; Ngie, A.; Darwish, A.; Ahmed, M.; Arafat, S.; Ahmed, F. Assessment of urban heat island using remotely sensed imagery over Greater Cairo, Egypt. *Adv. Remote Sens.* **2015**, *04*, 35–47. [CrossRef]
6. Shishegar, N. Street Design and Urban Microclimate: Analyzing the Effects of Street Geometry and Orientation on Airflow and Solar Access in Urban Canyons. *J. Clean Energy Technol.* **2013**, *1*, 52–56. [CrossRef]
7. Kotharkar, R.; Bagade, A.; Ramesh, A. Assessing urban drivers of canopy layer urban heat island: A numerical modeling approach. *Landsc. Urban Plan.* **2019**, *190*, 103586. [CrossRef]
8. Rodríguez-Algeciras, J.; Tablada, A.; Chaos-Yeras, M.; De la Paz, G.; Matzarakis, A. Influence of aspect ratio and orientation on large courtyard thermal conditions in the historical centre of Camagüey-Cuba. *Renew. Energy* **2018**, *125*, 840–856. [CrossRef]
9. Morakinyo, T.E.; Kong, L.; Lau, K.K.-L.; Yuan, C.; Ng, E. A study on the impact of shadow-cast and tree species on in-canyon and neighborhood's thermal comfort. *Build. Environ.* **2017**, *115*, 1–17. [CrossRef]
10. Ketterer, C.; Matzarakis, A. Human-biometeorological assessment of heat stress reduction by replanning measures in Stuttgart, Germany. *Landsc. Urban Plan.* **2014**, *122*, 78–88. [CrossRef]
11. Qaid, A.; Ossen, D.R. Effect of asymmetrical street aspect ratios on microclimates in hot, humid regions. *Int. J. Biometeorol.* **2015**, *59*, 657–677. [CrossRef]
12. Yola, L.; Siong, H.C.; Djaja, K. Climatically responsive urban configuration in residential area: Research gaps. In *The 4th International Tropical Renewable Energy Conference (i-TREC 2019)*; AIP Publishing LLC: Bali, Indonesia, 2020.
13. Wang, Y.; Berardi, U.; Akbari, H. Comparing the effects of urban heat island mitigation strategies for Toronto, Canada. *Energy Build.* **2016**, *114*, 2–19. [CrossRef]
14. Andreou, E. Thermal comfort in outdoor spaces and urban canyon microclimate. *Renew. Energy* **2013**, *55*, 182–188. [CrossRef]

15. Matzarakis, A.; Amelung, B. Physiological equivalent temperature as indicator for impacts of climate change on thermal comfort of humans. In *Seasonal Forecasts, Climatic Change and Human Health: Health and Climate*; Springer: Dordrecht, The Netherlands, 2008; pp. 161–172.
16. Morakinyo, T.E.; Lam, Y.F. Simulation study on the impact of tree-configuration, planting pattern and wind condition on street-canyon's micro-climate and thermal comfort. *Build. Environ.* **2016**, *103*, 262–275. [CrossRef]
17. Abdollahzadeh, N.; Biloria, N. Outdoor thermal comfort: Analyzing the impact of urban configurations on the thermal performance of street canyons in the humid subtropical climate of Sydney. *Front. Archit. Res.* **2021**, *10*, 394–409. [CrossRef]
18. De, B.; Mukherjee, M. Optimisation of canyon orientation and aspect ratio in warm-humid climate: Case of Rajarhat Newtown, India. *Urban Clim.* **2018**, *24*, 887–920. [CrossRef]
19. Memon, R.A.; Leung, D.Y.C.; Liu, C.-H. Effects of building aspect ratio and wind speed on air temperatures in urban-like street canyons. *Build. Environ.* **2010**, *45*, 176–188. [CrossRef]
20. Jamei, E.; Ossen, D.R.; Seyedmahmoudian, M.; Sandanayake, M.; Stojcevski, A.; Horan, B. Urban design parameters for heat mitigation in tropics. *Renew. Sustain. Energy Rev.* **2020**, *134*, 110362. [CrossRef]
21. Kolokotsa, D.; Lilli, K.; Gobakis, K.; Mavrigiannaki, A.; Haddad, S.; Garshasbi, S.; Mohajer, H.R.H.; Paolini, R.; Vasilakopoulou, K.; Bartesaghi, C.; et al. Analyzing the Impact of Urban Planning and Building Typologies in Urban Heat Island Mitigation. *Buildings* **2022**, *12*, 537. [CrossRef]
22. Balany, F.; Ng, A.W.; Muttil, N.; Muthukumaran, S.; Wong, M.S. Green Infrastructure as an Urban Heat Island Mitigation Strategy—A Review. *Water* **2020**, *12*, 3577. [CrossRef]
23. Takebayashi, H.; Moriyama, M. Relationships between the properties of an urban street canyon and its radiant environment: Introduction of appropriate urban heat island mitigation technologies. *Sol. Energy* **2012**, *86*, 2255–2262. [CrossRef]
24. Lan, H.; Lau, K.K.-L.; Shi, Y.; Ren, C. Improved urban heat island mitigation using bioclimatic redevelopment along an urban waterfront at Victoria Dockside, Hong Kong. *Sustain. Cities Soc.* **2021**, *74*, 103172. [CrossRef]
25. Hamdi, R.; Schayes, G. Sensitivity study of the urban heat island intensity to urban characteristics. *Int. J. Climatol. A J. R. Meteorol. Soc.* **2008**, *28*, 973–982. [CrossRef]
26. De Lieto Vollaro, A.; De Simone, G.; Romagnoli, R.; Vallati, A.; Botillo, S. Numerical Study of Urban Canyon Microclimate Related to Geometrical Parameters. *Sustainability* **2014**, *6*, 7894–7905. [CrossRef]
27. Pioppi, B.; Pigliautile, I.; Pisello, A.L. Human-centric microclimate analysis of Urban Heat Island: Wearable sensing and data-driven techniques for identifying mitigation strategies in New York City. *Urban Clim.* **2020**, *34*, 100716. [CrossRef]
28. Andreou, E.; Axarli, K. Investigation of urban canyon microclimate in traditional and contemporary environment. Experimental investigation and parametric analysis. *Renew. Energy* **2012**, *43*, 354–363. [CrossRef]
29. Aboelata, A. Vegetation in different street orientations of aspect ratio (H/W 1: 1) to mitigate UHI and reduce buildings' energy in arid climate. *Build. Environ.* **2020**, *172*, 106712. [CrossRef]
30. Lobaccaro, G.; Acero, J.A.; Sanchez Martinez, G.; Padro, A.; Laburu, T.; Fernandez, G. Effects of Orientations, Aspect Ratios, Pavement Materials and Vegetation Elements on Thermal Stress inside Typical Urban Canyons. *Int. J. Environ. Res. Public Health* **2019**, *16*, 3574. [CrossRef]
31. Vuckovic, M.; Kiesel, K.; Mahdavi, A. Trees and the microclimate of the urban canyon: A case study. In Proceedings of the 2nd ICAUD International Conference in Architecture and Urban Design, Tirana, Albania, 8–10 May 2014.
32. Mohamed, E. Analysis of urban growth at Cairo, Egypt using remote sensing and GIS. *Nat. Sci.* **2012**, *4*, 355–361.
33. Hassan, A.A.M. Dynamic expansion and urbanization of greater Cairo metropolis, Egypt. In Proceedings of the 6th International Conference on Urban Planning—Essen and Regional Development in the Information Society, Essen, Germany, 18–20 May 2011.
34. Zied, E., Vialard, A. Syntactic Stitching: Towards a Better Integration of Cairo's Urban Fabric. In Proceedings of the 11th International Space Syntax Symposium, Instituto Superior Técnico—Universidade de Lisboa, Lisbon, Portugal, 3–7 July 2017.
35. Elbardisy, W.M.; Salheen, M.A.; Fahmy, M. Solar Irradiance Reduction Using Optimized Green Infrastructure in Arid Hot Regions: A Case Study in El-Nozha District, Cairo, Egypt. *Sustainability* **2021**, *13*, 9617. [CrossRef]
36. Osman, R.; Ferrari, E.; McDonald, S. Water scarcity and irrigation efficiency in Egypt. *Water Econ. Policy* **2016**, *2*, 1650009. [CrossRef]
37. Envi-Board. Envi-Met Support Center. Envi-Met. Available online: http://www.envi-hq.com/ (accessed on 30 April 2023).
38. Middel, A.; Chhetri, N.; Quay, R. Urban forestry and cool roofs: Assessment of heat mitigation strategies in Phoenix residential neighborhoods. *Urban For. Urban Green.* **2015**, *14*, 178–186. [CrossRef]
39. Shahidan, M.F.; Shariff, M.K.M.; Jones, P.; Salleh, E.; Abdullah, A.M. A comparison of *Mesua ferrea* L. and *Hura crepitans* L. for shade creation and radiation modification in improving thermal comfort. *Landsc. Urban Plan.* **2010**, *97*, 168–181. [CrossRef]
40. Weather and Climate. Available online: https://weather-and-climate.com/Cairo-July-averages (accessed on 10 October 2022).
41. Time and Date. Available online: https://www.timeanddate.com/weather/egypt/cairo/historic?month=7&year=2022 (accessed on 10 October 2022).
42. Karimimoshaver, M.; Khalvandi, R.; Khalvandi, M. The effect of urban morphology on heat accumulation in urban street canyons and mitigation approach. *Sustain. Cities Soc.* **2021**, *73*, 103127. [CrossRef]

Disclaimer/Publisher's Note: The statements, opinions and data contained in all publications are solely those of the individual author(s) and contributor(s) and not of MDPI and/or the editor(s). MDPI and/or the editor(s) disclaim responsibility for any injury to people or property resulting from any ideas, methods, instructions or products referred to in the content.

land

MDPI

Article

The Integrated Analysis of Territorial Transformations in Inland Areas of Italy: The Link between Natural, Social, and Economic Capitals Using the Ecosystem Service Approach

Davide Marino, Antonio Barone, Angelo Marucci, Silvia Pili and Margherita Palmieri *

Department of Biosciences and Territory, University of Molise, 86090 Pesche, Italy; dmarino@unimol.it (D.M.); antonio.barone@unimol.it (A.B.); angelo.marucci@unimol.it (A.M.); silvia.pili@unimol.it (S.P.)
* Correspondence: margherita.palmieri@unimol.it

Abstract: This paper examines how spatial dynamics have impacted natural capital and the provision of ecosystem services. The units outlined by the National Strategy for Inland Areas (SNAI) have been used as the territorial units for this study. The SNAI is a public policy focused on enhancing the quality of services to citizens (such as transportation, healthcare, and education). It proposes the economic revitalization of inland areas undergoing processes of marginalization. Our focus on inland areas stems from two primary reasons: first, no previous studies in Italy have analyzed the changes in ecosystem services in SNAI areas; and second, SNAI areas are well-suited to providing ecosystem services that are in demand by urban areas. Although this study does not cover all aspects inherent to the topic, it represents a starting point aimed at understanding the links between environmental and socio-economic dynamics and ecosystem service changes. This is essential for both current and future generations. By analyzing the processes of permanence and transformation, modifications in the supply–demand balance have hereby been studied, as well as the economic variations in ecosystem services. The period considered runs from 1990 to 2018. These findings could help governmental institutions in developing sustainable governance models, in line with spatial policies and strategies.

Keywords: supply–demand balance ecosystem services; land use and land cover change; socio-economic dynamics; inland and rural areas

1. Introduction

1.1. Background

The National Strategy for Inland Areas (SNAI) in Italy is a territorial policy supported by European funds (ERDF, ESF, and EAFRD) and national programs. It aims to enhance the quality of services provided to citizens (such as transportation, health, and education) and fostering economic development in inland areas at risk of marginalization. In the document titled the "National Strategy for Inland Areas", produced by the Ministry for Territorial Cohesion and the South [1], inland areas are defined in terms of their distance from major urban centers and their communication infrastructures. The SNAI was initially introduced in the National Reform Program (NRP) in 2014 and further defined in the 2014–2020 Partnership Agreement. The SNAI classifies Italian municipalities into four clusters: belt areas, intermediate areas, peripheral areas, and outermost areas. Based on these classifications, the strategy identifies inland areas as places that, although distant from services, are rich in environmental and cultural resources as a result of long processes of human influence on the environment. These territories are characterized by a high biodiversity and heterogeneous ecosystems and are able to provide significant ecosystem services (ESs) to the community, such as drinking water supply, food supply, and carbon sequestration.

1.2. Territorial Socio-Economic Transformations and Ecosystem Service Balance

Rural abandonment and rapid urbanization, associated with agricultural intensification and climate change, have been key drivers in transforming terrestrial landscapes. This has had a significant impact on the quality of the environment and the ESs provided for human well-being [2]. It is clear that socio-economic factors have influenced and will continue to influence ES provision [3]. Pressure from human activities has altered the landscape, leading to a decreased supply and increased demand for ESs in urban areas [4]. The demand for ESs is increasing due to various factors such as population growth and urban expansion [5]. This increased demand can lead to ecosystem degradation by impacting the supply of goods and services, harming current and future generations [6]. It is important to analyze the demand for ESs at different scales [7]. For example, the demand for services like food supply and air purification is higher in urban areas than in rural areas due to human and industrial activities [4,8,9]. Conversely, regulatory services such as protection from erosion and hydrogeological disruption are analyzed at a local scale [10]. In recent decades, the gap between supply and social demand for ESs has increased [11]. Anthropogenic factors have caused an increase in demand, resulting in changes in land use and land cover, impacting natural capital and ES provision [12]. In recent decades, inland areas of Italy have experienced a significant migration of people to urban areas. This movement is attributed to several factors, including agricultural mechanization, which has reduced the need for agricultural labor, economic crises in local communities, the declining quality of public services, and environmental hazards [13]. The depopulation of rural and mountainous areas has led to changes in land use, altering the landscape. The abandonment of cultivation practices is one of the main causes of land use changes in Europe, particularly affecting mountainous areas [14]. The increase in forested areas has occurred at the expense of land intended for agriculture and grazing, resulting in a change in the landscape structure and changes in the ES supply [15]. These changes have led to a reduction in the supply of provisioning ESs and an increase in regulation ESs. Furthermore, land use and land cover changes have influenced both social and economic dynamics [16]. According to McPhearson [17], the complex interplay of socio-ecological systems (SESs) makes it challenging to identify long-term drivers. Changes in land use lead to variations in ESs on spatial and temporal scales [18–24] and also affect relationships among ESs [25]. The impacts of changes in land use and cover have highlighted the importance of identifying tools to monitor natural capital and ESs. Mendoza and colleagues [26] highlight the importance of quantifying these changes to understand their impacts on natural capital and human activities. One essential analysis tool used is the mapping of ESs, which helps identify the areas which need intervention for ecosystem maintenance and restoration. Mapping is crucial for assessing and visualizing ESs [27]. It allows us to identify where ES are available and where ES demand and supply are high [28]. The "European Commission's Mapping and Assessment of Ecosystems and their Services" initiative aims to improve the understanding of ecosystems and their services to support the implementation of Objective 2 of the Biodiversity Strategy by 2020. Various methods are currently used to map ESs, including models, proxy indicators, remote sensing data, and expert-based matrices. To effectively manage ESs, it is crucial to assess them from both a biophysical and economic perspective [29,30]. Biophysical evaluation helps one understand the qualitative and quantitative variations in the natural capital, while economic assessment allows one to assign an economic value to the benefits provided by ESs. The benefit transfer method is often used to estimate the benefits of ESs by transferring information (especially values) from studies conducted in another location and/or context [31]. In recent times, databases such as the Ecosystem Service Valuation Database (ESDV) [32] have been developed and made available online. These databases are useful for organizing economic information on ESs at different levels of detail, such as biomes.

1.3. The Research Objective

This paper aims to understand how demographic and economic dynamics have influenced changes in land use and land cover, thus affecting ES provision in SNAI areas. To achieve this, we analyzed the changes in soil and land cover in SNAI areas from 1990 to 2018, building upon a previous study by Marino et al. (2023) [31]. Using the ecosystem service matrix developed by Burkhard (2014) [28], we qualitatively evaluated the supply, demand, and balance of ESs. Inland areas are the focus of this study, as they are environmentally and socio-economically fragile and vulnerable territories. These areas require policies and interventions to support investment in innovation, protect the environment, and promote the unique features of the territory. This paper provides an initial assessment of the qualitative changes in the potential balance of ESs in these areas.

2. Materials and Methods

The research follows a methodological framework composed of 4 stages (Figure 1). Starting from a previous work [31], we represented the processes of transformations and permanence at the level of SNAI areas (step 1) and subsequently calculated the variation in the qualitative balance between the demand and supply of ESs due to these processes (step 2). To understand the impacts of these processes, economic changes in ES (step 3) and some social and economic transformations (step 4) were analyzed.

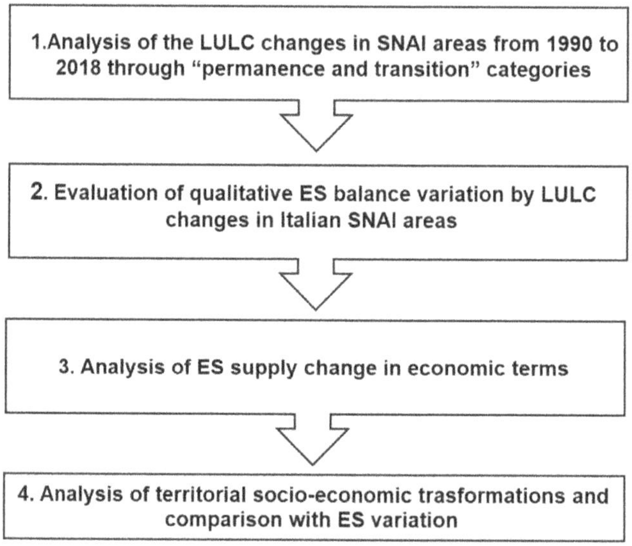

Figure 1. Methodological steps adopted in this work.

2.1. Study Area

The National Strategy for Inland Areas (SNAI) identifies 59 per cent (18 million ha) of Italy's territory as inland areas and defines the rest of the territory as central areas. Underlying this territorial subdivision is a criterion of distance from essential services (hospitals, railway stations, educational systems, etc.) measured in minutes of travel time by car.

Each Italian municipality (Figure 2) has been labeled by the SNAI as a hub (A), intermediate hub (B), or belt (C) municipality in the case of central areas or as intermediate (D), peripheral (E), or ultraperipheral (F) in the case of inland areas. In terms of the extent of these classes, intermediate municipalities (belonging to inland areas) are the most extensive (87 thousand km^2), followed by belt municipalities (84 thousand km^2).

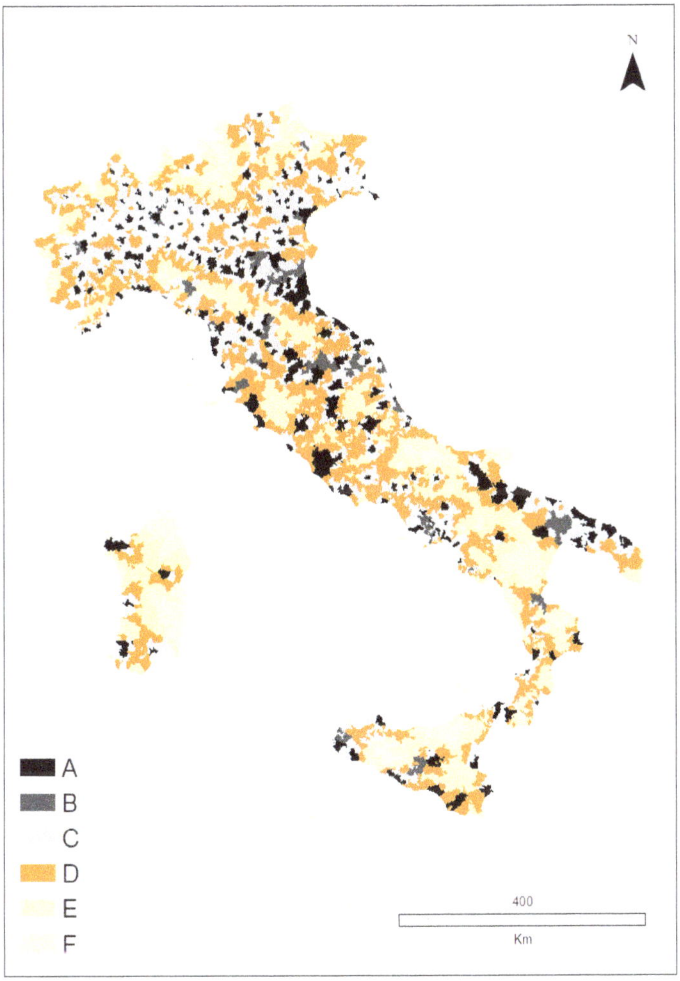

Figure 2. SNAI geographic distribution throughout the national territory. The letters identify the different SNAI areas or rather the central areas—hubs (A), intermediate hubs (B), and belt (C)—and internal areas, i.e., intermediate (D), peripheral (E), and ultraperipheral (F).

Regarding land use, Table 1 shows the relative percentage composition of each SNAI area in 2018 [CLC 2018 data]. Hub areas (A) are the areas with the highest relative concentration of urban areas (15%), while intermediate hubs (B) are those with the highest concentration of agricultural areas (66%), and ultraperipheral areas (F) are those whose area is predominantly occupied by forests (68%). Note that, in general, the core areas (A, B, C) are those whose relative composition is predominantly agricultural; in contrast, the interior areas E and F have a relative composition of predominantly forested areas. Again, in relative terms, the highest percentages of urbanization are found in the central areas (A, B, and C).

Table 1. Land composition (absolute values in km^2 and percentage) of Italian SNAI areas for the year 2018 (source: Corine Land Cover). For each SNAI area, the percentage identifies the weight of each surface type. The intensity of the light-blue color increases as the percentage increases.

SNAI Description	SNAI Code	Urban Areas		Agricultural Areas		Wooded and Semi-Natural Areas		Wetlands and Water Bodies		Total
		km^2	%	km^2	%	km^2	%	km^2	%	km^2
Hubs	A	4424	15%	18,706	65%	5135	18%	592	2%	28,857
Intermediate hubs	B	661	8%	5782	66%	2062	24%	242	3%	8746
Belt	C	7060	8%	53,602	63%	22,905	27%	1037	1%	84,605
Intermediate	D	2801	3%	44,168	50%	39,448	45%	1288	1%	87,705
Peripheral	E	1219	2%	29,230	40%	41,340	57%	608	1%	72,397
Ultraperipheral	F	279	1%	5825	30%	13,079	68%	112	1%	19,296
Total		16,444	5%	157,313	52%	123,969	41%	3879	1%	301,605

2.2. Analysis of the LULC Changes in SNAI Areas from 1990 to 2018 through Permanence and Transition Categories (Step 1)

The land use and cover change that occurred in Italian SNAI areas between 1990 and 2018 were analyzed in a GIS environment using the two respective shapefiles derived from the Corine Land Cover (CLC) project [33]. The shapefiles were then associated with each administrative unit, which, in turn, were associated, by ISTAT [34], with one of six SNAI categories. Next, the intersection function between the two land use shapefiles of 1990 and 2018 was used to obtain a third shapefile containing the polygons of the LULC changes that occurred during the period considered. Then, through the joint function, the types of LULC changes that occurred were named through the categories of transition and permanence already used in a previous work [31] (see Figure A1 in Appendix A). Finally, the shapefile attribute table was exported to an Excel file, and the surfaces for each transition and permanence type and SNAI area were summed. This elaboration provided information on the amount of area (absolute and relative) affected by each type of transition and permanence within each SNAI category.

2.3. Evaluation of ES Balance Variation by LULC Changes in Italian SNAI Areas (Step 2)

One of the most widely used approaches in the field of ES mapping is the one devised by Burkhard et al. [28,35,36], commonly known as the 'matrix approach' [37]. Specifically, this approach involves the association of qualitative values (where 0 identifies the minimum and 5 the maximum) of supply, demand, and balance (i.e., supply minus demand) of ESs to each CLC class at the third level of detail.

In our work, we obtained ES balance values for each CLC class using the ES potential supply and demand matrix contained in the work of Burkhard et al., from 2014 [28]. Specifically, for each CLC class, we subtracted the demand values from the potential supply values of ESs. The ES balance values obtained identified, for each CLC class, the net capacity to provide (when balance values > 0) or demand (when balance values < 0) ESs. The ES balance values obtained thus far were associated with third-level Corine land cover classes. We grouped the third-level classes and their associated ES balance values into macro classes (mostly aggregated according to the second Corine level) to allow for comparison (see also Section 2.3 below) with the results of our previous work, which was based on their aggregation into macro classes [31]. In particular, a weighted average for the area of the classes (III CLC level) contained in the macro class (II CLC level) was used to relate the ES balance values from the classes to the macro classes (Table 2).

Table 2. ES balance values associated with the LULC macro classes used in this work. The values are elaborated from Burkhard et al. (2014) [28]. The negative values (gradation of red) indicate that the demand for the ES of a specific macro class is higher than the relative supply. The positive values (gradations of green), on the contrary, indicate that the supply is greater than the demand.

Macroclass Description	Macroclass Code	Global Climate Regulation	Air Quality Regulation	Water Flow Regulation	Water Purification	Erosion Regulation	Natural Hazard Regulation	Crops	Fodder	Timber	Wild Food and Resources
Urban areas	100	−3.3	−4.7	−4.4	−4.1	0.1	−4.2	−3.0	−1.6	−3.2	−3.8
Arable lands	210	−1.1	0.0	−0.1	−0.2	−3.0	−1.0	5.0	4.9	0.0	1.0
Permanent crops	220	0.2	0.2	0.2	−1.7	−0.8	−2.6	4.0	0.0	1.1	0.1
Pastures	230	−1.0	−1.0	0.0	−2.0	0.0	−1.0	0.0	0.0	−1.0	2.0
Heterogeneous agricultural areas	240	0.5	0.5	0.5	−1.1	0.6	−0.6	3.5	2.1	0.5	1.5
Forests	310	5.0	5.0	3.0	5.0	5.0	4.0	0.0	1.0	5.0	5.0
Shrub and/or herbaceous vegetation	320	3.2	0.5	1.1	1.9	2.6	1.1	0.0	1.4	0.8	2.6
Non and sparsely vegetated areas	330	0.1	0.0	0.7	0.9	1.3	1.3	0.0	0.0	0.0	0.4
Wetland and water bodies	500	0.9	0.0	3.9	2.0	0.2	3.3	0.0	0.4	0.0	3.5

Using the qualitative coefficients calculated above, we then calculated the ES balance change that occurred from 1990 to 2018 in Italian SNAI areas. For this purpose, we calculated the ES unit balance variation values produced by each variation dynamic between the LULC macro classes (e.g., 210 to 100; see previous paragraph) and multiplied these by the surface area affected by the specific LULC variation dynamic. Finally, for each ES, we normalized the ES balance variation value (see equation below).

$$\Delta BDG_{ESy} = \Sigma\ (BDG_{ESyMx2018} - BDG_{ESyMz1990}) \times (A_{dinzx} \times 100/A_{tot})$$

where

ΔBDG_{ESy} = change in ES_y balance between 1990 and 2018;

$BDG_{ESyMx2018}$ = value of the balance of ES_y by macro class M_x at 2018;

$BDG_{ESyMz1990}$ = value of the balance of ES_y by macro class M_z at 1990;

A_{dinzx} = area affected by land use change dynamics (i.e., an area that was a macro class z in 1990 and changed to a macro class x in 2018);

A_{tot} = total area of the study area.

The results obtained were aggregated in terms of both LULC transition categories and SNAI areas.

2.4. Economic Evaluation of the ES Supply in SNAI Areas (Step 3)

To achieve the economic evaluation of ES supply variations in SNAI areas, we used the benefit transfer approach [38–40] and, in particular, the unitary economic coefficients (€/ha) collected in our previous work. We then associated, in a GIS environment, the unitary economic coefficients to the LULC macro class and aggregated them for each SNAI zone in 1990 and 2018. We obtained the economic value of the ES supply for each SNAI area for the years 1990 and 2018 and the relative difference.

2.5. Analysis of Territorial Socio-Economic Transformations (Step 4)

We examined the changes in territories based on socio-economic indicators. Specifically, we calculated (i) the population variation in municipalities from 1990 to 2018 [41] as a social indicator and (ii) employees and companies number variation from 2012 to 2018 [41]. Both changes were represented at SNAI level. While we understand that more indicators are needed for a comprehensive analysis of territorial changes, we chose these indicators because representative basic data were available at the municipal level. For this reason, the economic data had a different reference period.

3. Results

3.1. Main LULC Changes in SNAI Areas

The analysis of the land use changes in SNAI areas from 1990 to 2018 (Table 3) reveals trends that are common to all SNAI areas, as well as ones which are specific to each area type. In particular, by relating the extent of individual land use transitions to the total area of each SNAI zone, several key dynamics emerged (Table 3, Table A1). In all SNAI areas, permanence, representing approximately 90 percent by weight, is predominant over transitions. In central areas, permanence is primarily related to the continuation of agricultural land, particularly arable land. In inland areas, the primary permanence relates to the continuation of forested land. Conversely, agricultural intensification processes represent the most significant transition in all SNAI areas, with a range of 4.3% in hub areas to 5.3% in inter-municipal hubs. The process of renaturation is particularly prominent in the outermost areas, and, in general, it is greater in inland areas than in central areas. Urbanization has opposite dynamics: it has greater weight in central areas (with a maximum of 2.8 percent in hub areas) than in interior areas (with a minimum of 0.4 percent in peripheral and outermost areas).

Table 3. Changes in the LULC of SNAI areas between 1990 and 2018 summarized by transition categories [15]. Note: For each SNAI area (columns), we report the area of the transitions expressed as a percentage of the total area of the relevant SNAI area. Gradations of blue indicate increasing percentage values (light to dark).

LULC Permanence and Transition Categories	A	B	C	D	E	F
Agr. Extensification	0.3%	0.5%	0.3%	0.3%	0.2%	0.1%
Transformation to complex system	2.5%	2.2%	2.5%	2.6%	2.1%	1.5%
Agr. Intensification	4.3%	5.3%	4.5%	4.9%	4.9%	4.4%
Forest extension	0.9%	1.1%	0.7%	1.2%	1.6%	2.9%
Urbanization	2.8%	1.6%	2.0%	0.8%	0.4%	0.4%
Other	0.6%	0.4%	0.5%	0.3%	0.2%	0.3%
Permanence	88.7%	88.9%	89.5%	89.8%	90.6%	90.4%

3.2. Qualitative ES Balance Potential Variation

First, ES balance variation was analyzed for each type of land use change dynamic, regardless of the surface area covered by each dynamic (Table A2). As can be observed in Table A2, urbanization is the process of land use change that can produce the most significant decreases in the balance of ESs: this is due to a substantial increase in demand and, conversely, a notable decline in the supply of all ESs. In this regard, the soil sealing of forested lands represents the most detrimental dynamic. When forests undergo such a conversion, the air purification ES is most adversely affected (−10 points). In general, the loss of forests due to urbanization and agricultural intensification processes is among the most deleterious changes.

Then, to evaluate the ES balance variation considering the specific area affected by each dynamic of land use change, the qualitative values commented on in the previous paragraph were weighted (see Section 2). The radar chart below (Figure 3) presents the overall weighted ES balance qualitative value variation determined by each LULC dynamic from (i) arable land, (ii), permanent crops, (iii) pasture, (iv) heterogeneous agricultural areas, (v) forests, (vi) shrub and herbaceous vegetation, and (vii) open spaces with little or no vegetation.

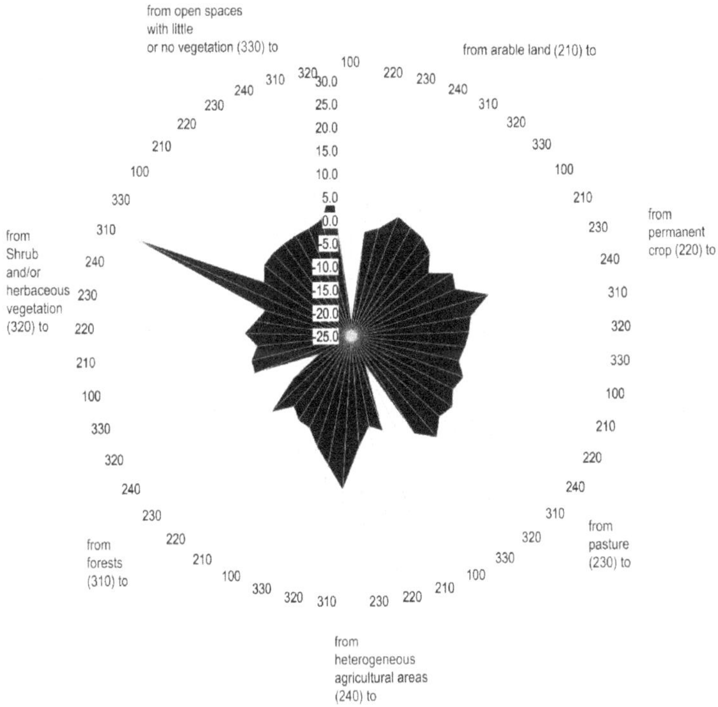

Figure 3. Radar chart: sum of ES balance changes for each dynamic, considering the respective surfaces.

Negative radar peaks (Figure 3) identify the LULC dynamics of change that produce the greatest detrimental impacts on ES balance variation. Specifically, the urbanization of arable land (210100) produces −21 points, while the homogenization of heterogeneous agricultural zones (240100) produces −18 points. The internal trends that define forest classes and, namely, the replacement of woodland with shrubland (310320) also produce a high overall decrease in ESs (−17 points).

These negative changes in the ES balance between 1990 and 2018 are weakly offset by the positive changes in forest extension. In particular, land use changes from shrubland (320) and heterogeneous agricultural areas (240) to forests (310) generated, respectively, +28 and +8 points.

By delving deeper into the details of the effects of LULC changes on balance variation for each ES (Table A3), it is possible to comment that the greatest negative and positive variations were found, on one hand, in (i) the erosion regulation ES as a consequence of the agricultural intensification processes of heterogeneous agricultural areas being transformed into arable zones (−5.0) and, on the other hand, in (ii) the air quality regulation ES, as a response to forest extension at the expense of shrublands (+5.4).

The effective changes in the ES balance generated by the territorial LULC dynamics (Table A3) were then grouped by LULC transition categories and SNAI area types (see Table 4 below).

Table 4. Total and mean ES qualitative balance variation at the SNAI area level. The colour gradations are proportional to the intensity of the change (negative in red, positive in green) in the ES balance.

SNAI Code	Global Climate Regulation	Air Quality Regulation	Water Flow Regulation	Water Purification	Erosion Regulation	Natural Hazard Regulation	Crops	Fodder	Timber	Wild Food and Resources	TOT	MEAN
A	−0.8	−1.2	−1.1	−0.9	0.2	−0.8	−1.5	−1.1	−0.8	−1.2	−9.3	−0.9
B	−0.1	−0.2	−0.2	−0.1	−0.1	−0.1	−0.2	−0.1	−0.1	−0.2	−1.4	−0.1
C	−1.6	−2.7	−2.4	−2.5	0.3	−2.4	−2.9	−2.7	−1.9	−2.9	−21.7	−2.2
D	−0.3	−0.2	−0.7	−0.6	0	−0.8	0	0	0	−0.5	−3.1	−0.3
E	−0.7	0.4	−0.2	−0.3	−0.7	−0.3	0.9	0.7	0.3	0	0.1	0.01
F	0.1	0.2	0.1	0.2	0.1	0.2	−0.1	0.1	0.2	0.2	1.3	0.1
TOT	−3.5	−3.7	−4.5	−4.1	−0.2	−4.3	−3.8	−3.2	−2.2	−4.7	−34.1	−3.4

The results reveal that the belt municipalities (C) experienced the most pronounced negative variations in the total ES balance (−2.7) among all SNAI areas. These changes were attributable to a drastic decline especially in the provision of crops (−2.9), wild food and resources (−2.7), and forage (−2.9).

In belt areas, the primary driver of this change was the urbanization of arable land (from 211 to 100) and heterogeneous agricultural areas (240 to 100), which, respectively, resulted in a loss of ES balance of −11.0 and −7.7 points. Urbanization, in fact, generated an increase in ES demand and a drastic decline in ES supply. Further negative variations, smaller in magnitude than the preceding ones (between −3 and −2.5), could be observed in relation to (i) internal transformation within forest classes (from forests 310 to shrublands 320) and (ii) processes of agricultural intensification, specifically from forests (310) to permanent crops (220) and heterogeneous agricultural areas (240). Conversely, in terms of positive ES balance variations, we observed a +3.8 in changes within the forestry classes, i.e., from 320 to 310.

The results of this analysis allow for the identification of two distinct patterns, which exert a differential influence on central and internal SNAI areas. In particular, in central areas, the process that most negatively affects ES balance is the urbanization of agricultural areas (especially arable land). In contrast, in internal areas, the worst process in terms of variation in the ES balance has been identified to include LULC transitions within forest classes, and, in particular the transitions from forests to shrublands (310320). Furthermore, the order of magnitude of negative fluctuations in central areas is significantly higher than in interior ones. Concerning positive ES balance variations, the most influential LULC change dynamic in both interior and central SNAI areas is the conversion of shrublands into woodlands (320310). This dynamic is responsible for an increase in the overall ES supply.

The land use changes that took place between 1990 and 2018 resulted in a total average variation in the ES balance that was markedly negative (see Table 5 below), with a final value of −3.41, out of a possible range ranging from −5 to +5. Going into more detail, the land use changes that occurred between 1990 and 2018 resulted, on average, in a fairly marked decrease in the overall balance of ESs in belt (−2.17) and hub (−0.93) areas. Less marked but still negative were the changes that occurred in the intermediate (−0.31) and inter-municipal hubs (−0.14). Slight increases, on the other hand, were recorded for the more inland areas such as the peripheral (+0.01) and ultraperipheral (+0.13) areas. These changes in the ES balance were driven by land use and land cover transitions that occurred in each SNAI area. Almost 90% of the urbanization that occurred at the Italian national level occurred, in fact, in the belt (46%), hub (21%), and intermediate (19%) areas, causing, especially in the case of the belt and hub areas, the greatest relative decrease in SEs (−2.22 and −1.05, respectively). Also, agricultural intensification led to decreases in the ES balance, especially in intermediate (−1.02), belt (−1.00), and peripheral (−0.94) areas. Regarding the positive changes in the ES balance, reforestation was the process that, on average, led to the most marked increase in service (+2.08), followed by the processes of complex system evolution (+0.59) and agricultural extensification (+0.11). As it is possible to note, permanencies and the other categories also led to positive increases. In particular, the permanence category also included internal forest transitions (320310) that led to ES balance growth. In general, reforestation processes resulted in point increases in all SNAI areas. However, in the case of peripheral and ultraperipheral areas, these processes were not counterbalanced by the processes of urbanization and agricultural intensification, leading to a final negative balance variation.

Table 5. Qualitative ES average balance variation by LULC transitions and permanence categories. The colour gradations are proportional to the intensity of the change (negative in red, positive in green) in the ES balance.

Transitions and Permanence	A	B	C	D	E	F	TOT by Transitions and Permanence
Agr. Extensification	0.01	0.01	0.03	0.04	0.02	0	0.11
Transformation to complex system	0.07	0.02	0.23	0.2	0.06	0.01	0.59
Agr. Intensification	−0.26	−0.08	−1	−1.02	−0.94	−0.25	−3.55
Forest extension	0.13	0.06	0.4	0.63	0.61	0.25	2.08
Urbanization	−1.05	−0.19	−2.22	−0.93	−0.43	−0.11	−4.94
Other	0.16	0.02	0.36	0.24	0.13	0.04	0.96
Permanence	0.01	0.01	0.03	0.54	0.56	0.19	1.34
TOT by SNAI areas	−0.93	−0.14	−2.17	−0.31	0.01	0.13	−3.41

3.3. Economic Evaluation of the ES Supply

The results of the economic analysis of the supply of ESs are presented in Table 6 below. The results obtained are the total and average (per hectare) economic value of the ES supply in each SNAI zone for the years 1990 and 2018, as well as the differences in the supply between the two periods. In terms of the average change per hectare, all central areas experienced a significant decline, including the intermediate (D) zone. This was particularly evident in hubs and belts. Conversely, peripheral (E) and ultraperipheral (F) areas experienced a slight increase. With regard to the total economic value, the belt (C) appeared to be the zone most affected by land use changes. The total variation amounted to −692 million EUR (74% of the total variation at the national level). As was also the case for the SNAI central areas, specifically hubs, decreases were the results of the loss of agricultural areas (in particular arable land). Notably, these results in terms of economic changes in the ES supply were in line with the findings of our qualitative balance evaluation (Table 4). Also concerning the qualitative balance, SNAI areas A, B, C, and D exhibited a reduction over time, while areas E and F demonstrated slight increases.

Table 6. Economic valuation of ES supply in SNAI areas and its change from 1990 to 2018. The colour gradations are proportional to the intensity of the change (negative in red, positive in green) in the ES supply economic value.

SNAI Code	km^2	1990		2018		VAR 1990−2018	
		MLN €	EUR/ha	MLN EUR	EUR/ha	Var EUR/ha	Var MLN EUR
A	28,857	9702	3362	9383	3252	−110	−319
B	8746	3125	3573	3102	3546	−27	−24
C	84,605	29,582	3496	28,889	3415	−82	−692
D	87,705	28,418	3240	28,331	3230	−10	−87
E	72,397	22,162	3061	22,300	3080	19	137
F	19,296	5518	2860	5576	2890	30	58
TOT	301,605	98,508	3266	97,581	3235	−30	−927

3.4. Analysis of Socio-Economic Transformations and Comparison with the Variation in the ES

Table 7 below provides a summary of the indicators of social and economic dynamics in relation to the economic and qualitative variation in ESs. The areas exhibiting the most pronounced decline in population between 1990 and 2018 are the peripheral (−5.7%) and ultraperipheral ones (F) (−9.4%). This trend confirms the long-term depopulation dynamics in the most marginal and isolated municipalities. This phenomenon contributed to the activation of the National Strategy for Internal Areas during the same period, when municipalities in the belt category experienced an increase of 16.4%. In terms of economic dynamics, indicators related to the number of employees and companies for the period 2012–2018 describe a negative trend in the number of companies, particularly in the intermediate (−3.9%), peripheral (−3.7%), and ultraperipheral (−3.2%) municipalities.

Table 7. Comparison of variations in territorial socio-economic indicators, qualitative balance, and economic value of ESs. (Note: The changes in the socio-economic indicators are assessed in the period of 2012–2018, while the changes in the ES values are for the period of 1990–2018.).

SNAI Code	Population (var %)	Number of Companies (var %)	Number of Employees (var %)	ES Average Qualitative Balance Variation (−5, +5)	ES Mean Economic Value (var EUR/ha)	ES Total Economic Value (var MLN EUR)	ES Economic Value per Inhabitant (var EUR/Inhabitant)
A	−0.9	0.9	6.9	−0.9	−110.5	−318.8	−10.7
B	3.6	−1.3	2.2	−0.1	−27.2	−23.8	−45.5
C	16.4	−3.0	1.1	−2.2	−81.8	−692.5	−248.1
D	6.2	−3.9	−1.0	−0.3	−9.9	−86.9	−213.9
E	−5.7	−3.7	−0.9	0.01	19.0	137.4	379.2
F	−9.4	−3.2	0.5	0.1	30.1	58.0	827.9

These data indicate weaker job creation in peripheral municipalities, consistent with the decrease in the number of employees in the intermediate and peripheral municipalities. Comparing the economic and social data with the estimate of ESs shows a negative economic value variation in the more central municipalities, but a positive variation in the peripheral and ultraperipheral municipalities. Importantly, data on the economic variation in ESs confirm that inland areas play a central role in the delivery of benefits. The economic value per hectare and inhabitant are higher in the peripheral and outermost areas, where depopulation and economic decline have been more marked. In the central areas, in fact, there is a greater demand for goods and services than in the inland areas, as shown by the qualitative values of the balance.

4. Discussion

Today, there is considerable interest in inland areas in institutional and scientific circles due to awareness of their great potential and the possible new settlement scenarios that could emerge in these areas [42]. The SNAI strategy aims to enhance the economic value of territories and improve the provision of essential public services [43].

The innovative contribution of our work was to provide a study of the dynamics of land use change and ESs provision in the territories within SNAI areas.

In Italy, scientific research on SNAI areas has been carried out by several authors [42,44,45], with only a few of them focusing on ESs [46]. Our investigation further examines the impact of socio-economic transformations on land use changes and ES provision. Some authors have demonstrated how spatial transformation processes linked to socio-economic factors, have impacted land use change and the provision of ESS (see, for example, [12,47]). The spatial heterogeneity of ES and their functional relationships are indeed influenced by natural and socio-economic factors [48,49].

In this paper, we highlight the significant impact of spatial dynamics on land use and the provision of ESS. Depopulation and economic decline in mountainous and inland areas have led to land abandonment in marginal areas, resulting in natural reforestation.

Conversely, agricultural intensification has increased in easily accessible areas such as hubs and intermediate hubs. Another crucial spatial dynamic is urbanization, which has significantly impacted the belt areas over the period under investigation (1990–2018). The aforementioned spatial dynamics have significantly impacted the balance between supply and demand. There is an increased demand for goods and services, especially in urbanization and intensification processes, while reforestation has resulted in a surplus of supply. However, reforestation has also resulted in an increase in regulation ESs (e.g., CO_2 absorption) at the expense of provisioning ESs.

To fully understand the various paths of territorial development, it is essential to analyze the characteristics and dynamics of the socio-economic structure of the areas in question. This analysis provides insights into the interaction of natural, social, and economic capital within socio-ecological systems (SESs) [50]. When natural capital contributes to ESs, its interaction with social and economic capital affects the overall economic well-being of the community [18]. The interactions between natural, social, and economic capital are influenced by various complex mechanisms and depend on several factors. Natural capital comprises ecosystems that form without human intervention and provide the resources that support economic activities [18,51]. Economic capital consists of human-produced capital, including consumer goods, intermediate goods used for production, and financial capital. Social capital, including human capital, encompasses social networks, norms, culture, and institutions that facilitate cooperation [52,53]. The combination of these capitals generates various human welfare benefits. Analyzing these interactions can offer insights into achieving environmental, economic, and social sustainability goals. The use of indicators to monitor the achievement of sustainability goals is a valuable tool for such analyses. In the present study, we selected some indicators that are most representative of the phenomenon under investigation. We chose three indicators to analyze the potential relationships between the three forms of capital: the economic value of ESs (natural capital), the resident population (social capital), and the number of employees (economic capital).

The indicators were selected on the basis of the availability of data for the specified reference period. Although they could be improved, they provide a framework for interpreting the phenomena that have occurred in the country's inland areas in recent decades.

The results demonstrate that the territorial changes in the SNAI areas resulted in a positive increase in the economic capital in the hub areas (+1.00), while, at the same time, there was a decrease in the natural capital (−0.46) in the same areas. Additionally, social capital experienced significant growth, particularly in the belt areas (+1.00), which appeared to be the final destination for the population moving from the hubs and inner areas (Figure 4).

It is worth noting that the social and economic factors affecting these areas have not caused a significant increase in the supply of ESs. Natural capital shows weakly positive values in peripheral and ultraperipheral areas. The provision of ESs depends, in fact, on the interaction of various forms of capital and the proper maintenance of the territory. This, in turn, affects the provision of essential goods and services for society, such as food and fodder supply [18,50].

The combination of different capital forms is indispensable in the integration of social, economic, and ecosystem perspectives and the fusion of these concepts to better understand and manage ESs [54]. It is therefore crucial to consider strategies that balance economic development with the conservation of natural capital, for example, by identifying and preserving, at least, critical natural capital [55,56] in the most urbanized areas.

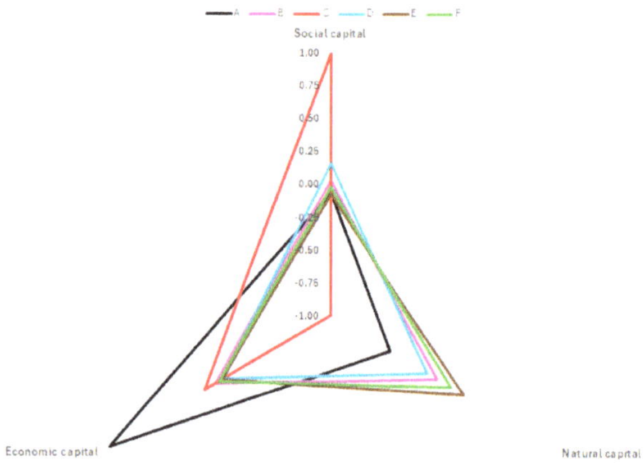

Figure 4. Relation between natural capital, economic capital, and social capital in SNAI areas. The values shown in the graph have been normalized on a scale from −1 to +1.

Further developments in our research will have to focus on overcoming the current limitations, refining the approach presented here, and applying it to other contexts. Future investigations could involve analyzing the relationships among different ESs through the concept of "bundles". Bundles are defined by Raudsepp-Hearne et al. (2010) [57] as a "set of associated ES that repeatedly appear together in time or space". In effect, bundle analysis allows for the evaluation of potential synergies and trade-offs, thereby supporting appropriate landscape planning and management [19]. Additionally, it would be undoubtedly beneficial to incorporate a quantitative assessment of the demand for ESs. This would enable a quantitative assessment of the relationship between supply and demand and, thus, also of the balance of the ESs over time. The advantages of economic evaluation are manifold and include the possibility of comparing variables and indicators commonly used in cost–benefit analyses that underlie, for example, the ex ante evaluation of public policies, possible public–private agreements, and the activation of potential payment for ecosystem service (PES) schemes. Another line of future research can be identified in the application of the ES balance approach in other specific territorial areas such as protected areas or metropolitan cities. This would facilitate the analysis of the effects of different forms of spatial planning on the dynamics of ES production and consumption over time.

5. Conclusions

Our research stands out from others conducted in Italy [58–60] within SNAI areas due to our focus on analyzing the changes in ESs in connection to social and economic factors from 1990 to 2018. This was achieved by combining transition and permanence matrices with a qualitative ES matrix and using coefficients to estimate economic changes. Although our approach is experimental, it has allowed us to demonstrate how spatial changes have impacted natural capital and the provision of ESs and identify which SNAI areas have been most affected by these changes. The LULC analysis revealed that agricultural intensification represents the most significant transition across all SNAI areas, while urbanization has a greater impact on central areas compared to inland areas. Urbanization processes have caused a considerable increase in demand and a decrease in the supply of ESs, bringing the ES balance into deficit. These spatial dynamics have influenced both the qualitative change in demand and supply and the economic value of ESs. On average, during the period studied, all central areas experienced significant decreases in terms of ES economic value per hectare. These results suggest the need for government agencies to develop sustainability governance models to support the implementation of territorial policies and strategies. Furthermore, the research results provide a tool for mapping the data, which allows for the identification of areas requiring prompt intervention for the purposes of improved planning and management. Additionally, the economic estimation of ESs highlights their areas of surplus and deficit and identifies the sustainable limits of natural resource consumption. This approach can also support the implementation of SEEA Ecosystem Accounting (SEEA EA) [61], which is designed to account for global and local governance. The findings of this study are pertinent to the implementation of Act No. 22, from 28 December 2015, which focuses on environmental measures designed to promote a green economy and reduce the overuse of natural resources. This law introduced specific obligations for municipalities to contribute to the conservation and protection of natural capital. It also requires municipalities to develop a preliminary strategy for accounting for natural capital and ESs, as well as implementing measures for their valorization, which may include payments for ESs. One of the key objectives of SNAI is to enhance productive land use and address demographic decline to strengthen the overall cohesion of the national territory. In this context, the ESs provided by these territories should be at the core of development policies that take into account their value in community well-being, as also indicated by the recent Kunming–Montreal Global Biodiversity Framework (COP 15).

Author Contributions: Conceptualization, D.M. and M.P.; methodology, D.M., M.P., A.M. and A.B.; software, A.B. and S.P.; formal analysis, A.B. and S.P.; investigation, A.B., A.M. and M.P.; data curation, A.B. and A.M.; writing—original draft preparation, A.B., M.P. and A.M.; writing—review and editing, M.P., A.M., A.B. and S.P.; visualization, A.B. and S.P.; and supervision, D.M., A.M. and M.P. All authors have read and agreed to the published version of the manuscript.

Funding: This research received no external funding.

Data Availability Statement: The data presented in this study are available upon request from the corresponding author.

Conflicts of Interest: The authors declare no conflicts of interest.

Appendix A

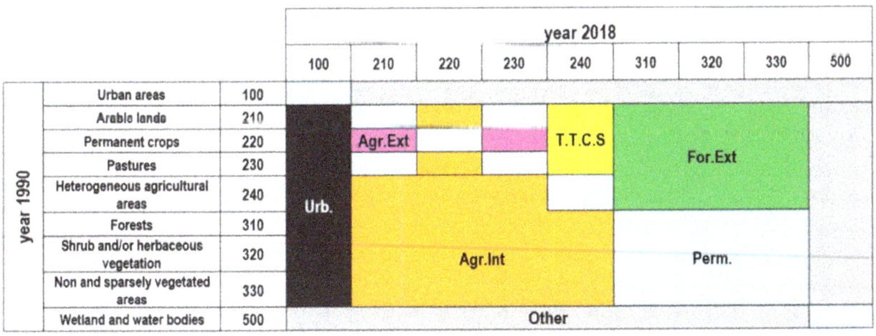

Figure A1. Transition categories associated with LULC changes between macro classes. Based on [31]. Urb = urbanization; Agr. Ext = agricultural extensification; Agr. Int = agricultural intensification; T.T.C.S = transformation to complex system; Perm = permanence of the same type of LULC; and other = transitions not considered in this analysis.

Table A1. Land use changes aggregated at the SNAI classes level (surface area and percentage coverage).

A	**28,857**		B	**8746**		C	**84,605**	
	km^2	%		**km^2**	%		**km^2**	%
Perm (210210)	9838	34	Perm (210210)	3114	36	Perm (210210)	29,016	34
Perm (240240)	4275	15	Perm (310310)	1599	18	Perm (310310)	15,622	18
Perm (100100)	3615	13	Perm (240240)	1183	14	Perm (240240)	11,662	14
Perm (310310)	3164	11	Perm (220220)	718	8	Perm (220220)	5954	7
Perm (220220)	2250	8	Perm (100100)	518	6	Perm (100100)	5340	6
Perm (320320)	1212	4	Perm (500500)	227	3	Perm (320320)	4064	5
Perm (500500)	552	2	Perm (320320)	226	3	Etcs (210240)	1200	1
Int (240210)	532	2	Urb (240210)	223	3	Int (240210)	1165	1
Etcs (210240)	438	2	Etcs (210240)	89	1	Perm (330330)	1099	1
Urb (210100)	408	1	Int (240220)	86	1	Perm (500500)	956	1
Urb (240100)	272	1	Etcs (220240)	76	1	Urb (210100)	886	1
Etcs (220240)	232	1	Urb (210100)	74	1	Int (240220)	810	1
Int (240220)	211	1	Int (210220)	68	1	Etcs (220240)	725	1
Perm (320310)	153	1	Perm (330330)	54	1	Perm (230230)	709	1
			Urb (240100)	46	1	Urb (240100)	580	1
						Perm (320310)	506	1
						Int (210220)	492	1
D	**87,705**		E	**72,397**		F	**19,296**	
	km^2	%		**km^2**	%		**km^2**	%
Perm (310310)	25,010	29	Perm (310310)	23,292	32	Perm (310310)	5325	28
Perm (210210)	19,592	22	Perm (210210)	12,468	17	Perm (320320)	4761	25
Perm (240240)	11,121	13	Perm (320320)	10,317	14	Perm (210210)	2230	12
Perm (320320)	7854	9	Perm (240240)	7448	10	Perm (240240)	1797	9
Perm (220220)	5555	6	Perm (330330)	3361	5	Perm (330330)	1249	6
Perm (330330)	2605	3	Perm (220220)	2990	4	Perm (320310)	508	3
Perm (100100)	2099	2	Perm (320310)	1340	2	Perm (220220)	468	2
Perm (500500)	1239	1	Perm (230230)	1031	1	Perm (310320)	297	2
Int (240210)	1203	1	Int (240210)	917	1	For Ext (240320)	284	1
Perm (320310)	1145	1	Perm (100100)	895	1	Perm (330320)	231	1
Etcs (210240)	993	1	Etcs (320240)	818	1	Etcs (320240)	216	1
Etcs (220240)	988	1	Perm (330320)	758	1	Int (240210)	204	1
Perm (230230)	892	1	Perm (310320)	755	1	Perm (100100)	198	1
Perm (330320)	803	1	Etcs (210240)	583	1	Int (320210)	163	1
Int (240220)	669	1	Perm (500500)	564	1	Etcs (210240)	128	1
Perm (310320)	651	1	Etcs (220240)	557	1	Perm (230230)	123	1
Etcs (320240)	628	1	Int (320210)	488	1	Etcs (220240)	116	1
Int (210220)	554	1	For Ext (240320)	443	1	Perm (500500)	98	1

Table A2. LULC change dynamics and consequences on the potential balance sheet of each ES.

Transition and Permanence	Change	Global Climate Regulation	Air Quality Regulation	Water Flow Regulation	Water Purifcation	Erosion Regulation	Natural Hazard Regulation	Crops	Fodder	Timber	Wild Food and Resources	Mean
Urb	210100	−2	−5	−4	−4	3	−3	−8	−6	−3	−5	−3
Int	210220	1	0	0	−2	2	−2	−1	−5	1	−1	0
Perm	210230	0	−1	0	−2	3	0	−5	−5	−1	1	−1
Etcs	210240	2	0	1	−1	4	0	−2	−3	1	1	0
For Ext	210310	6	5	3	5	8	5	−5	−4	5	4	2
For Ext	210320	4	1	1	2	6	2	−5	−3	1	2	1
For Ext	210330	1	0	1	1	4	2	−5	−5	0	−1	0
Urb	220100	−4	−5	−5	−2	1	−2	−7	−2	−4	−4	−2
Agr Ext	220210	−1	0	0	2	−2	2	1	5	−1	1	0
Agr Ext	220230	−1	−1	0	0	1	2	−4	0	−2	2	0
Etcs	220240	0	0	0	1	1	2	0	2	−1	1	1

Table A2. *Cont.*

Transition and Permanence	Change	Global Climate Regulation	Air Quality Regulation	Water Flow Regulation	Water Purifcation	Erosion Regulation	Natural Hazard Regulation	Crops	Fodder	Timber	Wild Food and Resources	Mean
For Ext	220310	5	5	3	7	6	7	−4	1	4	5	3
For Ext	220320	3	0	1	4	3	4	−4	1	0	3	1
For Ext	220330	0	0	1	3	2	4	−4	0	−1	0	0
Urb	230100	−2	−4	−4	−2	0	−3	−3	−2	−2	−6	−2
Int	230210	0	1	0	2	−3	0	5	5	1	−1	1
Int	230220	1	1	0	0	−1	−2	4	0	2	−2	0
Etcs	230240	1	1	0	1	1	0	3	2	2	−1	1
For Ext	230310	6	6	3	7	5	5	0	1	6	3	3
For Ext	230320	4	2	1	4	3	2	0	1	2	1	1
For Ext	230330	1	1	1	3	1	2	0	0	1	−2	1
Urb	240100	−4	−5	−5	−3	0	−4	−6	−4	−4	−5	−3
Int	240210	−2	0	−1	1	−4	0	2	3	−1	−1	0
Int	240220	0	0	0	−1	−1	−2	0	−2	1	−1	−1
Int	240230	−1	−1	0	−1	−1	0	−3	−2	−2	1	−1
For Ext	240310	5	5	3	6	4	5	−3	−1	4	4	2
For Ext	240320	3	0	1	3	2	2	−3	−1	0	1	1
For Ext	240330	0	0	0	2	1	2	−3	−2	−1	−1	0
Urb	310100	−8	−10	−7	−9	−5	−8	−3	−3	−8	−9	−5
Int	310210	−6	−5	−3	−5	−8	−5	5	4	−5	−4	−2
Int	310220	−5	−5	−3	−7	−6	−7	4	−1	−4	−5	−3
Int	310230	−6	−6	−3	−7	−5	−5	0	−1	−6	−3	−3
Etcs	310240	−5	−5	−3	−6	−4	−5	3	1	−4	−4	−2
Perm	310320	−2	−4	−2	−3	−2	−3	0	0	−4	−2	−2
Perm	310330	−5	−5	−2	−4	−4	−3	0	−1	−5	−5	−2
Urb	320100	−7	−5	−5	−6	−3	−5	−3	−3	−4	−6	−3
Int	320210	−4	−1	−1	−2	−6	−2	5	3	−1	−2	−1
Int	320220	−3	0	−1	−4	−3	−4	4	−1	0	−3	−1
Int	320230	−4	−2	−1	−4	−3	−2	0	−1	−2	−1	−1
Etcs	320240	−3	0	−1	−3	−2	−2	3	1	0	−1	−1
Perm	320310	2	4	2	3	2	3	0	0	4	2	2
Perm	320330	−3	−1	0	−1	−1	0	0	−1	−1	−2	−1
Urb	330100	−3	−5	−5	−5	−1	−5	−3	−2	−3	−4	−3
Int	330210	−1	0	−1	−1	−4	−2	5	5	0	1	0
Int	330220	0	0	−1	−3	−2	−4	4	0	1	0	0
Int	330230	−1	−1	−1	−3	−1	−2	0	0	−1	2	−1
Etcs	330240	0	0	0	−2	−1	−2	3	2	1	1	0
Perm	330310	5	5	2	4	4	3	0	1	5	5	2
Perm	330320	3	1	0	1	1	0	0	1	1	2	1
Perm Macro	–	0	0	0	0	0	0	0	0	0	0	0
Other	–	1	1	1	1	0	1	1	1	1	1	0
	tot	−3	−4	−5	−4	0	−4	−4	−3	−2	−5	−2

Table A3. Changes in the area-weighted balance of each LULC change dynamic.

Transition and Permanence	Change	km²	Global Climate Regulation	Air Quality Regulation	Water Flow Regulation	Water Purification	Erosion Regulation	Natural Hazard Regulation	Crops	Fodder	Timber	Wild Food and Resources	Sum
Urb	210100	1725	−1.3	−2.7	−2.5	−2.2	1.8	−1.8	−4.6	−3.7	−1.8	−2.7	−21.5
Int	210220	1496	0.6	0.1	0.2	−0.8	1.1	−0.8	−0.5	−2.4	0.5	−0.5	−2.4
Perm	210230	299	0.0	−0.1	0.0	−0.2	0.3	0.0	−0.5	−0.5	−0.1	0.1	−0.9
Etcs	210240	3430	1.8	0.5	0.7	−1.0	4.0	0.5	−1.7	−3.1	0.6	0.6	2.8
For Ext	210310	173	0.3	0.3	0.2	0.3	0.5	0.3	−0.3	−0.2	0.3	0.2	1.9
For Ext	210320	398	0.6	0.1	0.2	0.3	0.7	0.3	−0.7	−0.5	0.1	0.2	1.3
For Ext	210330	46	0.0	–	0.0	0.0	0.1	0.0	−0.1	−0.1	–	−0.0	−0.0
Urb	220100	294	−0.3	−0.5	−0.5	−0.2	0.1	−0.2	−0.7	−0.2	−0.4	−0.4	−3.2
Agr Ext	220210	695	−0.3	−0.1	−0.1	0.4	−0.5	0.4	0.2	1.1	−0.2	0.2	1.1
Agr Ext	220230	20	−0.0	−0.0	−0.0	−0.0	0.0	0.0	−0.0	−0.0	−0.0	0.0	−0.0
Etcs	220240	2694	0.2	0.2	0.2	0.6	1.2	1.7	−0.4	1.9	−0.5	1.3	6.5
For Ext	220310	126	0.2	0.2	0.1	0.3	0.2	0.3	−0.2	0.0	0.2	0.2	1.6
For Ext	220320	174	0.2	0.0	0.0	0.2	0.2	0.2	−0.2	0.1	−0.0	0.1	0.8
For Ext	220330	33	−0.0	−0.0	0.0	0.0	0.0	0.0	−0.0	−0.0	−0.0	0.0	0.0
Urb	230100	89	−0.1	−0.1	−0.1	−0.1	0.0	−0.1	−0.1	−0.0	−0.1	−0.2	−0.8
Int	230210	357	−0.0	0.1	−0.0	0.2	−0.4	−0.0	0.6	0.6	0.1	−0.1	1.1
Int	230220	47	0.0	0.0	0.0	0.0	−0.0	−0.0	0.1	0.0	0.0	−0.0	0.1
Etcs	230240	463	0.2	0.2	0.1	0.1	0.1	0.1	0.5	0.3	0.2	−0.1	1.8
For Ext	230310	67	0.1	0.1	0.1	0.2	0.1	0.1	–	0.0	0.1	0.1	0.9
For Ext	230320	587	0.8	0.3	0.2	0.8	0.5	0.4	–	0.3	0.4	0.1	3.8
For Ext	230330	10	0.0	0.0	0.0	0.0	0.0	0.0	–	–	0.0	−0.0	0.0

Table A3. *Cont.*

Transition and Permanence	Change	km^2	Global Climate Regulation	Air Quality Regulation	Water Flow Regulation	Water Purification	Erosion Regulation	Natural Hazard Regulation	Crops	Fodder	Timber	Wild Food and Resources	Sum
Urb	240100	1357	−1.7	−2.3	−2.2	−1.4	−0.2	−1.6	−2.9	−1.7	−1.7	−2.4	−18.0
Int	240210	4245	−2.2	−0.7	−0.9	1.2	−5.0	−0.6	2.1	3.9	−0.8	−0.7	−3.5
Int	240220	2158	−0.2	−0.2	−0.2	−0.5	−1.0	−1.4	0.3	−1.5	0.4	−1.0	−5.2
Int	240230	259	−0.1	−0.1	−0.0	−0.1	−0.0	−0.0	−0.3	−0.2	−0.1	0.0	−1.0
For Ext	240310	819	1.2	1.2	0.7	1.6	1.2	1.3	−0.9	−0.3	1.2	1.0	8.2
For Ext	240320	1320	1.2	0.0	0.3	1.3	0.9	0.7	−1.5	−0.3	0.1	0.5	3.2
For Ext	240330	75	−0.0	−0.0	0.0	0.0	0.0	0.0	−0.1	−0.1	−0.0	−0.0	−0.1
Urb	310100	137	−0.4	−0.4	−0.3	−0.4	−0.2	−0.4	−0.1	−0.1	−0.4	−0.4	−3.2
Int	310210	315	−0.6	−0.5	−0.3	−0.5	−0.8	−0.5	0.5	0.4	−0.5	−0.4	−3.4
Int	310220	415	−0.7	−0.7	−0.4	−0.9	−0.8	−0.9	0.5	−0.1	−0.5	−0.7	−5.1
Int	310230	59	−0.1	−0.1	−0.1	−0.1	−0.1	−0.1	–	−0.0	−0.1	−0.1	−0.8
Etcs	310240	928	−1.4	−1.4	−0.8	−1.9	−1.4	−1.4	1.1	0.3	−1.4	−1.1	−9.3
Perm	310320	2219	−1.3	−3.3	−1.4	−2.3	−1.7	−2.2	–	0.3	−3.1	−1.7	−16.7
Perm	310330	177	−0.3	−0.3	−0.1	−0.2	−0.2	−0.2	–	−0.1	−0.3	−0.3	−1.9
Urb	320100	136	−0.3	−0.2	−0.2	−0.3	−0.1	−0.2	−0.1	−0.1	−0.2	−0.3	−2.1
Int	320210	1190	−1.7	−0.2	−0.5	−0.8	−2.2	−0.8	2.0	1.4	−0.3	−0.7	−3.9
Int	320220	243	−0.2	−0.0	−0.1	−0.3	−0.3	−0.3	0.3	−0.1	0.0	−0.2	−1.2
Int	320230	543	−0.8	−0.3	−0.2	−0.7	−0.5	−0.4	–	−0.3	−0.3	−0.1	−3.5
Etcs	320240	2037	−1.9	−0.1	−0.4	−2.0	−1.4	−1.1	2.4	0.5	−0.2	−0.8	−5.0
Perm	320310	3692	2.2	5.4	2.4	3.8	2.9	3.6	–	−0.5	5.1	2.9	27.8
Perm	320330	646	−0.7	−0.1	−0.1	−0.2	−0.3	0.0	–	−0.3	−0.2	−0.5	−2.3
Urb	330100	18	−0.0	−0.0	−0.0	−0.0	−0.0	−0.0	−0.0	−0.0	−0.0	−0.0	−0.2
Int	330210	60	−0.0	–	−0.0	−0.0	−0.1	−0.0	0.1	0.1	–	0.0	0.0
Int	330220	17	0.0	0.0	−0.0	−0.0	−0.0	−0.0	0.0	0.0	0.0	−0.0	−0.0
Int	330230	8	−0.0	−0.0	−0.0	−0.0	−0.0	−0.0	–	–	−0.0	0.0	−0.0
Etcs	330240	99	0.0	0.0	−0.0	−0.1	−0.0	−0.1	0.1	0.1	0.0	0.0	0.1
Perm	330310	108	0.2	0.2	0.1	0.1	0.1	0.1	–	0.0	0.2	0.2	1.2
Perm	330320	2200	2.3	0.4	0.3	0.7	1.0	−0.1	–	1.0	0.6	1.6	7.7
Perm Macro	-	261,823	–	–	–	–	–	–	–	–	–	–	−0.0
Other	-	1080	0.9	1.2	1.2	0.9	−0.0	0.9	1.3	0.8	0.8	1.2	9.0
	tot	301,605	−3.5	−3.7	−4.5	−4.1	−0.2	−4.3	−3.8	−3.2	−2.2	−4.7	−34.1

Table A4. Overall changes relative to each dynamic within each SNAI zone. Very weak changes have not been represented in this table (values between −0.5 and +0.5).

A	−9.3	B	−1.4	C	−21.7	D	−3.1	E	0.1	F	1.3
210100	−5.1	210100	−0.9	210100	−11.0	310320	−4.9	310320	−5.7	310320	−2.2
240100	−3.6	240100	−0.6	240100	−7.7	240100	−3.7	310240	−2.6	310240	−0.5
310320	−0.7			310320	−3.0	210100	−3.2	320240	−2.0	320210	−0.5
220100	−0.6			310220	−2.6	310240	−2.9	320230	−1.9	320240	−0.5
310100	−0.6			310240	−2.5	240220	−1.6	240100	−1.9	240100	−0.5
240220	−0.5			240220	−2.0	310220	−1.6	320210	−1.6	240310	0.6
310240	−0.5			310210	−1.5	320240	−1.5	210100	−1.1	240320	0.7
320100	−0.5			220100	−1.3	320210	−1.1	320330	−0.9	330320	0.8
240310	0.5			310100	−1.0	240210	−1.0	240210	−0.8	320310	3.8
220240	0.6			240210	−1.0	320230	−0.9	240220	−0.7		
320310	1.1			210220	−0.8	210220	−0.9	310330	−0.5		
				320100	−0.6	310100	−0.9	310210	−0.5		
				320240	−0.6	310210	−0.9	310100	−0.5		
				310330	−0.5	220100	−0.8	210240	0.5		
				230320	0.9	320330	−0.5	230240	0.6		
				210240	1.0	330310	0.5	230320	0.7		
				330320	1.2	210310	0.5	240320	1.1		
				220240	1.8	230240	0.6	220240	1.3		
				240310	1.8	240320	0.8	240310	2.6		
				320310	3.8	210240	0.8	330320	2.7		
						230320	1.5	320310	10.1		
						220240	2.4				
						240310	2.5				
						330320	2.8				
						320310	8.6				

References

1. Accordo di Partenariato 2014–2020, Strategia nazionale per le Aree interne: Definizione, obiettivi, strumenti e governance. *Aedon Riv. di Arti e Dirit. Line* **2018**, *3*, 1–70. [CrossRef]
2. Vatitsi, K.; Ioannidou, N.; Mirli, A.; Siachalou, S.; Kagalou, I.; Latinopoulos, D.; Mallinis, G. LULC Change Effects on Environmental Quality and Ecosystem Services Using EO Data in Two Rural River Basins in Thrace, Greece. *Land* **2023**, *12*, 1140. [CrossRef]
3. Egarter Vigl, L.; Tasser, E.; Schirpke, U.; Tappeiner, U. Using Land Use/Land Cover Trajectories to Uncover Ecosystem Service Patterns across the Alps. *Reg. Environ. Chang.* **2017**, *17*, 2237–2250. [CrossRef]
4. Fang, Z.; Ding, T.; Chen, J.; Xue, S.; Zhou, Q.; Wang, Y.; Wang, Y.; Huang, Z.; Yang, S. Impacts of Land Use/Land Cover Changes on Ecosystem Services in Ecologically Fragile Regions. *Sci. Total Environ.* **2022**, *831*, 154967. [CrossRef]
5. Kuang, W.; Du, G.; Lu, D.; Dou, Y.; Li, X.; Zhang, S.; Chi, W.; Dong, J.; Chen, G.; Yin, Z.; et al. Global Observation of Urban Expansion and Land-Cover Dynamics Using Satellite Big-Data. *Sci. Bull.* **2021**, *66*, 297–300. [CrossRef]
6. Buytaert, W.; Zulkafli, Z.; Grainger, S.; Acosta, L.; Alemie, T.C.; Bastiaensen, J.; De Bièvre, B.; Bhusal, J.; Clark, J.; Dewulf, A.; et al. Citizen Science in Hydrology and Water Resources: Opportunities for Knowledge Generation, Ecosystem Service Management, and Sustainable Development. *Front. Earth Sci.* **2014**, *2*, 26. [CrossRef]
7. Geijzendorffer, I.R.; Roche, P.K. The Relevant Scales of Ecosystem Services Demand. *Ecosyst. Serv.* **2014**, *10*, 49–51. [CrossRef]
8. Meisch, C.; Schirpke, U.; Huber, L.; Rüdisser, J.; Tappeiner, U. Assessing Freshwater Provision and Consumption in the Alpine Space Applying the Ecosystem Service Concept. *Sustainability* **2019**, *11*, 1131. [CrossRef]
9. Zhang, Z.; Liu, Y.; Wang, Y.; Liu, Y.; Zhang, Y.; Zhang, Y. What Factors Affect the Synergy and Tradeoff between Ecosystem Services, and How, from a Geospatial Perspective? *J. Clean. Prod.* **2020**, *257*, 120454. [CrossRef]
10. Zhang, L.; Huang, Q.; He, C.; Yue, H.; Zhao, Q. Assessing the Dynamics of Sustainability for Social-Ecological Systems Based on the Adaptive Cycle Framework: A Case Study in the Beijing-Tianjin-Hebei Urban Agglomeration. *Sustain. Cities Soc.* **2021**, *70*, 102899. [CrossRef]
11. Wei, H.; Fan, W.; Wang, X.; Lu, N.; Dong, X.; Zhao, Y.; Ya, X.; Zhao, Y. Integrating Supply and Social Demand in Ecosystem Services Assessment: A Review. *Ecosyst. Serv.* **2017**, *25*, 15–27. [CrossRef]
12. Schirpke, U.; Tasser, E.; Borsky, S.; Braun, M.; Eitzinger, J.; Gaube, V.; Getzner, M.; Glatzel, S.; Gschwantner, T.; Kirchner, M.; et al. Past and Future Impacts of Land-Use Changes on Ecosystem Services in Austria. *J. Environ. Manag.* **2023**, *345*, 118728. [CrossRef]
13. Oppio, A. Migrants and Italian Inner Areas for an Anti-Fragility Strategy. *Valori Valutazioni* **2021**, *2021*, 93–102. [CrossRef]
14. Dax, T.; Schroll, K.; Machold, I.; Derszniak-Noirjean, M.; Schuh, B.; Gaupp-Berghausen, M. Land Abandonment in Mountain Areas of the EU: An Inevitable Side Effect of Farming Modernization and Neglected Threat to Sustainable Land Use. *Land* **2021**, *10*, 591. [CrossRef]
15. Marino, D.; Palmieri, M.; Marucci, A.; Pili, S. Long-Term Land Cover Changes and Ecosystem Services Variation: Have the Anthropogenic Transformations Degraded Human Well-Being in Italy? *Ital. Rev. Agric. Econ.* **2022**, *77*, 7–23. [CrossRef]
16. Esengulova, N.; Balena, P.; De Lucia, C.; Lopolito, A.; Pazienza, P. Key Drivers of Land Use Changes in the Rural Area of Gargano (South Italy) and Their Implications for the Local Sustainable Development. *Land* **2024**, *13*, 166. [CrossRef]
17. McPhearson, T.; Cook, E.M.; Berbés-Blázquez, M.; Cheng, C.; Grimm, N.B.; Andersson, E.; Barbosa, O.; Chandler, D.G.; Chang, H.; Chester, M.V.; et al. A Social-Ecological-Technological Systems Framework for Urban Ecosystem Services. *One Earth* **2022**, *5*, 505–518. [CrossRef]
18. Costanza, R.; de Groot, R.; Sutton, P.; van der Ploeg, S.; Anderson, S.J.; Kubiszewski, I.; Farber, S.; Turner, R.K. Changes in the Global Value of Ecosystem Services. *Glob. Environ. Chang.* **2014**, *26*, 152–158. [CrossRef]
19. de Groot, R.; Brander, L.; van der Ploeg, S.; Costanza, R.; Bernard, F.; Braat, L.; Christie, M.; Crossman, N.; Ghermandi, A.; Hein, L.; et al. Global Estimates of the Value of Ecosystems and Their Services in Monetary Units. *Ecosyst. Serv.* **2012**, *1*, 50–61. [CrossRef]
20. Blumstein, M.; Thompson, J.R. Land-Use Impacts on the Quantity and Configuration of Ecosystem Service Provisioning in Massachusetts, USA. *J. Appl. Ecol.* **2015**, *52*, 1009–1019. [CrossRef]
21. Schirpke, U.; Tscholl, S.; Tasser, E. Spatio-Temporal Changes in Ecosystem Service Values: Effects of Land-Use Changes from Past to Future (1860–2100). *J. Environ. Manag.* **2020**, *272*, 111068. [CrossRef] [PubMed]
22. García-Nieto, A.P.; Geijzendorffer, I.R.; Baró, F.; Roche, P.K.; Bondeau, A.; Cramer, W. Impacts of Urbanization around Mediterranean Cities: Changes in Ecosystem Service Supply. *Ecol. Indic.* **2018**, *91*, 589–606. [CrossRef]
23. Gomes, E.; Inácio, M.; Bogdzevič, K.; Kalinauskas, M.; Karnauskaitė, D.; Pereira, P. Future Land-Use Changes and Its Impacts on Terrestrial Ecosystem Services: A Review. *Sci. Total. Environ.* **2021**, *781*, 146716. [CrossRef] [PubMed]
24. Marucci, A.; Marino, D.; Palmieri, M.; Pili, S. The Role of Agroforestry Areas in the Potential Provision of Ecosystem Services: The Case of the Molise Region. *L'Ital. For. Mont.* **2022**, *77*, 153–163. [CrossRef]
25. Han, Z.; Song, W.; Deng, X.; Xu, X. Trade-Offs and Synergies in Ecosystem Service within the Three-Rivers Headwater Region, China. *Water* **2017**, *9*, 588. [CrossRef]
26. Mendoza-González, G.; Martínez, M.L.; Lithgow, D.; Pérez-Maqueo, O.; Simonin, P. Land Use Change and Its Effects on the Value of Ecosystem Services along the Coast of the Gulf of Mexico. *Ecol. Econ.* **2012**, *82*, 23–32. [CrossRef]
27. Burkhard, B.; Santos-Martin, F.; Nedkov, S.; Maes, J. An Operational Framework for Integrated Mapping and Assessment of Ecosystems and Their Services (MAES). *One Ecosyst.* **2018**, *3*, e22831. [CrossRef]

28. Burkhard, B.; Kandziora, M.; Hou, Y.; Müller, F. Ecosystem Service Potentials, Flows and Demands-Concepts for Spatial Localisation, Indication and Quantification. *Landsc. Online* **2014**, *34*, 1–32. [CrossRef]
29. Pandeya, B.; Buytaert, W.; Zulkafli, Z.; Karpouzoglou, T.; Mao, F.; Hannah, D.M. A Comparative Analysis of Ecosystem Services Valuation Approaches for Application at the Local Scale and in Data Scarce Regions. *Ecosyst. Serv.* **2016**, *22*, 250–259. [CrossRef]
30. Schild, J.E.M.; Vermaat, J.E.; van Bodegom, P.M. Differential Effects of Valuation Method and Ecosystem Type on the Monetary Valuation of Dryland Ecosystem Services: A Quantitative Analysis. *J. Arid Environ.* **2018**, *159*, 11–21. [CrossRef]
31. Marino, D.; Barone, A.; Marucci, A.; Pili, S.; Palmieri, M. Impact of Land Use Changes on Ecosystem Services Supply: A Meta Analysis of the Italian Context. *Land* **2023**, *12*, 2173. [CrossRef]
32. Ecosystem Services Valuation Database. Available online: https://www.esvd.info/ (accessed on 8 August 2023).
33. SINAnet Groupware. Available online: https://groupware.sinanet.isprambiente.it/uso-copertura-e-consumo-di-suolo/library/copertura-del-suolo/corine-land-cover (accessed on 25 March 2024).
34. Istat Aree Interne. Available online: https://www.istat.it/notizia/la-geografia-delle-aree-interne-nel-2020-vasti-territori-tra-potenzialita-e-debolezze/ (accessed on 25 March 2024).
35. Burkhard, B.; Kroll, F.; Müller, F.; Windhorst, W. Landscapes' Capacities to Provide Ecosystem Services—A Concept for Land-Cover Based Assessments. *Landsc. Online* **2009**, *15*, 1–22. [CrossRef]
36. Burkhard, B.; Kroll, F.; Nedkov, S.; Müller, F. Mapping Ecosystem Service Supply, Demand and Budgets. *Ecol. Indic.* **2012**, *21*, 17–29. [CrossRef]
37. Jacobs, S.; Burkhard, B.; Van Daele, T.; Staes, J.; Schneiders, A. 'The Matrix Reloaded': A Review of Expert Knowledge Use for Mapping Ecosystem Services. *Ecol. Model.* **2015**, *295*, 21–30. [CrossRef]
38. Bateman, I.J.; Harwood, A.R.; Mace, G.M.; Watson, R.T.; Abson, D.J.; Andrews, B.; Binner, A.; Crowe, A.; Day, B.H.; Dugdale, S.; et al. Bringing Ecosystem Services into Economic Decision-Making: Land Use in the United Kingdom. *Science* **2013**, *341*, 45–50. [CrossRef]
39. Boutwell, J.L.; Westra, J.V. Benefit Transfer: A Review of Methodologies and Challenges. *Resources* **2013**, *2*, 517–527. [CrossRef]
40. Richardson, L.; Loomis, J.; Kroeger, T.; Casey, F. The Role of Benefit Transfer in Ecosystem Service Valuation. *Ecol. Econ.* **2015**, *115*, 51–58. [CrossRef]
41. ISTAT ASC. Available online: https://asc.istat.it/ASC/ (accessed on 1 June 2024).
42. D'Andria, E.; Fiore, P. The RI.P.R.O.VA.RE. Project for the Regeneration of Inland Areas: A Focus on the Ufita Area in the Campania Region (Italy). *Buildings* **2023**, *13*, 336. [CrossRef]
43. Tantillo, F. Una Strategia per Le Aree Interne. In *Borgofuturo+: Un Progetto Locale per le Aree Interne;* Quodlibet: Macerata, Italy, 2022; pp. 57–62. [CrossRef]
44. Fenu, N.; Giaccaria, P. La Scommessa della SNAI. La Strategia Nazionale delle Aree Interne. 2023. Available online: https://www.documentigeografici.it/index.php/docugeo/article/view/476 (accessed on 1 June 2024).
45. Lucia, C.; Annunziata, P. Spatial Analysis for Territorial Marginality Mapping. GIS-Based Application to the Case of Southern and Insular Italy. In Proceedings of the SUPTM 2022: 1st Conference on Future Challenges in Sustainable Urban Planning & Territorial Management, Online, 17–19 January 2022; pp. 1–4. [CrossRef]
46. Pilogallo, A.; Scorza, F.; Murgante, B. Ecosystem Services-Based City Ranking in Italy: A Tool to Enhance Sustainable Thinking in Regeneration Strategies. *Land* **2024**, *13*, 891. [CrossRef]
47. Wang, Y.; Rao, Y.; Zhu, H. Analyzing the Land Use and Cover Change Inside and Outside China's Ecological Function Area. *Land* **2023**, *12*, 1447. [CrossRef]
48. Shen, J.; Li, S.; Liu, L.; Liang, Z.; Wang, Y.; Wang, H.; Wu, S. Uncovering the Relationships between Ecosystem Services and Social-Ecological Drivers at Different Spatial Scales in the Beijing-Tianjin-Hebei Region. *J. Clean. Prod.* **2021**, *290*, 125193. [CrossRef]
49. Luo, Y.; Yang, D.; O'Connor, P.; Wu, T.; Ma, W.; Xu, L.; Guo, R.; Lin, J. Dynamic Characteristics and Synergistic Effects of Ecosystem Services under Climate Change Scenarios on the Qinghai–Tibet Plateau. *Sci. Rep.* **2022**, *12*, 2540. [CrossRef] [PubMed]
50. Marino, D.; Palmieri, M.; Marucci, A.; Tufano, M. Comparison between Demand and Supply of Some Ecosystem Services in National Parks: A Spatial Analysis Conducted Using Italian Case Studies. *Conservation* **2021**, *1*, 36–57. [CrossRef]
51. Pearce, D.W.; Turner, R.K. *Economics of Natural Resources and the Environment*; Harvester Wheatsheaf: Birmingham, UK, 1990; ISBN 0 7450 0225 0.
52. Marino, D.; Palmieri, M. Investing in nature: Working with public expenditure and private payments for a new governance model. In *Reconnecting Natural and Cultural Capital: Contributions from Science and Policy*; Paracchini, M.L., Zingari, P.C., Blasi, C., Eds.; European Commission, Joint Research Centre: Brussels, Belgium, 2018; ISBN 9789279599491.
53. Putnam, R.D. Tuning in, turning out: The strange disappearance of social capital in America. *Political Sci. Politics* **1995**, *27*, 64–83.
54. You, C.; Qu, H.; Wang, C.; Feng, C.C.; Guo, L. Trade-off and Synergistic of Ecosystem Services Supply and Demand Based on Socio-Ecological System (SES) in Typical Hilly Regions of South China. *Ecol. Indic.* **2024**, *160*, 111749. [CrossRef]
55. Ekins, P. Identifying Critical Natural Capital Conclusions about Critical Natural Capital. *Ecol. Econ.* **2003**, *44*, 277–292. [CrossRef]
56. Chaplin-Kramer, R.; Neugarten, R.A.; Sharp, R.P.; Collins, P.M.; Polasky, S.; Hole, D.; Schuster, R.; Strimas-Mackey, M.; Mulligan, M.; Brandon, C.; et al. Mapping the Planet's Critical Natural Assets. *Nat. Ecol. Evol.* **2023**, *7*, 51–61. [CrossRef]
57. Raudsepp-Hearne, C.; Peterson, G.D.; Bennett, E.M. Ecosystem Service Bundles for Analyzing Tradeoffs in Diverse Landscapes. *Proc. Natl. Acad. Sci. USA* **2010**, *107*, 5242–5247. [CrossRef] [PubMed]

58. Rossitti, M.; Dell'ovo, M.; Oppio, A.; Torrieri, F. The Italian National Strategy for Inner Areas (Snai): A Critical Analysis of the Indicator Grid. *Sustainability* **2021**, *13*, 6927. [CrossRef]
59. Porta, A.; Segre, G.; Arauzo-Carod, J.M. The Italian National Strategy for Inner Areas: First Insights from Regions' Specialization in Cultural and Creative Industries. In *Clusters in Times of Uncertainty*; Edward Elgar Publishing: Cheltenham, UK, 2024; pp. 240–261. [CrossRef]
60. Pettenella, D.; Romano, R. Le Foreste Delle Aree Interne: Potenzialità Di Sviluppo Dell'economia Locale e Di Offerta Di Beni Pubblici. *Agriregionieuropa* **2016**, *12*, 50–55.
61. Obst, C.; Hein, L.; Edens, B. National Accounting and the Valuation of Ecosystem Assets and Their Services. *Environ. Resour. Econ.* **2016**, *64*, 1–23. [CrossRef]

Disclaimer/Publisher's Note: The statements, opinions and data contained in all publications are solely those of the individual author(s) and contributor(s) and not of MDPI and/or the editor(s). MDPI and/or the editor(s) disclaim responsibility for any injury to people or property resulting from any ideas, methods, instructions or products referred to in the content.

land

MDPI

Article

Integrating Gender Perspectives in Participation to Guide Changes in Urban Planning in Serbia

Nataša Čolić Marković and Nataša Danilović Hristić *

Institute of Architecture and Urban & Spatial Planning of Serbia, 11000 Beograd, Serbia; natasac@iaus.ac.rs
* Correspondence: natasadh@iaus.ac.rs or ndanilovichristic@gmail.com

Abstract: Participation is an essential instrument for the expression of diverse knowledge and interests in urban planning. It ultimately influences outcomes in the built environment, fostering inclusive, safe, and accessible spaces and environments. However, urban planning is often viewed, by the public and practitioners alike, as an expert-driven, administrative, and semi-closed system, focusing on the provision of resources that are fit for all, seldom targeting the participation of gender expertise in decision-making. Drawing on empirical qualitative research conducted with planning practitioners in Belgrade in 2023 within the UN Women project "Safety of Women and Girls in Public Spaces", this paper investigates if and how gender perspectives are integrated into urban planning policy and practice in Serbia, and the role of knowledge creation and participation in informing planning about diverse needs at the local level. The research draws on practical insights from practitioners in the fields of urban and spatial planning, traffic, and urban safety. The findings reveal how limitations to modifying the national legislative framework in terms of gender integration could be overcome at the level of planning policy and practice. The importance of including the participation of gender expertise alongside achieving gender parity in knowledge creation is highlighted.

Keywords: participation; gender perspectives; urban development; planning practice; UN Women; Serbia

1. Introduction

Addressing gender inequalities has been recognized as crucial in the formation, implementation, and evaluation of public policy for sustainable urban development [1–3]. Planning and governance practices are expected to incorporate gender perspectives into knowledge creation, considering how new planning solutions impact the most vulnerable, ensuring that no one is left behind [4–7].

As a strategy to systematically integrate gender perspectives into all stages of policy development, planning, and decision-making across sectors, gender mainstreaming was established in the late 1990s by the United Nations to advance equality, justice, inclusivity, and sensitivity to the diverse gender-related needs [8,9]. The goal of gender mainstreaming is to ensure equal opportunities across various spheres of everyday life for all genders, to address their needs, including those related to the care economy, which are often unrecognized, and to guarantee accessibility both in decision-making processes and in the planning of urban resources, including public spaces and services [1,10,11]. Integrating these goals into national urban planning frameworks requires prioritizing safety, affordability, accessibility, and a sense of belonging in urban areas by identifying the needs regarding the provision, access, and usability of public resources that significantly impact the daily lives of women and men [12,13]. While gender mainstreaming has been considered to be a political and organizational strategy across the EU, the prevalence and effectiveness of gender perspectives in urban development across countries diverge with regard to legal, organizational, and cultural aspects [14]. According to Zibell et al. [14] the integration of gender perspectives in urban development in Europe is led by initiatives and programs of international and EU research. However, its implementation at the local level depends on the openness of institutions, with one of the most successful examples in Europe being the city of Vienna [11,14].

Research in the field of gender and planning suggests that, at the local level, gender perspectives should be integrated through the substantive representation of women and gender expertise in planning processes to challenge drivers of inequality [10,11,15–19]. This perspective is supported by the communicative planning model in planning theory, which involves fostering participation and communication to acquire knowledge representing diverse consumption decisions, work responsibilities, social practices, and economic habits within communities [18,19]. Participation helps to shape the knowledge base among process participants, including practitioners, the general public, private sector representatives and decision-makers, to guide more just planning decisions and outcomes [1,20,21]. It operates on the premise that social groups and individuals learn from one another during the process, which influences their decision-making behaviors [22]. Knowledge is considered the most significant outcome of a dynamic, collaborative approach to planning, providing a base for decision-makers' and professionals' actions that reflect the diverse meanings of a good quality of life [22–25]. However, planning processes can remain in a semi-closed setting due to the established "rules of the game" and the weight of resources, which may influence what counts as knowledge [26,27]. A fuller picture of power relations in planning, including those related to gender perspectives, thus requires constant updating [28]. While the literature in this field on more developed democratic societies has evolved [1–3,10–20], there is a limited understanding of how gender perspectives are integrated into the hybrid planning and participation of transitional and post-socialist societies, such as Serbia. This research problem is addressed in this paper.

Participation in planning practice, among other forms of representative democracy, is widely viewed as a means to uncover and address inequalities, challenging the assumption of homogeneity among urban populations and their needs [24,29,30]. This perspective aligns with institutional frameworks in many developed democracies, where the state actively supports the realization of essential community interests by promoting the exercise of individual human, civil, and political rights through participatory practices [31,32]. Still, in some planning practices, the integration of gender perspectives in knowledge production has been marginalized for decades [33–35]. Urban planning in Europe has long been dominated by technology-driven models and investment-oriented priorities [34]. This approach often reduces complex socio-spatial challenges—such as public infrastructure, healthcare, sanitation, public space hygiene, and affordable housing—to technical or engineering solutions [35,36]. Urban planning still struggles to integrate the gender perspectives of diverse daily needs and experiences in planning proposals due to "...the lack of tools to clearly highlight women's needs to decision-makers, practitioners, technicians and the local community" [1] (p. 2). Furthermore, in some post-socialist countries transitioning to market-based economies, societal human rights were significantly more developed than political human rights during the era of state socialism. Current planning systems have thus retained elements of the unitary and collectivist principles established during that period through norms at the level of planning policy and institutions [37–40]. At the same time, substantive aspects of planning, such as practitioners' values, relationships, and experiences of working under the uncertainty brought by transition, enhance our understanding of how acting in little ways may help achieve more inclusive planning and participation [41].

In Serbia, the citizens' rights to participate in urban planning processes have been safeguarded by law since 1949 [42]. The planning system has traditionally relied on technical rules and normative planning standards as the most accurate reflection of socially framed standards to organize public spaces and services [38,39]. Until 2014, the legal requirements for public participation in the plan formation process entailed informing the public about participation options via local newspapers, media and/or municipality websites, public inquiry on the draft plan for 30 days in the premises of the local self-governance unit, and reporting responses to the comments and complaints received during the participation process during public discussion. The transition to an urban governance framework, alongside Serbia's ongoing process of European integration, has introduced changes to the participatory practices [41]. These include the formalization of two-tiered public participation processes from 2014 by conducting an early public inquiry on the concept plan for 15 days in addition to the traditional public inquiry on the draft plan. Testing innovative participatory methods primarily takes place within new governance initiatives, such as the formation processes of sustainable and integrated urban and territorial strategies in Serbian cities [43]. Moreover, the importance of participatory planning has been highlighted by increasingly vocal citizen groups and grassroots movements responding to the usurpation of public space and services, as well as the threats posed to the environment and cultural heritage by new developments in Belgrade [44].

Along with the process of EU integration, Serbia adopted the Strategy for Gender Equality in 2016 [45,46], and the Gender Equality Law [47]. This framework made local authorities responsible for the integration of gender perspectives through gender mainstreaming and gender-budgeting in the planning and implementation of policies financed from the national, regional, and local budget. However, in 2024, the Constitutional Court of Serbia temporarily suspended the implementation of the Gender Equality Law, citing concerns that it might not be aligned with the Serbian Constitution [48]. These circumstances, along with the lack of participation from the planning and construction sector in the working group for the development of the national strategy for gender equality, may challenge the integration of gender perspectives into urban planning policies and their implementation [49,50].

Introducing gender perspectives into planning legislation and practice in order to address the needs, roles, identities, and limitations of different societal and cultural groups requires adjustments in both the normative institutional framework and, substantively, within the planning community [15,19,40]. Using the conceptual framework of duality between the normative and substantive aspects of planning [40,51], this research explores the potential for integrating gender perspectives into hybrid planning systems in transitional and post-socialist societies, such as Serbia, through the following questions. From a normative perspective, it examines how gender perspectives are integrated into Serbian policy in the fields of planning, participation, and governance as a planning norm, and how coherence is achieved between the respective policy goals. From a substantive perspective, the research considers how gender perspectives are used as a standard in practitioners' work. It emphasizes "niches" and small-scale changes in planning and participation that foster knowledge creation, aiming to better meet diverse local needs.

Researching planning practice from the practitioners' viewpoint allows the opening of the "black box" of formal processes where planning agendas are set [52–55]. The paper will demonstrate that, although the gender equality framework is not significantly integrated into planning policy and practice in Serbia, there are areas of planning policy and practice where gendered participatory action may be able to improve urban development.

2. Materials and Methods

The research methodology is based on the assumption of critical pragmatism that joint work towards transformation and acting in "little ways" may enable the realization of different outcomes in a given political, historical, and socio-economic context [56,57]. Creating experiences through collaboration may allow for the re-evaluation of previously held beliefs [22–24]. The research design thus considers the experiential and practical knowledge of practitioners as essential for reflecting on and co-producing innovative, gender-responsive practices through small, pragmatic steps [57,58].

This empirical research of gender perspectives in participation and planning in Serbia is part of the UN Women project "Safety of Women and Girls in Public Space" that was implemented in Belgrade, Serbia, between April and September 2023. The project aimed to improve gender responsiveness in planning legislation, practice, and research by promoting collaboration among practitioners in urban planning and related fields. The research was conducted in a live fashion throughout the five workshops with practitioners and policy makers situated in the project process (Figure 1), with the overt position of researchers. The main research methods are the content analysis of recent Serbian legislation and policy in the field of planning, participation, and governance, and ethnography with empirical qualitative research methods during the workshops with practitioners [56].

Figure 1. Workshops with practitioners within the project timeline.

The content analysis of the legislation and policy was conducted to address the first research question regarding the normative aspects of integrating a gender perspective in planning and participation. To address the second research question, the qualitative research included data collection via monitoring conversations at workshops, taking notes, "observing behavior, listening to what is said in conversations both between others and with the fieldworkers, and asking questions" [56] (p. 432). This was followed by anonymous questionnaires completed during the final workshop to assess the effectiveness of the project process towards building capacities to integrate gender perspectives in planning. The questionnaire used a Likert scale with three levels: "agree", "partially agree", and "disagree". It also included open-ended questions, enabling participants to express their views on the topics discussed during the workshops.

The original notes, questions, and responses derived from the workshop discussions and the questionnaire were in Serbian and have been translated into English. At the linear level of analysis, the coding of the narratives used a thematic approach [56], where practitioners' inputs were examined through identified normative and substantive aspects of gender perspectives in planning and participation. At the analytical level, the narratives were contextualized in relation to the results obtained from the questionnaire to explore the relevance of gender perspectives in practitioners' work, their willingness to apply the knowledge about gender-differentiated needs in urban spaces, and the recognition of their role in creating opportunities for more inclusive public participation.

The methodology for conducting "real-time" research within the project implementation process is grounded in Patton's assertion that engaging in and monitoring open communication and the reflections of participants during the working process enriches researchers' understanding of the research phenomena through direct interpersonal engagement [58]. The data analysis and interpretation decode the robust, specific, and tacit experiences of practitioners and policy makers, providing a contextual understanding of planning processes and the opportunities to change usual practices. Understanding how practitioners subjectively experience and interpret gender perspectives in urban planning and participation can encourage the reflection on their behaviors, practices, ethics, and relationships, generate context-dependent knowledge, and contribute to gradual change [59,60].

The practitioners who participated in the project workshops are the main subjects of the research—they include urban and spatial planners, traffic planning practitioners, urban safety experts, policy makers, representatives of universities, urban enterprises, international organizations, local associations, NGOs, and individuals. The sample of practitioners was chosen using a purposive sampling strategy [56], where researchers deliberately selected participants that were most likely to provide valuable insights to address the research questions. The practitioners were invited to participate in the workshops through contact with the legal representatives of the enterprises. The pool of practitioners was expanded through a snowball sampling approach, where the initial contacts pointed out other practitioners working in the institutions and fields relevant to the project aim [56]. A total of 96 individuals attended the workshops, 86% of whom were women. The high female representation can be attributed to women being asked by their supervisors to attend the workshops due to the project topic.

Purposeful sampling implies that the participants possess an intimate knowledge of the local context in which the researched phenomenon is shaped and articulated. Therefore, several preconditions were necessary in choosing the sample:

- All practitioners had to have experience in working on the planning of public resources (public land use and services of general economic interest) at the city or neighborhood level.
- All practitioners had to have experience in facilitating public participation and communicating with stakeholders of different interests, i.e., the general public, private investors, and the representatives of other public enterprises.
- At least some practitioners had to have experience in the application of gender equality provisions from the Gender Equality Strategy in planning practice.
- At least some practitioners had to have experience in the field of urban safety.
- At least some practitioners had to have experience in working on reevaluating planning policy or are currently in the position of policy makers who are in charge of it.

To protect participant anonymity, the researchers did not disclose their names or affiliations. The inputs are presented based on participants' sectors of employment (Table 1).

The workshops were designed to introduce participants to the topic of gender in urban planning, provide practical examples of gender-just participation and co-creation in planning and governance practices in EU countries, and leverage their collective professional expertise to generate ideas on how to integrate gender perspectives into policy and daily planning practices at the local level. The duration of each of the workshops was around four hours, which included brief presentations followed by group work and joint discussion. The different thematic scopes of the workshops meant that practitioners from different backgrounds were invited to each event.

The aim of the first workshop was to increase the knowledge capacities of participants from all sectors on gender approaches in urban planning and participation. A total of 50 participants were present (about 78% of the total invited).

The aim of the second workshop with planning practitioners and third workshop with experts from the field of communal space maintenance and urban safety was to reevaluate the urban planning methodology and procedures. These workshops included a joint analysis of four public micro-locations in Belgrade through focus group discussions. A total of 22 participants were present at the second workshop and 20 participants were present at the third workshop (about 80% of the total invited).

Table 1. Research sample.

Sector	Institutions	Code
Public—national level	Gender Equality Coordination Body of the Republic of Serbia Ministry of Construction, Transport and Infrastructure of the Republic of Serbia Agency for Spatial Planning of the Republic of Serbia	PNL
Public—authority, city level	City of Belgrade Office of the Chief City Architect Secretariat of Public City Transport Police Department for the City of Belgrade Municipality of Vračar	PCL
Public enterprises and institutes	City of Belgrade—Public Transport Greenmarkets Fortress of Belgrade Belgrade's Subway and Train Public Lighting Transportation Institute CIP Urban Planning Institute of Belgrade Institute of Protection of the Heritage of Belgrade	PE
Professional associations	Serbian Chamber of Engineers National Association of Architects National Association of Urban Planners National Association of Spatial Planners	PA
Education and research	University of Belgrade—Faculty of Geography, Department of Spatial Planning University of Belgrade—Faculty of Transport Institute of Architecture and Urban and Spatial Planning of Serbia	ER
Non-governmental	Academy for Female Leadership Standing Conference of Towns and Municipalities	NGO
International organizations	UN Women British Embassy Belgrade	INT

The aim of the fourth thematic workshop involving practitioners and policy makers was to facilitate discussions on the proposed measures to integrate gender perspectives in urban planning policy. A total of 19 participants responded to the invitation to participate at the workshop (about 73% of the total invited). Before the discussion on the measures for the integration of gender perspectives in planning policy, a content analysis of the policy and legislation supporting participation and governance was presented to the practitioners.

The aim of the fifth workshop was to gather all-sector participants in a joint discussion on the possibilities of modifying their practice-as-usual and implementing measures to increase gender responsiveness in planning. After the workshop, an anonymous questionnaire was distributed to the practitioners. There were 42 participants present at the final workshop (about 65% of the total invited), while 36 participants filled in the questionnaire.

At the workshops, practitioners were asked to analyze four public micro-locations in Belgrade through group work in four teams. Most of the participants were already familiar with these locations, as they live and work in Belgrade. The plans of the areas were the main tool for analysis, coupled with the insights from female users (practitioners) who took part in the workshops. Each team elaborated issues through mapping and proposed targeted measures to facilitate inclusive participation and planning standards that address the needs in public spaces from a gender perspective (Figure 2). The teams presented their proposed measures to the other groups, followed by a joint discussion. The cases selected represent typical neighborhoods in Belgrade, serving as critical examples that allow for learning and replicating knowledge [56] in the planning of urban areas across Serbia:

- the central city residential area of Vračar Municipality surrounding a busy green market, representing traditional mixed-use development;
- block 70 in New Belgrade, an open block neighborhood developed under socialist-modernist planning principles;
- the Kaluđerica neighborhood on the city's outskirts, characterized by informal, individually built houses in an underdeveloped area lacking infrastructure; and
- Ušće Park, a large-scale open public space along the central city riverbank.

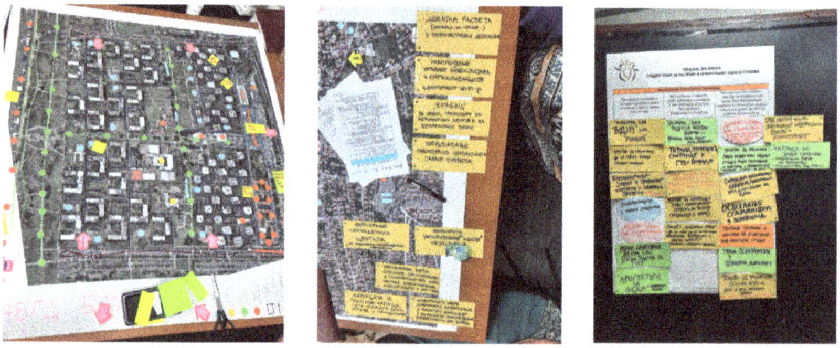

(a) (b) (c)

Figure 2. Case study work during workshops 2 and 3: (**a**) analysis of the open block socialist neighborhood of New Belgrade; (**b**) analysis of the Kaluđerica neighborhood containing informal housing; and (**c**) recommendations to integrate gender perspectives in the planning process and built environment.

3. Findings

The findings relating to the qualitative analysis of practitioners' inputs were contextualized with the content analysis of the legislation and policy in the fields of planning, participation, and governance to answer the first research question of how gender perspectives represent a norm in planning. To answer the second research question relating to the integration of gender perspectives as a substantive standard in the work of practitioners, the results are presented on the basis of practitioners' practical experiences, suggestions, and insights during the case study work, leading to an understanding of their underlying values and actions, which may enable the realization of different outcomes in planning.

3.1. Gender Perspectives as a Planning Norm

During the workshops, practitioners noted that, despite the adoption of a gender equality policy in 2009, the current formal planning policy remains gender neutral. This means that it does not prioritize or specifically address the needs, interests, or experiences of any gender in particular, assuming that their needs can be met without special consideration. One of the practitioners noted that the term "gender" is not represented in both the procedural participation elements and land use standards of the Planning and Construction Law [61]—a key normative framework for planning and development in Serbia. Furthermore, the Draft Spatial Plan of the Republic of Serbia [62] does not distinguish between the needs of women and men in the use of space but emphasizes the importance of considering the needs of all societal groups (including vulnerable groups) when defining priorities for physical development, such as the accessibility to services of general economic interest [62]. These findings suggest that the integration of gender perspectives in formal planning legislation at the national level remains infrequent. Still, one of the practitioners at the workshops recognized the following:

> "...reaching more differentiated data on needs through decision-making can be achieved by making extra efforts and small steps in informing those who are often excluded from decisions about their living environment, even if this is not mandatory by formal planning law" (PS, workshop 2).

Thus, despite the notion that formal planning legislation does not explicitly integrate gender perspectives in planning and participation, some practitioners spoke about the possibility of including them through informal, voluntary practice. The informal practices of planning professionals were previously registered in smaller municipalities in Serbia, where planning initiatives are announced through local radio, TV stations, and the municipality's social media platforms. In other cases, wider citizen participation is carried out at local festivals, which attract a diverse public, alongside more traditional methods such as newspaper ads and official websites [63], but they remain at the level of individual initiatives.

Broader political support to widen the scope of participatory practice in policy and legislation in field of governance can be traced to recent initiatives ratified by the government in line with the EU accession Chapter 22 on Regional Policy and the coordination of structural instruments [64] and the testing of governance initiatives in Serbia. In 2018, the government signed the Additional Protocol to the European Charter of Local Self-Government [65], which requires consultative processes, local referendums, and other tailored methods of participation for local decision-making in self-government units, particularly for residents of larger geographical areas, including rural regions. The need for broader geographical discussions on gender justice was raised by the practitioners at the workshops, as well as the following:

> "We need to decentralize the discourse on gender in planning from Belgrade to smaller urban and rural areas, where women tend to be unpaid household members who spend most of their time engaged in caregiving activities... in villages they are rarely homeowners, and also have limited accessibility to public services" (PS, workshop 5).

Besides the need to reevaluate the spatial scope of participation, another novel legislation in the field of governance, the Planning System Law of 2018 [66], stipulates the broader involvement of diverse citizen associations and civil society organizations in public policy formation. The Law provides a basis for the broader inclusion of gender expertise from the non-governmental sector, which resonated with practitioners at workshops:

> "There is a need for recognition and involvement of diverse groups of participants in decision-making processes, primarily the NGO sector that is making breakthroughs in the domain of gender equality in Serbia" (NGO, workshop 2).

Strengthening effective partnerships and dialogue between different sectors requires the support of subsidiarity in urban planning through an openness to other forms of knowledge. In Serbia, planning has traditionally been viewed as a public sector activity. However, the demands for democratization and governance highlight the need to reassess the traditional hierarchical roles between the public, private, and non-governmental sectors in planning.

The Local Self-Government Law of 2021 [67] aligns with the above-mentioned legislation and policy in the domain of governance and participation, emphasizing mandatory public hearings during the preparation of strategic, spatial, and urban plans. This article concerning this law is important because it outlines the need to tailor the participation process to different levels and types of planning documents—both formal and traditional plans, as well as strategies supporting governance approaches—which are typically considered to be less binding. Currently, all of the planning instruments follow the same procedure and have the same duration, regardless of their scale or territorial coverage.

The participants who attended workshop 4, which was dedicated to proposing measures to integrate gender perspectives in urban planning policy, highlighted that the Sustainable Urban Development Strategy of Serbia until 2030, adopted in 2019 [68], explicitly calls for the consideration of women's needs in its measures to provide accessible public spaces and services in Strategic Goal 3: Societal Wellbeing. The strategy also outlines a set of measures to implement multilevel and multi-stakeholder governance initiatives through participation in Strategic Goal 5: Urban Governance. Measure 5.1.5, titled the "Application of an integrated approach in urban development, synchronizing "new" strategic planning for urban settlements with traditional urban planning", implicitly supports the integration of gender perspectives in participation through both strategic and traditional planning.

Some of the initiatives to expand the formal framework for participation in planning presented at the workshops were recognized by the practitioners as "...small doors for the integration of gender perspectives in participation processes" (INT, workshop 1). One of the practitioners correctly noted that the institutionalization of such processes requires individual and institutional capacity development:

> "I thoroughly enjoyed the workshops and gained valuable insights into women's perspectives in the context of urban planning —something I had never considered before, even as a woman and a planner [...]. This highlights the need for institutional capacity development within local government to integrate gender perspectives into decisions regarding space. However, it also requires a shift in our mentality. For decades, we have planned public spaces and services under the assumption that they suit everyone, while simultaneously prioritizing private developments and aligning them carefully with investors' interests." (ER, workshop 5).

Capacity development and widening the scope of traditional participatory planning requires giving stronger support to urban governance, sharing responsibilities, and considering different forms of expertise and knowledge outside of the public sector. Additionally, the practitioners emphasized the need for broader geographical discussions on gender justice, where both gender and planning policies should address gender-specific issues, such as land management and land ownership patterns in Serbia, where women are significantly underrepresented [69,70].

The above findings reveal a deficiency in the integration of gender perspectives in formal planning legislation. Although the policy framework and legislation in the field of governance supporting inclusive participatory action, aligned with the EU Cohesion Policy for 2021–2027 [71], creates opportunities to enhance gender representation and expertise in urban planning, some practitioners highlighted the challenges related to institutional inertia. They expressed concerns about the lack of guarantees that the provisions outlined in the legal framework would be effectively implemented without strong political commitment at the local level, improved institutional capacities, and sufficient funding to support the institutionalization of rare initiatives, such as the project discussed in this paper, into standard practice.

3.2. Gender Perspectives as a Substantive Standard in Practitioners' Work

To investigate how gender perspectives could become a substantive standard reflecting practitioners' values in urban planning, we triangulated the quantitative data obtained from the case study work at the workshops with the questionnaire results. Figure 3 presents the results from the practitioners' responses to the anonymous questionnaire, highlighting aspects such as the relevance of the workshop themes—including the re-evaluation of planning for public spaces and services, as well as participatory planning—in their work to achieve more equitable urban planning processes and outcomes (1); practitioners' willingness to apply knowledge about gender-differentiated needs in urban spaces gained through the workshops in their ongoing practice (2); and professionals' recognition of their role in integrating gender perspectives as a professional standard in their work (3).

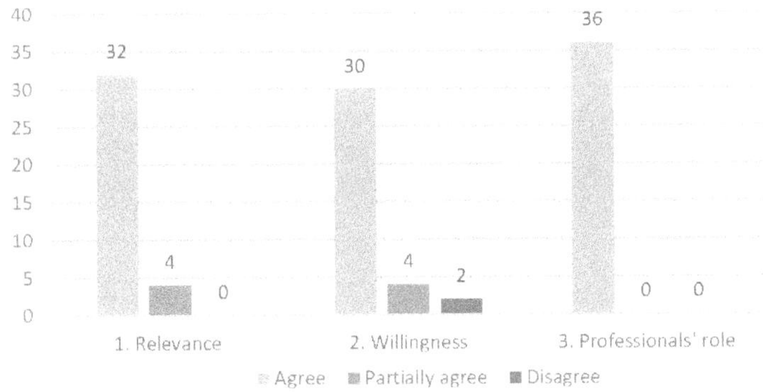

Figure 3. Practitioners' marks from the questionnaire.

Most practitioners gave the highest rating to the relevance of the workshop themes for their practice, while 11.1% partly agreed. During the workshops, they noted that the "case study approach was particularly useful for identifying different barriers and issues in space that women may encounter more frequently than men" (PLC, workshop 5). The method of analyzing the accessibility to public resources in four plans representing different types of neighborhoods in Belgrade, combined with joint discussions on the diverse issues and opportunities encountered by women and men in urban spaces, deepened practitioners' understanding of their varying needs related to consumption, transportation, and caregiving. After the analysis, each group proposed a set of measures to integrate gender perspectives in planning norms and standards, as well as to include gender expertise in plan formation processes through the recognition of diverse stakeholders and strategies to reach them. The exercise generated a large amount of data, but here, we focus only on a few key findings.

From the aspect of care, the practitioners observed that neighborhoods in the "open block" urban typology, with larger public open spaces typical of a socialist development in New Belgrade, provide a safer and more comfortable environment for women pedestrians and women with children compared to the high-density neighborhoods in the inner-city Vračar Municipality. From the aspect of mobility and transportation, the practitioners observed that public transportation stops were denser and more accessible by foot to the inhabitants of the New Belgrade, Vračar and Ušće neighborhoods, than in the informal neighborhood of Kaluđerica located on the outskirts of Belgrade. The professionals highlighted the issues of urban public safety, low level of mobility and accessibility to services of general interest for women in informal development neighborhoods, such as the Kaluđerica neighborhood. The theme of informal development is specific to the local planning context in Serbia where, according to the database of illegally built dwellings without a construction permit in Serbia, there were 2.05 million dwellings, or 43.52% of the total number of dwellings in 2017 [72]. The mapping of unsafe and inaccessible areas and services for women in Kaluđerica spurred the discussion at workshops 2 and 3 about the importance of planning standards in cases where inconsistent law enforcement contributed to unregulated construction on green and open public space and agricultural areas on the periphery of cities [73]. As integral components of urban plans, planning standards directly impact quality of life at the micro level [39,73]. Violating these standards in the built environment of informal settlements leads to a reduced accessibility to caregiving services, long distances to public transport stops, the absence of open green spaces, and a lack of safe and comfortable pedestrian areas. These deficiencies particularly affect women, especially caregivers. At the same time, it can be argued, from a gender perspective, that the government's decision to legalize illegally constructed housing en masse—built to address the basic housing needs of internally displaced persons and those unable to afford housing after the collapse of socialism—could be viewed in a positive light.

The high relevance of gender perspectives in planning and participation was agreed on by most of the participants, predominantly in the domains of practical application and the reinvention of planning standards, such as building regulations for public resources. The discussion around the possibility of influencing more inclusive participation by applying knowledge about gender perspectives in their usual practice was highly marked as well by 83.3% of practitioners. Some of them shared practical experiences in participation where gender parity was not achieved during the plan formation process:

> "In that particular case, it was important to involve as many diverse groups of citizens as possible, as this new plan represents the basis for urban and rural development of the municipality for the next 10 years. We advertised the participation process on the municipality website. However, at the public hearing there were only about twenty men present, and one woman who came from a larger city. As heads of family and landowners, we usually see men most often on such occasions. I wish we approached this process differently to reach more women and understand their perspective regarding the issues of mobility, agricultural land use, transportation needs and public services use, which are key to their wellbeing" (ER, workshop 2).

The practitioner observed that women did not participate in the public inquiry procedure, except for one who was present, and expressed a willingness to change their usual practices. In such contexts, targeted and tailored actions are necessary to involve gender experts from local action groups, gatekeepers representing individuals with care responsibilities, and self-organized actors in urban development [74]. The methods applied for informing, consultation, and active participation should consider that the diverse public may not be accustomed to engaging in planning processes or may be unaware of the impact of the plans [41,63].

All of the practitioners who filled in the questionnaire recognized their role in integrating gender perspectives as a professional standard in their ongoing practice. The results confirm previous research where planning practitioners in Belgrade see themselves as "protectors of the public interest" by adhering to planning standards for public land use and services [39,73,75] and through facilitating participation [41,63,76]. Some of them expressed a willingness to apply small steps to improve collaborative endeavours:

> "We should have enough time and resources to conduct in-depth stakeholder analysis before initiating participatory planning processes, with particular at-

> tention to those often excluded from participation, such as individuals with disabilities, women with children, youth, and the elderly. Additionally, we need to collaborate more closely with urban sociologists and demographers early in the planning process [...]. As a planning practitioner, I was especially pleased that these workshops provided an opportunity to discuss spatial issues with a representative from the police department. This interaction offered me a broader perspective and a new dimension for understanding neighborhoods' spatial scale. Multidisciplinary collaboration was standard practice during the 1970s planning era but has since been lost in the transition to a market economy and the push to make planning processes more efficient and shorter." (PS, workshop 2).

Multidisciplinary and multilevel collaboration in planning requires fostering coordination to integrate gender expertise into decision-making across various areas, including gender equality, social wellbeing, urban safety, the care economy, climate adaptation, mobility, and transportation. Having multidisciplinary teams working on case studies demonstrated that gender perspectives play a multifaceted role in planning and the use of public space. For instance, the insights provided by police officers about "blind spots" in the case study neighborhoods were considered to be crucial by traffic planners and urban designers in the initial stages of the planning process. They acknowledged that the pressure to speed up planning and construction processes requires pragmatic steps to counter-balance development imperatives, which is in line with the research of participatory processes in other European countries [77].

4. Discussion and Conclusions

This research examined the potential for integrating gender perspectives into urban planning in Serbia, using a conceptual framework that highlights the duality between normative and substantive aspects of participation and planning [40,51]. From a normative perspective, the empirical research conducted with practitioners in Belgrade, combined with a content analysis of the relevant legislation and policies, indicates a lack of gender references in the formal planning framework. Exceptions are found in the novel governance framework supported by the programs and initiatives of international and EU organizations [68,78–80]. At the national level, gender perspectives are explicitly incorporated in measures promoting equitable public space design and service provision, as outlined in Serbia's Sustainable Urban Development Strategy [68]. At the local level, these perspectives are reflected in the set of indicators used to monitor the implementation of territorial strategies, influenced by EU policies and the Sustainable Development Goals [80]. However, it is important to note that practitioners often view territorial and urban development strategies as informal planning instruments that operate alongside, rather than fully within, the formal framework of urban and spatial plans, despite their formal recognition under the Planning and Construction Law [61].

The Serbian practitioners involved in this research acknowledged the broader political support for inclusive participation driven by recent legislation in the field of governance initiatives. These initiatives are viewed as beneficial for acquiring new knowledge and strengthening institutional capacities to secure funding for integrated urban development projects aligned with the EU integration process. However, their impact on the practitioners' routine practices has been limited [53]. The interplay between formal and informal planning frameworks—where the latter ones are often subordinate—combined with the shortening of planning procedures and the institutional hierarchy, makes integrating gender perspectives into planning a complex task that requires more inclusiveness in the methods used to organize participation [14]. These findings highlight the efforts needed to establish a continuous loop in the dialogue between different sectors and levels of governance that have the same goals, i.e., societal wellbeing, but have different roles in governance regimes [14,74].

A question that is briefly addressed in this research but merits further exploration concerns the influence of socialist path dependency and unitary collectivist approaches on the integration of gender perspectives in planning and participation from a normative perspective. Participation, as a right to engage in decision-making and urban governance, is a political human right. Along with political rights, the corpus of human rights includes civil, social, economic, and cultural rights based on the principle of solidarity, which encompass the right to development, environmental protection, and housing [37]. The affirmation and implementation of these rights is supported by the United Nations, as well as the European Union, where the impact of international policy lies in its ability to transcend national jurisdictions. This is crucial for Serbia's efforts to integrate a gender perspective into planning, since, in former socialist countries, societal human rights were more developed, while political human rights were less emphasized [37,81].

From a substantive perspective, the research investigated how gender issues are reflected as a standard for practitioners' work by focusing on small-scale changes in planning and participation that foster knowledge creation, aiming to better meet diverse local needs. According to Healey [82], this process can be facilitated by creating experiences that allow for the re-evaluation of previously held beliefs during collaborative efforts through small, pragmatic steps, which was exemplified through joint work on the case studies of four Belgrade neighborhoods.

Integrating gender perspectives in case study work through the lens of other practitioners, including gender experts, in the traditionally technical plan formation process, challenged the established frameworks through new, alternative ideas that encourage change. Reflecting on the gendered experiences of urban space was described among practitioners as an important exercise at the workshops, highlighting their professional responsibility as "protectors of the public interest" by adhering to standards in participation and the planning of public land uses and services. They acknowledged the need to reinvent planning standards towards diverse users' needs and geographical scopes, including, but not limited to, the aspects of new transport options, shorter distances, the proximity of public services and mixed land use through a gender perspective lens, which resonates with the international research in counties beyond the EU framework [83–85]. New standards impacting households should be re-evaluated to determine whether they generate new economic possibilities for both women and men, or impose additional responsibilities on caregivers [13,14,19]. Participatory planning and governance practices, shaped by socially constructed criteria in addition to technical criteria, should thus guide the organization of human environments through participation by taking into account the "multiplicity of life courses or patterns for women and men" [12] (p. 91).

The research setting revealed a collective effort to transform the "rules of the game" [86], such as some of the established norms and processes of urban planning and governance in Serbia. Additionally, the question of changing the local mentality to be more responsive towards gender perspectives, as a substantive aspect of planning practice, was also raised. However, the reframing process is inherently complex, influenced by factors beyond the planning process, including the country's economic and human development, resource availability, power imbalances, income inequality, participation capacity, and overarching societal values and principles, coupled with the low participation of women in political, economic, and social spheres [87]. The transition from one socio-economic and political regime to another may have significantly impacted how gender perspectives are expressed in planning, leading to the emergence of hybrid models distinct from those in developed democratic contexts.

Author Contributions: Conceptualization, N.Č.M. and N.D.H.; methodology, N.Č.M. and N.D.H.; software, N.Č.M.; validation, N.Č.M. and N.D.H.; formal analysis, N.Č.M. and N.D.H.; investigation, N.Č.M. and N.D.H.; resources, N.Č.M. and N.D.H.; data curation, N.Č.M.; writing—original draft preparation, N.Č.M.; writing—review and editing, N.Č.M. and N.D.H.; visualization, N.Č.M.; supervision, N.D.H. and N.Č.M.; project administration, N.D.H.; funding acquisition, N.D.H. All authors have read and agreed to the published version of the manuscript.

Funding: The research presented in this paper was undertaken under the Ministry of Science, Technological Development and Innovation of the Republic of Serbia Grant number 451-03-66/2024-03/200006. The research is supported by the UN Women project "Safety of Women and Girls in Public Space" Grant number 40005537, which was implemented in Belgrade, Serbia, from May to September 2023. This initiative was led by the British Embassy in Belgrade and the Government of the Republic of Serbia, through the Gender Equality Coordination Body, Ministry of Labor, Employment, Social and Veterans Affairs.

Data Availability Statement: The raw qualitative data obtained from the workshops are unavailable due to ethical reasons. Ethical approval for publishing the undertaken research is available.

Conflicts of Interest: The authors declare no conflicts of interest. The funders had no role in the design of the study; in the collection, analyses, or interpretation of data; in the writing of the manuscript; or in the decision to publish the results.

References

1. Carpio-Pinedo, J.; De Gregorio Hurtado, S.; Sánchez De Madariaga, I. Gender Mainstreaming in Urban Planning: The Potential of Geographic Information Systems and Open Data Sources. *Plan. Theory Pract.* **2019**, *20*, 221–240. [CrossRef]
2. Rodríguez-García, M.J.; Donati, F. European Integral Urban Policies from a Gender perspective. Gender-Sensitive Measures, Transversality and Gender Approaches. *Sustainability* **2021**, *13*, 9543. [CrossRef]
3. Damyanovic, D.; Zibell, B. Brief historical review of gendered approaches in spatial development and planning. In *Gendered Approaches to Spatial Development in Europe: Perspectives, Similarities, Differences*; Zibell, B., Damyanovic, D., Sturm, U., Eds.; Routledge: London, UK, 2019; pp. 25–36.
4. UN. *Transforming Our World: The 2030 Agenda for Sustainable Development*; UN: New York, NY, USA, 2015.
5. European Commission. *Urban Agenda for the EU: Empowering Cities to Deliver the Sustainable Development Goals*; European Commission: Brussels, Belgium, 2016.
6. European Commission. *Territorial Agenda for the EU Until 2030: A Future for All Places*; European Commission: Brussels, Belgium, 2020.
7. European Commission. *New Leipzig Charter: The Transformational Power of Cities for the Common Good*; European Commission: Brussels, Belgium, 2020.
8. United Nations. *Gender Mainstreaming: An Operational Strategy for the UN System*; UN: New York, NY, USA, 2002.
9. European Union. *Treaty of Amsterdam Amending the Treaty on European Union, the Treaties Establishing the European Communities and Certain Related Acts*; European Union: Brussels, Belgium, 1999.
10. Umweltbundesamt. *Gender und Klimawandel: Ein Überblick über Forschung und Politik, Umweltbundesamt*; Umweltbundesamt: Dessau-Roßlau, Germany, 2018. Available online: https://www.umweltbundesamt.de/sites/default/files/medien/1410/publikationen/2018-03-15_texte_23-2018_gender-klima.pdf (accessed on 13 December 2024).
11. Damyanovic, D.; Reinwald, F.; Weikmann, A. *Manual for Gender Mainstreaming in Urban Planning and Urban Development*; Urban Development and Planning: Vienna, Austria, 2013.
12. Wotha, B. Urban governance and gender-aware planning. In *Fair Shared Cities: The Impact of Gender Planning*; Sanchez de Madriaga, I., Roberts, M., Eds.; Routledge: London, UK, 2016; pp. 91–106.
13. Horelli, L. Engendering urban planning in different contexts—Successes, constraints and consequences. *Eur. Plan. Stud.* **2017**, *25*, 1779–1796. [CrossRef]
14. Zibell, B.; Damyanovic, D.; Strum, U. *Gendered Approaches to Spatial Development in Europe*; Routledge: New York, NY, USA, 2019.
15. Escalante, S.O.; Valdivia, B.G. Planning from below: Using feminist participatory methods to increase women's participation in urban planning. *Gender Dev.* **2015**, *23*, 113–126. [CrossRef]
16. Horwood, K.; Lock, A.B.; Manns, S.; Morphet, C.; Palit, N. The substantive and descriptive representation of women in planning: Analysis from practice and academia. *Town Plann. Rev.* **2022**, *93*, 657–675. [CrossRef]
17. Kronsell, A.; Dymén, C.; Rosqvist, L.S.; Stepanova, O.; Hiselius, L.W. Critical masculine and feminine norms in sustainable municipal transport policies and planning. *J. Environ. Policy Plann.* **2023**, *25*, 663–675. [CrossRef]

18. Zibell, B. The model of the European city in the light of gender planning and sustainable development. In *Fair Shared Cities: The Impact of Gender Planning*; Sanchez de Madriaga, I., Roberts, M., Eds.; Routledge: London, UK, 2016; pp. 75–88.
19. Spitzner, M. Transformative Klimapolitik—Genderwirkungsprüfung und emanzipative Suffizienz-Rechte. *Forum Wiss.* **2020**, *3*, 17–21. Available online: https://www.bdwi.de/forum/archiv/archiv/10876264.html (accessed on 13 December 2024).
20. Cabo, M.A. Poder y género en la lucha ambiental. Casos de estudio en el norte de Cataluña. *Investig. Femin.* **2010**, *1*, 9–26.
21. Mattila, H. Habermas revisited: Resurrecting the contested roots of communicative planning theory. *Prog. Plan.* **2019**, *141*, 100431. [CrossRef]
22. Innes, J.E. Consensus building: Clarifications for the critics. *Plann. Theory* **2004**, *3*, 5–20. [CrossRef]
23. Forester, J. *The Deliberative Practitioner*; MIT Press: Cambridge, MA, USA, 1999.
24. Healey, P. Collaborative planning in perspective. *Plann. Theory* **2003**, *2*, 101–124. [CrossRef]
25. Mendonça, R.F.; Ercan, S.A.; Asenbaum, H. More than words: A multidimensional approach to deliberative democracy. *Political Stud.* **2022**, *70*, 153–172. [CrossRef]
26. Lukes, S. *Power: A Radical View*, 2nd ed.; Palgrave Macmillan: Basingstoke, UK, 2005.
27. Aitken, M. A three-dimensional view of public participation in Scottish land-use planning: Empowerment or social control? *Plan. Theory* **2010**, *9*, 248–264. [CrossRef]
28. Westin, M. The framing of power in communicative planning theory: Analysing the work of John Forester, Patsy Healey and Judith Innes. *Plan. Theory* **2022**, *21*, 132–154. [CrossRef]
29. Beauvais, E.; Warren, M.E. What can deliberative mini-publics contribute to democratic systems? *Eur. J. Political Res.* **2019**, *58*, 893–914. [CrossRef]
30. Healey, P. *Collaborative Planning: Shaping Places in Fragmented Society*; Macmillan Press: London, UK, 1997.
31. Perry, J.L.; Wise, L.R. The motivational bases of public service. *Public Adm. Rev.* **1990**, *50*, 367–373. [CrossRef]
32. Bengs, C. Planning theory for the naïve? *Eur. J. Spat. Dev.* **2005**, *3*, 1–12.
33. Katz, C. Vagabond capitalism and the necessity of social reproduction. *Antipode* **2001**, *33*, 709–728. [CrossRef]
34. Carrasco, C. Beyond equality: Towards a system of non-androcentric indicators. In *Gender and Well-Being in Europe: Historical and Contemporary Perspectives*; Harris, B., Gálvez, L., Machado, H., Eds.; Routledge: London, UK, 2016; pp. 185–201.
35. Schubert, D. Urban hygiene and slum clearance as catalysts: The emergence of the sanitary city and town planning. In *European Planning History in the 20th Century: A Continent of Urban Planning*; Guerra, M.W., Abarkan, A., Castrillo Romón, M.A., Pekár, M., Grau, V., Eds.; Routledge: New York, NY, USA; London, UK, 2023; pp. 27–38.
36. Malaza, N.; Todes, A.; Williamson, A. *Gender in Planning and Urban Development*; Discussion Paper 7; Commonwealth Secretariat: London, UK, 2009.
37. Dimitrijević, V.; Paunović, M. *Ljudska Prava (Human Rights)*; Beogradski Centar za Ljudska Prava: Beograd, Serbia, 1997.
38. Hirt, S. Planning during post-socialism. *Int. Encycl. Soc. Behav. Sci.* **2015**, *18*, 187–192.
39. Čolić, N.; Nedović-Budić, Z. Public interest as a basis for planning standards in urban development: State-socialist and post-socialist cases in Serbia. *J. Plann. Educ. Res.* **2024**, *44*, 1288–1302. [CrossRef]
40. Calderon, C.; Westin, M. Understanding context and its influence on collaborative planning processes: A contribution to communicative planning theory. *Int. Plan. Stud.* **2021**, *26*, 14–27. [CrossRef]
41. Čolić, N.; Dželebdžić, O.; Čolić, R. Building on recent experiences and participatory planning in Serbia: Toward a new normal. In *The 'New Normal' in Planning, Governance and Participation*; Lissandrello, E., Sørensen, J., Olesen, K., Steffansen, R.N., Eds.; Springer: Cham, Switzerland, 2023; pp. 41–56. [CrossRef]
42. National Assembly of the Federal People's Republic of Yugoslavia. Basic Resolution on General Urban Plan. In *Official Gazette of the Federal People's Republic of Yugoslavia, No. 78/49*; National Assembly of the Federal People's Republic of Yugoslavia: Belgrade, Yugoslavia, 1949.
43. Čolić, R. Integrated Urban Development Strategy as an Instrument for Supporting Urban Governance. *Serbian Archit. J.* **2015**, *7*, 317–342. [CrossRef]
44. Jovanović, P.R.; Stupar, A.B. The Emerging Community Planning in the Super-Blocks of New Belgrade. *Urban Des. Int.* **2022**, *27*, 275–287. [CrossRef]
45. National Assembly of the Republic of Serbia. Strategy for Gender Equality of the Republic of Serbia. 2016. Available online: https://issuu.com/ekoneckg/docs/strategija-rodna_ravnopravnost190_l (accessed on 13 December 2024).
46. National Assembly of the Republic of Serbia. Strategy for Gender Equality of the Republic of Serbia. 2021. Available online: https://www.rodnaravnopravnost.gov.rs/sr-Latn/node/732 (accessed on 13 December 2024).
47. National Assembly of the Republic of Serbia. Gender Equality Law. In *Official Gazette of the Republic of Serbia, No. 52/2021*; National Assembly of the Republic of Serbia: Belgrade, Serbia, 2021.
48. Constitutional Court of the Republic of Serbia. Available online: https://ustavni.sud.rs/sednice-suda/saopstenja-sa-sednice-suda/saopstenje-sa-8-sednice-ustavnog-suda-odrzane (accessed on 13 December 2024).

49. Pajvančić-Cizelj, A.; Hjuson, M. Urbanization and urban planning at the European semi-periphery: Unintended gender consequences. *Sociologija* **2018**, *60*, 275–287. [CrossRef]
50. Babović, M. Gender Equality in Serbia. In *Undermining Democracy: Processes and Institutions in Serbia 2010–2020*; Spasojević, D., Ed.; CRTA—Center for Research, Transparency, and Accountability: Belgrade, Serbia, 2021; pp. 149–164.
51. Fainstein, S.S. Planning Theory and the City. *J. Plan. Educ. Res.* **2005**, *25*, 121–130. [CrossRef]
52. Forester, J.; Kuitenbrouwer, M.; Laws, D. Enacting reflective and deliberative practices in action research. *Policy Stud.* **2019**, *40*, 456–475. [CrossRef]
53. Čolić, R.; Milić, Đ.; Petrić, J.; Čolić, N. Institutional Capacity Development within the National Urban Policy Formation Process–Participants' Views. *Environ. Plan. C Polit. Space* **2022**, *40*, 69–89. [CrossRef]
54. Mäntysalo, R.; Westin, M.; Mattila, H. Public planner—A deliberative authority. *Plan. Theory Pract.* **2023**, *24*, 11–29. [CrossRef]
55. Bua, A.; Escobar, O. Participatory-deliberative processes and public policy agendas: Lessons for policy and practice. *Policy Des. Pract.* **2018**, *1*, 126–140. [CrossRef]
56. Bryman, A. *Social Research Methods*; Oxford University Press: Oxford, UK, 2012.
57. Forester, J. On the evolution of a critical pragmatism. In *Encounters in Planning Thought*; Haselsberger, B., Ed.; Routledge: London, UK, 2017; pp. 298–314.
58. Patton, M.Q. Evaluation in the field: The need for site visit standards. *Am. J. Eval.* **2015**, *36*, 444–460. [CrossRef]
59. Flyvbjerg, B. Phronetic planning research: Theoretical and methodological Reflections. *Plan. Theory Pract.* **2004**, *5*, 283–306. [CrossRef]
60. Schram, S. Phronetic social science: An idea whose time has come. In *Real Social Science: Applied Phronesis*; Flyvbjerg, B., Landman, T., Sanford, S., Eds.; Cambridge Univeristy Press: New York, NY, USA, 2012; pp. 15–26.
61. National Assembly of the Republic of Serbia. Planning and Construction Law. 2023. Available online: https://www.paragraf.rs/propisi/zakon_o_planiranju_i_izgradnji.html (accessed on 13 December 2024).
62. Group of Institutions. *PPRS 2021–2035: Spatial Plan of the Republic of Serbia, 2021–2035*; (In Process of Adoption); National Assembly of the Republic of Serbia: Belgrade, Serbia, 2021.
63. Čolić, N.; Dželebdžić, O. Beyond formality: A contribution toward revising the participatory planning practice in Serbia. *Spatium Int. Rev.* **2018**, *39*, 17–25. [CrossRef]
64. Government of the Republic of Serbia, Ministry of European Integration. Action Plan for Chapter 22—Regional Policy and Coordination of Structural Instruments Adopted. Available online: https://www.mei.gov.rs/eng/news/908/more/w/0/action-plan-for-chapter-22-regional-policy-and-coordination-of-structural-instruments-adopted/ (accessed on 13 December 2024).
65. National Assembly of the Republic of Serbia. Additional Protocol to the European Charter of Local Self-Government on the right to participate in the affairs of a local authority. In *Official Gazette of the Republic of Serbia, No. 8/2018*; National Assembly of the Republic of Serbia: Belgrade, Serbia, 2018.
66. National Assembly of the Republic of Serbia. Planning System Law. In *Official Gazette of the Republic of Serbia, No. 30/2018*; National Assembly of the Republic of Serbia: Belgrade, Serbia, 2018.
67. National Assembly of the Republic of Serbia. Local Self Government Law. In *Official Gazette of the Republic of Serbia, No. 111/2021*; National Assembly of the Republic of Serbia: Belgrade, Serbia, 2021.
68. National Assembly of the Republic of Serbia. Sustainable Urban Development Strategy of the Republic of Serbia. In *Official Gazette of the Republic of Serbia, No. 47/2019*; National Assembly of the Republic of Serbia: Belgrade, Serbia, 2019.
69. Stadtmüller, H. Gender analysis for the GIZ-project "Strengthening Municipal Land Management" in Serbia. GIZ Internal Document 2015. *unpublished*.
70. Ranković Plazinić, B.; Jović, J. Women and Transportation Demands in Rural Serbia. *J. Rural Stud.* **2014**, *36*, 207–218. [CrossRef]
71. European Commission. *EU Cohesion Policy for the Period 2021–2027*; European Commission: Brussels, Belgium, 2021; Available online: https://ec.europa.eu/regional_policy/2021-2027_en (accessed on 22 January 2025).
72. Ministry of Construction, Transport and Infrastructure. Available online: https://www.mgsi.gov.rs/cir/dokumenti/baza-nezakonito-izgradjenih-objekata (accessed on 17 December 2024).
73. Petovar, K. Grad bez građana—Tako se gradi(lo) u Beogradu i Srbiji. *Republika* **2010**, *484–485*, 15–26.
74. Horelli, L.; Wallin, S. Civic engagement in urban planning and development. *Land* **2024**, *13*, 1446. [CrossRef]
75. Vujošević, M.; Petovar, K. Public interest vs. strategies of individual actors in urban and spatial planning. *Sociologija* **2006**, *48*, 356–382. [CrossRef]
76. Danilović Hristić, N.; Stefanović, N. The Role of Public Insight in Urban Planning Process: Increasing Efficiency and Effectiveness. *Spatium Int. Rev.* **2013**, *30*, 33–39. [CrossRef]
77. Calderon, C.; Mutter, A.; Westin, M.; Butler, A. Navigating swift and slow planning: Planners' balancing act in the design of participatory processes. *Eur. Plan. Stud.* **2024**, *32*, 390–409. [CrossRef]

78. Rikanović, M.; Zindović, M.; Đurić, N.; Lalić, M. Analiza Urbanističkog Planiranja i Projektovanja u Srbiji sa Stanovišta Rodne Ravnopravnosti. 2020. Available online: https://www.dropbox.com/scl/fi/9i02f3p2ybx5k5figyrej/Analiza-urbanog-planiranja-i-projektovanja-u-Srbiji-rodni-aspekti.pdf?rlkey=lvkdvgvh5rphlm3f4ut9kq6rq&e=1&dl=0 (accessed on 13 December 2024).
79. Zindovic, M. Početna analiza stanja u oblasti urbanističkog planiranja i projektovanja iz rodne perspektive, sa setom preporuka. In Proceedings of the Second Assembly of the Committee for Urban Planning, Housing and Construction of the Standing Conference of Towns and Municipalities, Srebrno Jezero, Serbia, 19 April 2017.
80. EU Local Development Program—EU PRO Plus. Available online: https://www.euproplus.org.rs/biblioteka/teritorijalne-strategije-biblioteka (accessed on 16 December 2024).
81. Zaharijević, A. Dissidents, Disloyal Citizens and Partisans of Emancipation: Feminist Citizenship in Yugoslavia and Post-Yugoslav Spaces. *Womens Stud. Int. Forum* **2015**, *49*, 93–100. [CrossRef]
82. Healey, P. Knowledge Flows, Spatial Strategy Making, and the Roles of Academics. *Environ. Plan. C Gov. Policy* **2008**, *26*, 861–881. [CrossRef]
83. Tian, M.; Li, Z.; Xia, Q.; Peng, Y.; Cao, T.; Du, T.; Xing, Z. Walking in China's Historical and Cultural Streets: The Factors Affecting Pedestrian Walking Behavior and Walking Experience. *Land* **2022**, *11*, 1491. [CrossRef]
84. Zorrilla-Muñoz, V.; Agulló-Tomás, M.S.; Rodríguez-Blázquez, C.; Ayala, A.; Fernandez-Mayoralas, G.; Forjaz, M.J. Ageing Perception as a Key Predictor of Self-Rated Health by Rural Older People—A Study with Gender and Inclusive Perspectives. *Land* **2022**, *11*, 323. [CrossRef]
85. Slayi, M.; Zhou, L.; Thamaga, K.H.; Nyambo, P. The Role of Social Inclusion in Restoring Communal Rangelands in Southern Africa: A Systematic Review of Approaches, Challenges, and Outcomes. *Land* **2024**, *13*, 1521. [CrossRef]
86. Moulaert, F.; Martinelli, F.; González, S.; Swyngedouw, E. Introduction: Social Innovation and Governance in European Cities. Urban Development Between Path Dependency and Radical Innovation. *Eur. Urban Reg. Stud.* **2007**, *14*, 195–209. [CrossRef]
87. European Parliament. Report on the Situation of Fundamental Rights in the European Union. 2012. A7-0136/2013. Available online: https://www.europarl.europa.eu/doceo/document/A-7-2013-0136_EN.html (accessed on 16 December 2024).

Disclaimer/Publisher's Note: The statements, opinions and data contained in all publications are solely those of the individual author(s) and contributor(s) and not of MDPI and/or the editor(s). MDPI and/or the editor(s) disclaim responsibility for any injury to people or property resulting from any ideas, methods, instructions or products referred to in the content.

land

MDPI

Article

Scale-Dependent Relationships Between Urban Morphology and Noise Perception: A Multi-Scale Spatiotemporal Analysis in New York City

Siting Chen [1], Bingjie Yu [1,*], Guang Shi [1], Yiping Cai [2], Yanyu Wang [1] and Pingge He [1,*]

1 School of Architecture, Southwest Jiaotong University, Chengdu 611756, China; chensting@swjtu.edu.cn (S.C.); guangsh@my.swjtu.edu.cn (G.S.); wangyy@swjtu.edu.cn (Y.W.)
2 School of Public Administration, Southwest Jiaotong University, Chengdu 611756, China; ykuer@my.swjtu.edu.cn
* Correspondence: bjyu@my.swjtu.edu.cn (B.Y.); hpg@my.swjtu.edu.cn (P.H.); Tel.: +86-183-8017-3778 (B.Y.); +86-158-9223-3795 (P.H.)

Abstract: Urban morphology significantly influences residents' noise perceptions, yet the impact across different spatial and temporal scales remains unclear. This study investigates the scale-dependent relationship between urban morphology and noise perception in New York City using noise complaint rates (NCR) as a proxy for perceived noise levels. A multi-scale analysis framework was applied, including four spatial scales (100 m, 200 m, 500 m, and 1000 m) and three temporal classifications (daytime/nighttime/dawn, weekdays/weekends, and seasonal divisions). Statistical analyses, including Spearman correlation, Moran's I test, and Geographically Weighted Regression (GWR), examined spatiotemporal heterogeneity. Results show: (1) NCR and urban morphology indicators vary significantly across spatial and temporal aggregations. (2) Correlations between NCR and urban morphology indicators generally strengthen with larger spatial units, revealing a scale effect. Temporal variations, e.g., residential land ratio (RES) and greenery percentage (SVI Green), show stronger correlations with NCR in summer than in winter. (3) The Moran's I index revealed significant spatial clustering at the 1000 m scale. Multi-temporal GWR analysis revealed spatial variations in urban morphology-noise relationships across different temporal contexts; in residential areas, building density exacerbates complaints more during non-working periods than during working hours. This study enhances understanding of urban sound environments, offering insights required for more precise urban planning policies.

Keywords: urban morphology; noise complaint; scale effect; spatiotemporal analysis; multi-scale analysis

1. Introduction

1.1. Background

Urban noise pollution has emerged as a critical environmental health challenge in cities worldwide, affecting residents' health and well-being [1]. Particularly in high-density urban areas, noise pollution has become a major threat to public health. Residents exposed to high noise levels over long periods are at a greater risk of experiencing hearing loss, depression, cardiovascular diseases, and other health problems. As cities continue to densify and expand, this environmental stressor continues to impact a vast population, making it a pressing concern for urban sustainability and public health.

The severity of urban noise exposure is inherently linked to the physical characteristics of the built environment. Urban morphology, encompassing elements such as building density, street layout, and land use patterns, plays a crucial role in determining how sound propagates through urban spaces [2–4]. Understanding these morphological influences on noise has become increasingly important for urban planners and policymakers seeking to create healthier urban environments.

Previous studies have made significant strides in examining the relationship between urban morphology and noise; however, a critical challenge in understanding these relationships lies in the scale at which they are studied. In recent years, there has been an increasing focus on scale effects in urban environmental research, as the choice of aggregation units at different scales can lead to different interpretations of spatial characteristics. This phenomenon is related to the Modifiable Areal Unit Problem (MAUP) [5]. Some studies have examined noise-morphology relationships at the neighborhood level, while others have focused on block-level or city-wide analyses, potentially resulting in inconsistent conclusions [6,7]. Furthermore, temporal variations in noise patterns—from daily fluctuations to seasonal changes—add another layer of complexity that has not been fully addressed in current research [8,9].

This gap in understanding the scale-dependent effects of urban morphology on noise perception poses significant challenges for urban planning and noise management. Urban planning interventions operate at various spatial scales, ranging from building-level renovations to neighborhood-wide revitalization efforts. Without a comprehensive understanding of how the relationships between noise and urban morphology vary across different scales and time periods, designing effective noise mitigation strategies becomes increasingly difficult.

Moreover, while objective noise measurements have been widely studied, residents' perceptions of noise, often reflected through complaints, introduces another crucial dimension to understanding urban noise problems. This subjective aspect of noise exposure can vary significantly across different urban contexts and time periods, yet few studies have systematically examined how these variations relate to urban morphological characteristics across different spatial and temporal scales.

These research gaps highlight the need for a comprehensive multi-scale analysis that considers both spatial and temporal dimensions in understanding how urban morphology influences residents' perception of noise. Such analysis would not only advance our theoretical understanding of urban noise issues but also provide insights for developing more precise and rational planning and regulatory policies, ultimately helping to improve urban sound environment quality and to promote public health.

1.2. Literature Review

An increasing number of studies have identified a correlation between urban morphology and the urban sound environment across different regions, aiming to provide detailed scientific recommendations for mitigating urban noise [4,10]. Researchers have leveraged vast amounts of urban big data to explore the role of urban morphology in alleviating noise pollution. However, these studies have focused on different regions, spatial scales, or temporal contexts, leading to diverse and even contradictory results regarding the relationship between urban morphology and the sound environment.

For instance, regarding road density, Yildirim et al. found a positive correlation between road density and urban noise pollution, indicating that areas with higher road density experience more severe noise pollution [11]. Conversely, Tang et al. argued that denser road networks and more densely packed intersections could lead to lower traffic volumes, thereby reducing noise pollution [12]. These discrepancies can be attributed to three fundamental components of urban morphology studies: scale, time, and region [13,14]. Therefore, the following parts of this section will review the existing literature through these three key aspects: scale effects, temporal variations, and spatial heterogeneity, to address the key findings and limitations in research on the relationship between urban morphology and the urban sound environment, as well as to identify existing research gaps.

1.2.1. Scale Effects

Scale represents the planning unit that integrates urban environmental elements, often calculated in terms of the average values of various urban environmental factors within blocks or grids during research [15]. Different planning units at various scales can significantly alter how urban morphology elements are represented [16]. For example, small-scale units (e.g., 100 m) may not adequately reflect the spatial arrangement patterns of buildings, whereas large-scale units (e.g., 1000 m) tend to average out the variations in the shapes of individual buildings. Therefore, scholars have extensively investigated scale effects in urban environmental research, revealing significant scale-dependent differences in the relationship between urban morphology and urban challenges, such as the urban heat island effect and air pollution [17–19].

However, only a limited number of studies have explored the scale effects on the relationship between urban morphology and the sound environment. For instance, Efstathios et al. considered the relationship between urban morphology and noise levels at three different spatial scales. They found that, at the macro scale, green space patterns, urban structure, and traffic noise levels were correlated with other morphological parameters. At the meso scale, an increase in internal road connectivity led to higher traffic noise levels. At the micro scale, different areas could have the same building coverage but exhibit different noise levels [6]. Similarly, Song et al. conducted a comparative analysis of the relationship between urban morphology and urban noise at three planning scales (300 m, 600 m, and 800 m), found that at scales below 600 m, green coverage positively impacted traffic noise level. In contrast, at scales of 600 m or above, the effect turned negative [20].

While these studies confirm and explore the scale effects between urban morphology and sound environment, enhancing our understanding of the MAUP issue in sound environments, they have not considered the temporal changes in scale effects. In contrast, the temporal variation in scale effects has been widely studied in other urban environmental fields [21,22].

1.2.2. Temporal Variations

The urban sound environment exhibits significant temporal distribution differences, primarily driven by urban activities composed of human activities [23]. This is because human activities vary significantly across different time periods, such as day and night, weekdays and weekends, and across seasons. Hong et al. confirmed that temporal factors play a key role in the relationship between urban morphology and the sound environment. Similarly, Guo et al. found that complaints about building noise were not significantly different from those caused by residential and commercial operations during the day, while complaints about building noise occurred with much higher frequency at night [24]. Wang et al. discovered that the built environment plays a crucial role in noise complaints, with suburban areas, characterized by fewer green spaces, being more likely to receive complaints from neighbors in the spring and summer months [25].

Although the above studies preliminarily examined the temporal changes in the sound environment, the differences in the relationship between the sound environment and specific urban morphological features across different time periods still require further investigation.

1.2.3. Spatial Heterogeneity

In addition to scale and time, the spatial heterogeneity in the relationship between urban morphology and sound environment also warrants attention. The spatial variation in the urban sound environment can be attributed to underlying urban surface characteristics, which result in various outcomes due to the effects of urban morphology across different regions [23]. For example, Zhou found that the distribution of high, medium, and low noise levels exhibits distinct spatial patterns across urban planning units, with low-noise levels demonstrating stronger spatial aggregation within each study unit. In these low-noise areas, all building layout indicators, except for average building height, showed a significant negative correlation with noise levels [26]. Zhao et al. revealed the spatial heterogeneity of the sound environment in Chicago, suggesting that this heterogeneity is closely associated with the spatial variations in park areas and population density [27]. Karapostoli et al. combined quantitative and qualitative sound and architectural data to analyze, describe, and design acoustic spaces. In the historic center of Thessaloniki, Greece, they selected five areas with unique sound characteristics and compared their architectural features [28]. He et al. argued that residents in relatively quiet suburban areas are more susceptible to noise [29]. Finally, Chen et al. found that complex building shapes in high-density urban areas are more effective at alleviating noise compared to low-density areas, thereby reducing noise complaints [3].

However, while these studies provide valuable insights into spatial heterogeneity, they primarily focus on spatial factors without integrating the temporal dimension. This oversight limits a comprehensive understanding of the spatiotemporal heterogeneity of the relationship between urban morphology and the sound environment.

In summary, previous studies have revealed a complex interplay between urban morphology and the noise environment, influenced by scale, time, and spatial heterogeneity. While many studies have established correlations between urban morphology and noise levels, the findings often vary due to differences in analytical scales, temporal contexts, and spatial heterogeneity. However, gaps still remain in understanding the comprehensive spatiotemporal dynamics of urban noise, particularly in integrating temporal changes into the analysis of scale effects. Further research employing multi-scale and multi-temporal approaches is needed to enhance our understanding of urban noise and inform effective urban noise mitigation strategies.

In addition, due to the difficulty of obtaining large-scale, real-time, high-precision environmental noise data [30], scholars began to measure the spatiotemporal distribution of urban sound environments by recording residents' noise complaint data [31]. Unlike physical sound decibel data [32], noise complaint data can, to some extent, eliminate perceptual differences in how individuals experience physical noise, providing a better measurement of urban sound environment quality [11,33].

1.3. Research Purpose

Building on the insights gained from the literature review, the primary purpose of this research is to investigate how urban morphological characteristics influence residents' perceptions of noise levels across different spatial and temporal contexts, with a specific focus on New York City. This study particularly emphasizes cyclical temporal patterns (daily, weekly, and seasonal) that are directly related to human activity patterns, as these temporal cycles may strongly influence how residents experience and respond to urban noise. Utilizing noise complaint rates as a measure of residents' perceptions of noise levels, this study establishes a multi-scale analysis framework to explore the relationship between urban morphology and noise complaints across different spatial scales and temporal divisions.

Ultimately, this research aims to address the following questions: (1) Do the distributions of urban morphological characteristics and noise complaint rates vary significantly across different spatial scales and temporal contexts? (2) How do the correlations between urban morphological characteristics and noise complaint rates change across varying spatial scales and temporal contexts? (3) How do the spatial patterns of the impact of urban morphological characteristics on noise complaint rates differ across different spatial scales and temporal contexts?

By exploring these questions, this study aims to provide scientific evidence for urban planning and sound environment management, and to promote more human-centered and health-oriented urban design.

2. Materials and Methods

2.1. Study Area

This study selects New York City as the research area, as shown in Figure 1. As one of the largest cities in the world, New York City is characterized by a highly dense urban morphology and a crowded population distribution. The city's concentrated built environment, coupled with its intensive human activities, has led to deteriorating sound environment over time. Moreover, New York City maintains a comprehensive open data policy, providing access to detailed noise complaint records through its 311 system. This rich dataset enables a thorough assessment of the urban sound environment and its spatiotemporal variations.

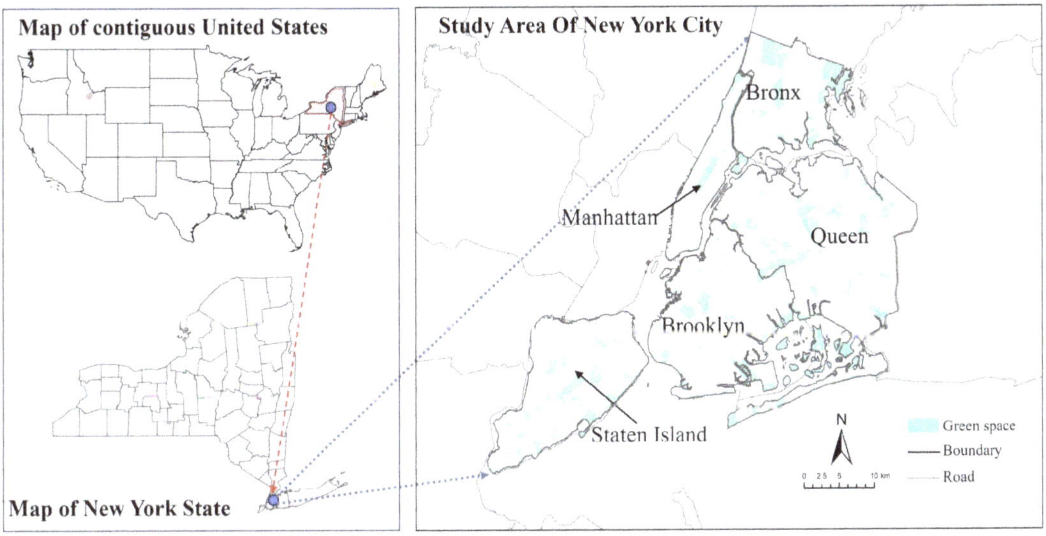

Figure 1. Study area.

2.2. Research Framework

To systematically investigate the scale-dependent relationship between urban morphology and residents' perceptions of noise levels in New York City, this study employs a multi-scale analysis framework that integrates diverse data sources and analytical techniques. Specifically, this research consists of four main parts:

(1) Indicator System Development and Data Preprocessing: This study utilizes noise complaint rate as a measure of residents' perceptions of noise levels in New York City and establishes an indicator system for urban morphology based on a literature review. This system allows for the derivation of a quantifiable representation

of urban morphological characteristics. Subsequently, four spatial scales are identified: 100 m, 200 m, 500 m, and 1000 m, corresponding to the sizes of different urban planning units. Additionally, three temporal classification schemes are defined: daytime/nighttime/dawn, weekdays/weekends, and seasonal divisions. Noise complaint data and urban morphology indicators are preprocessed and aggregated based on the average according to the defined spatial scales and temporal divisions for further analysis.

(2) Comparison of Variable Distribution across Different Spatial Scales and Temporal Divisions: To elucidate the scale effects of noise complaint rate and urban morphology indicators, this study compares their distributions across the identified spatial scales and temporal divisions. Statistical tests are employed to assess whether the means of the variables differ significantly when using different scales of aggregation units. This analysis aims to provide preliminary evidence for the existence of scale effects.

(3) Comparison of Correlation across Different Spatial Scales and Temporal Divisions: To investigate the differences in correlation patterns between urban morphology and noise complaint rates as influenced by spatial scales and temporal divisions, this study calculates and compares the Spearman correlation coefficients for urban morphology and noise complaint rates across the specified spatial scales and temporal divisions. This analysis aims to provide a comprehensive understanding of how the correlation between urban morphological characteristics and noise complaints varies depending on the selected spatial units and temporal divisions for data aggregation.

(4) Comparison of Spatial Autocorrelation and Spatiotemporal Heterogeneity: this study assesses the spatial autocorrelation of noise complaint rates across four spatial scales and three temporal divisions using Moran's I test. The spatial aggregation unit exhibiting the most significant clustered pattern is then selected for further analysis. Subsequently, Geographically Weighted Regression (GWR) models are constructed for each of the temporal divisions to explore the spatiotemporal heterogeneity in the relationship between urban morphology and noise complaint rates. This approach provides valuable insights into how these relationships vary across different spatial scales and temporal divisions.

The specific research framework is shown in Figure 2.

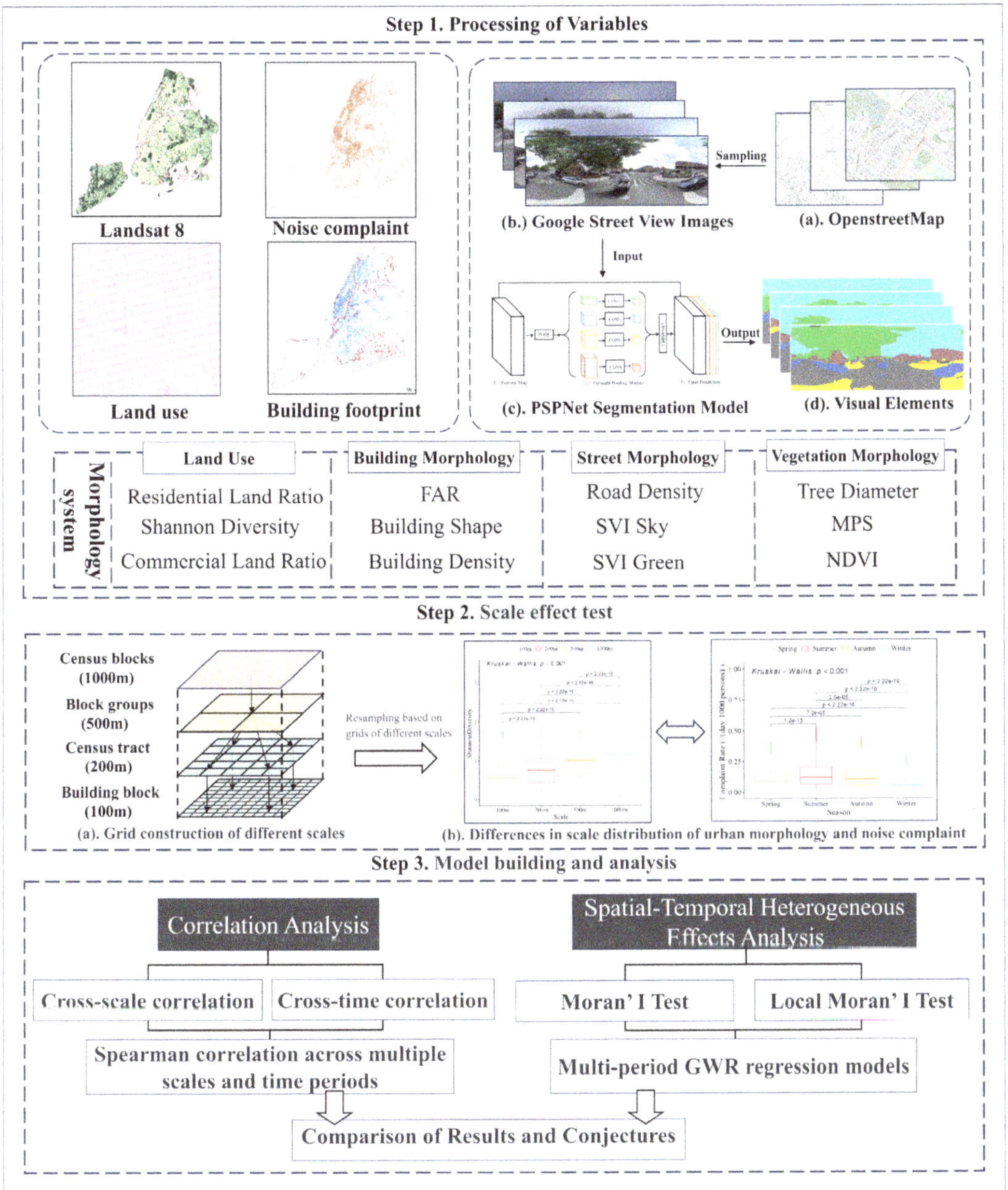

Figure 2. Research framework.

2.3. Indicators System

In developing the urban morphology indicator system, we integrate insights from previous research and consider the data availability. Consequently, this study establishes an urban morphology indicator system from four aspects: building morphology, vegetation morphology, street morphology, and land use [3,34–36].

The development of the urban morphology indicator system involves several steps. Initially, we selected a total of 30 indicators from existing related studies. Subsequently,

Pearson correlation tests are performed to assess the relationships among these indicators, thereby eliminating those with multicollinearity (defined as a Pearson coefficient greater than 0.7). As a result, the selected indicators for building morphology include building density, floor area ratio (FAR), and building shape index. For vegetation morphology, the indicators selected are the normalized difference vegetation index (NDVI), mean patch size (MPS), and tree diameter. For street morphology, the study incorporates road density, the proportion of sky elements in Street View images (SVI Sky), and the proportion of green elements in Street View images (SVI Green). Lastly, for land use, the indicators considered are the Shannon diversity index, commercial land ratio, and residential land ratio.

The selected indicators are designed to represent the morphological characteristics of buildings, vegetations, streets and land uses of a city area. A summary of the meaning and reference sources for each indicator is shown in Table 1.

Table 1. Variable definitions and data sources.

Type	Metrics	Description	Unit	Resource	Reference
Street Morphology	SVI Sky	The proportion of sky elements in Street View images	%	Calculated based on Google Street View Image	[37]
	SVI Green	The proportion of green elements in Street View images	%		
	Road Density	Road network density in each grid	km/km^2	Calculated based on road network data from Openstreet map (Building Footprints (Deprecated) ǀ NYC Open Data)	[38]
Building Morphology	FAR	Mean floor area ratio of buildings in each grid	%	Calculated based on building footprint dataset from NYC OpenData	[39]
	Building Shape	The mean of the building shape index in each grid	-		[3]
	Building Density	The building density in each grid	%		[40]
Land Use	Shannon Diversity	Shannon diversity index in each grid	-	Calculated based on MapPLUTO dataset from NYC Department of City Planning (DCP) (https://www.nyc.gov/site/planning/data-maps/open-data/dwn-pluto-mappluto.page)	[41]
	Com Land Ratio	The percentage of commercial land in each grid	%		[42]
	Res Land Ratio	The percentage of residential land in each grid	%		[43]
Vegetation Morphology	NDVI	Mean of normalized difference vegetation index in each grid	-	Calculated based on USGS Landsat 8 Collection 2 Tier 1 TOA Reflectance Dataset	[44]
	MPS	The mean patch size of vegetation in each grid	km^2		[34]
	Tree Diameter	The mean trunk diameter of trees in each grid	m	Forestry Tree Points dataset from NYC Department of Parks and Recreation (DPR) ((Forestry Tree Points ǀ NYC Open Data)	[3]

2.4. Data Preprocessing

2.4.1. Grid Construction at Specified Scales

The determination of an appropriate spatial scale for urban analysis is challenging due to the varying urban structural characteristics (e.g., compactness, topography). Previous studies have employed multiple spatial resolutions to address this complexity. For example, Song et al. measured the relationship between traffic noise and urban morphology in Dongguan, China, using three spatial resolutions: 300 m, 600 m, and 1200 m [20]. Similarly, Li et al. selected five spatial resolutions (30 m, 90 m, 300 m, 500 m, and 1000 m) to measure the urban morphology at different scales in Beijing [45].

In New York City, urban neighborhoods are traditionally classified using three administrative units: "census tract" (approximately 1000 m), "block groups" (approximately 500 m), and "census blocks" (approximately 200 m) [46]. Building upon these established scales, we incorporated building-level analysis (100 m) to create a four-tier spatial resolution framework: 100 m (building scale), 200 m (census block scale), 500 m (block group scale), and 1000 m (census tract scale) (Figure 3). These varying spatial scales reflect different contexts in which residents experience and respond to their urban environment, from immediate surroundings (100 m) to neighborhood-level effects (1000 m), enabling analysis across different levels of urban organization while maintaining practical relevance to existing planning and administrative frameworks.

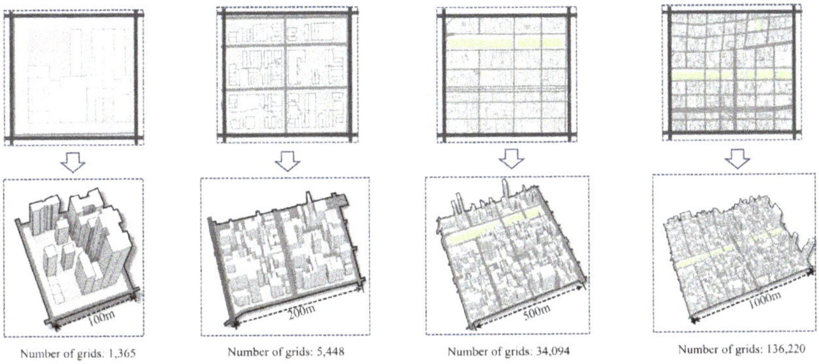

Figure 3. Differences in urban morphology at different spatial scales.

To facilitate data aggregation, we utilized the ArcGIS Pro 3.0 fishnet tool to generate standard grids corresponding to these four scale sizes (EPSG:3857).

2.4.2. Noise Complaint Data Preprocessing

We obtained 3,180,000 georeferenced noise complaint records from NYC's 311 platform (2018–2022). These complaints serve as a proxy for residents' perception of noise levels, capturing how people experience and respond to their sound environment within different urban morphological contexts. Each record includes temporal details, geographic coordinates, and complaint type, reflecting community noise disturbances.

Temporal patterns were analyzed using three classification schemes: daily (daytime: 06:00–17:59, nighttime: 18:00–23:59, dawn: 00:00–05:59), weekly (weekdays vs. weekends), and seasonal (spring, summer, autumn, winter).

The complaint records of each temporal division were spatially aggregated based on standard grids corresponding to the four established spatial scales (100 m, 200 m, 500 m, 1000 m). For the daily time division, we calculated hourly complaint frequencies within each cell. For weekly and seasonal analyses, we computed daily complaint frequencies. To account for demographic variations, we normalized the complaints by calculating complaint rates per 1000 residents within each grid cell, dividing the total number of complaints by the population size of the corresponding area. Complaints were normalized by population, calculating rates per 1000 residents within each grid cell. Additionally, outlier detection identified cells with unusually high rates, ensuring validity as genuine noise issues.

2.4.3. Urban Morphology Indicators

(1) Building morphology indicators

Building morphology indicators, including Floor Area Ratio (FAR), building shape index, and building coverage ratio, were computed using building footprint data from NYC OpenData.

The building shape index was calculated using the *momepy* Python package according to the following formula:

$$Building\ shape\ index = \frac{\frac{p-\sqrt{p^2-16a}}{4}}{\frac{p}{2}-\frac{p-\sqrt{p^2-16a}}{4}} \quad (1)$$

where a is the area of the object and p is its perimeter. The closer the index is to 1, the closer the building shape is to a rectangle, and the closer it is to 0, the more diversified the building shape is [47].

The building shape index was first calculated at the individual building level, then spatially aggregated by taking the mean value within each grid cell of the four identified spatial resolutions (100 m, 200 m, 500 m, and 1000 m).

Floor Area Ratio (FAR) was computed as the ratio of total floor area in a cell to the cell area [48], while building coverage ratio represents the proportion of land area occupied by building footprints within each grid cell.

(2) Vegetation morphology indicators

Normalized Difference Vegetation Index (NDVI) was computed using the USGS Landsat 8 Collection 2 Tier 1 TOA Reflectance dataset accessed through the Google Earth Engine. We selected satellite images captured between 1 January 2020, and 1 January 2021, applying a cloud cover threshold of less than 20% to ensure image quality. The median values of the filter images were used to minimize the influence of seasonal variations and atmospheric conditions. NDVI was calculated using the following formula:

$$NDVI\ index = \frac{NIR - R}{NIR + R} \quad (2)$$

where NIR is the near-infrared band (Band 5, 0.85–0.88 μm) and R is the red band (Band 4, 0.64–0.67 μm) of Landsat 8 imagery. The resulting NDVI map was then resampled by calculating the mean values within each cell of the standard grids corresponding to the four defined spatial scales (100 m, 200 m, 500 m, and 1000 m).

Mean Patch Size (MPS) was derived from the NDVI map to characterize the fragmentation of vegetated area. First, the NDVI map was reclassified into a binary map where pixels with NDVI values above 0.2 were classified as vegetation patches and those below 0.2 as non-vegetation. Connected vegetation pixels were then grouped into distinct patches. MPS was calculated based on following formula:

$$MPS = \frac{A}{n} \quad (3)$$

where A is the total area (m^2) of all vegetation patches in a grid cell; n is the number of vegetation parches within that grid cell. A larger MPS value indicates a more contiguous vegetated area, while smaller values suggest more fragmented vegetation patterns.

(3) Land use indicators

Land use indicators were calculated using the MapPLUTO dataset, which provides detailed land use information at the tax lot level. The NYC Department of City Planning (DCP) classified land use into 11 categories: One & Two Family Buildings (1), Multi-Family Walk-Up Buildings (2), Multi-Family Elevator Buildings (3), Mixed Residential & Commercial Buildings (4), Commercial & Office Buildings (5), Industrial & Manufacturing (6), Transportation & Utility (7), Public Facilities & Institutions (8), Open Space & Outdoor Recreation (9), Parking Facilities (10), and Vacant Land (11).

To quantify land use diversity, we calculated the Shannon Diversity Index using the following formula:

$$Shannon\ Diversity\ Index = -\sum p_i * \ln(p_i) \quad (4)$$

where p_i is the proportion of land use category i within each grid cell. The index value of 0 indicates homogeneous land use (single category), while higher values reflect greater land use diversity within the grid cell.

Additionally, we computed commercial and residential land use ratios for each grid cell. These ratios were calculated by summing the areas of all tax lots classified as commercial (category 4 and 5) or residential (categories 1, 2, and 3), respectively, and dividing the sum by the total area of the grid cell.

(4) Street morphology indicators

Road density was computed using OpenStreetMap (OSM) road data as the total street length in a grid cell divided by its area (km/km^2). For the Street View Index (SVI), 63,276 sampling points were generated along OSM streets in NYC at 100 m intervals using ArcGIS Pro. At each point, four Google Street View images (cardinal directions) were retrieved via API. Semantic segmentation using a pre-trained PSPNet model classified image pixels into categories like building, tree, or sky. The SVI was calculated as the mean proportion of sky and greenery pixels across the four images at each point, using the following formula [49]:

$$Street\ View\ index = \frac{\sum_{i=1}^{4} character\ Pixels}{\sum_{i=1}^{4} Total\ Pixels} \quad (5)$$

where *character pixels* is either sky or greenery pixels (calculated separately for SVI Sky and SVI Green), and *i* denotes the image number (1–4) for each sampling point.

The final SVI values for each grid cell were calculated by averaging the indices of all sampling points within the respective cell boundaries.

2.5. Analysis Methods

2.5.1. Scale-Dependent Variable Distribution Analysis

To examine scale effects on noise complaint rates and urban morphology indicators, this study analyzed their distributions across four spatial resolutions (100 m, 200 m, 500 m, 1000 m) using boxplots and non-parametric tests. Normality was assessed via Shapiro–Wilk and Kolmogorov–Smirnov tests, revealing violations of normality assumptions. Thus, the Kruskal–Wallis test was used for overall comparisons, with pairwise Wilcoxon tests for specific scale comparisons. Analyses and visualizations were conducted in R, utilizing *dplyr* and *tidyr* for data manipulation and *ggpubr* for statistical visualizations.

2.5.2. Scale-Dependent Correlation Analysis

To elucidate how the correlation between noise complaints and urban morphology varies across different spatial scales and temporal divisions, this study conducted correlation analysis between noise complaint rates and urban morphology indicators across four spatial scales and three temporal divisions. Given the violation of normality assumption indicated by the results from Shapiro–Wilk and Kolmogorov–Smirnov tests, this study employed the Spearman non-parametric test instead of the Pearson parametric test.

The correlation analysis and visualizations were performed in R, using the stats packages for computing Spearman correlations and the *ggplot2* package for generating correlation matrices and plots.

2.5.3. Multi-Scale and Temporal Spatial Pattern Analysis

Spatial pattern analyses, including spatial autocorrelation and clustering, were conducted to examine noise complaint rate (NCR) patterns across spatial scales and temporal divisions. Spatial autocorrelation is a spatial analysis method used to determine whether the observations at particular locations are significantly correlated with those at neighboring locations [50]. Global Moran's I was used to assess whether NCR distributions were clustered (Moran's I > 0), dispersed (Moran's I < 0), or random (Moran's I ≈ 0), highlighting scale-dependent spatial patterns. Local Moran's I identified clusters of high/low NCR values and spatial outliers. These analyses also revealed changes in spatial patterns with different aggregation scales. Analyses were performed in R using the sf package for spatial data, *spdep* for spatial statistics, and *tmap* for visualization.

2.5.4. Multi-Temporal Geographically Weighted Regression Analysis

Based on multi-scale spatial pattern analysis, we selected the spatial scale with the most clustered noise complaint rate (NCR) distribution to develop Geographically Weighted Regression (GWR) models for each temporal division. Unlike general linear models such as Ordinary Least Squares (OLS) [51], GWR models account for geographic differences by calculating local estimates for each unit, revealing spatial variations in relationships between NCR and urban morphology indicators [39,52].

The GWR model for this study is defined as follows:

$$yi = \beta_0(u_i, v_i) + \sum_k \beta_k(u_i, v_i)x_{ik} + \varepsilon_o \quad (6)$$

where yi is NCR in grid cell i, (u_i, v_i) are the geographic coordinates of grid cell i, $\beta_0(u_i, v_i)$ is the local intercept for grid cell i; $\beta_k(u_i, v_i)$ is the local coefficient for urban morphology indicator k; x_{ik} is the value of urban morphology indicator k at grid cell i; and ε_o is the random error term.

The GWR models are implemented in R, using the *sf* and *sp* packages for spatial data handling, the *GWRmodel* package for GWR modeling, and the *tmap* package for visualizing model outputs. We employed an adaptive bandwidth approach, which allows the bandwidth to vary based on the density of observations. The optimal bandwidth was determined using the *bw.gwr* function with AIC (Akaike Information Criterion) optimization, ensuring the best balance between model fit and complexity.

To be noted, GWR applies a single bandwidth to all variables and assumes local linear relationships between variables, which will potentially overlook the complex scale effect and nonlinear interactions between urban morphology and complaints. Nevertheless, its ability to reveal spatial heterogeneity aligns with our primary goal of understanding spatiotemporal variations in urban morphology-NCR relationships.

3. Results

3.1. Scale Effects on Variable Distributions

The distribution patterns of urban morphology indicators across four specified spatial scales (100 m, 200 m, 500 m, and 1000 m) are illustrated in Figure 4, with summary statistics presented in Table 2. The results of the Kruskal–Wallis tests revealed significant scale effects on all urban morphology indicators ($p < 0.001$), and Wilcoxon tests identified the distinct variable distributions between the pairwise scales ($p < 0.001$).

Moreover, the distributions of several indicators demonstrated consistent variation patterns across spatial scales. For example, the means of the Shannon diversity index and MPS increases with the size of aggregation units. In contrast, the means of road density, FAR, building density, residential land ratio, and tree diameter decrease as the size of aggregation units increases. The remaining indicators did not show a consistent pattern across spatial scales.

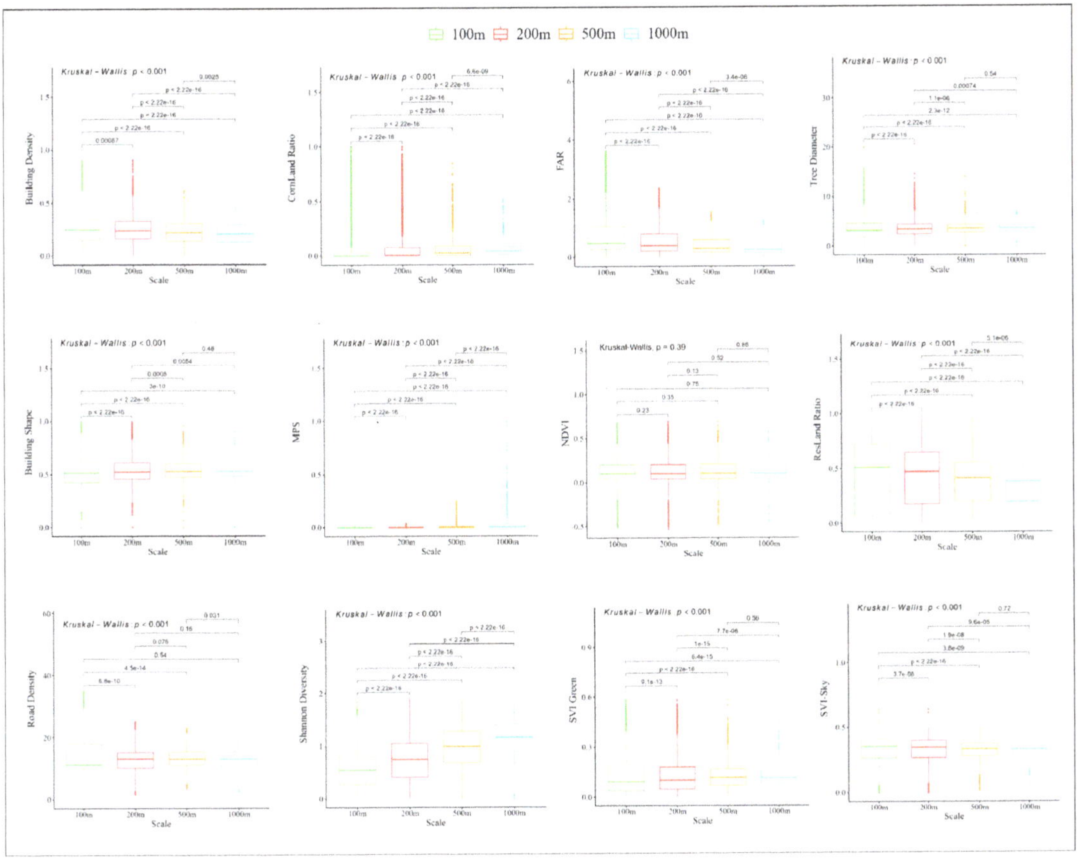

Figure 4. Distribution of urban morphology indicators across four specified spatial scales.

Table 2. Distribution of urban morphology indicators at different spatial scales.

	100 m			200 m			500 m			1 km		
	Mean	**Sd**	**KS. P**	**Mean**	**Sd**	**KS. P**	**Mean**	**Sd**	**SW. P**	**Mean**	**Sd**	**SW. P**
SVI Sky	0.324	0.111	***	0.326	0.099	***	0.328	0.073	***	0.330	0.062	***
SVI Green	0.124	0.107	***	0.125	0.098	***	0.127	0.075	***	0.126	0.064	***
Road Density	17.362	12.621	***	15.188	7.816	***	13.474	5.241	***	12.704	4.324	***
FAR	3.529	15.790	***	1.503	4.438	***	0.703	1.096	***	0.503	0.613	***
Building Shape	0.544	0.125	***	0.545	0.108	***	0.543	0.093	***	0.542	0.081	***
Building Density	0.275	0.138	***	0.256	0.124	***	0.227	0.114	***	0.211	0.109	***
Shannon Diversity	0.573	0.329	***	0.751	0.379	***	0.966	0.400	***	1.124	0.380	***
Com Land Ratio	0.060	0.139	***	0.066	0.122	***	0.064	0.094	***	0.062	0.074	***
Res Land Ratio	0.436	0.304	***	0.417	0.265	***	0.374	0.221	***	0.347	0.192	***
Tree Diameter	3.730	1.879	***	3.645	1.448	***	3.636	1.134	***	3.606	0.891	ns
NDVI	0.131	0.126	***	0.130	0.127	***	0.133	0.128	***	0.131	0.128	***
MPS	0.000	0.002	***	0.002	0.007	***	0.016	0.041	***	0.046	0.113	***

Note: *** $p < 0.01$, SW. P: p-value from Shapiro–Wilk test; KS. P. p-value from Kolmogorov–Smirnov tests.

These results preliminarily validate our hypothesis that urban morphology indicators exhibit scale-dependent distribution patterns, highlighting the importance of considering scale effects in urban morphological analysis.

Figure 5 illustrates the distribution of noise complaint rates (NCR) across different temporal divisions aggregated at four spatial scales (100 m, 200 m, 500 m, and 1000 m). The results of the Kruskal–Wallis tests and Wilcoxon tests demonstrated significant differences in the mean values of NCRs among temporal divisions within each of the three temporal classification schemes we defined ($p < 0.001$).

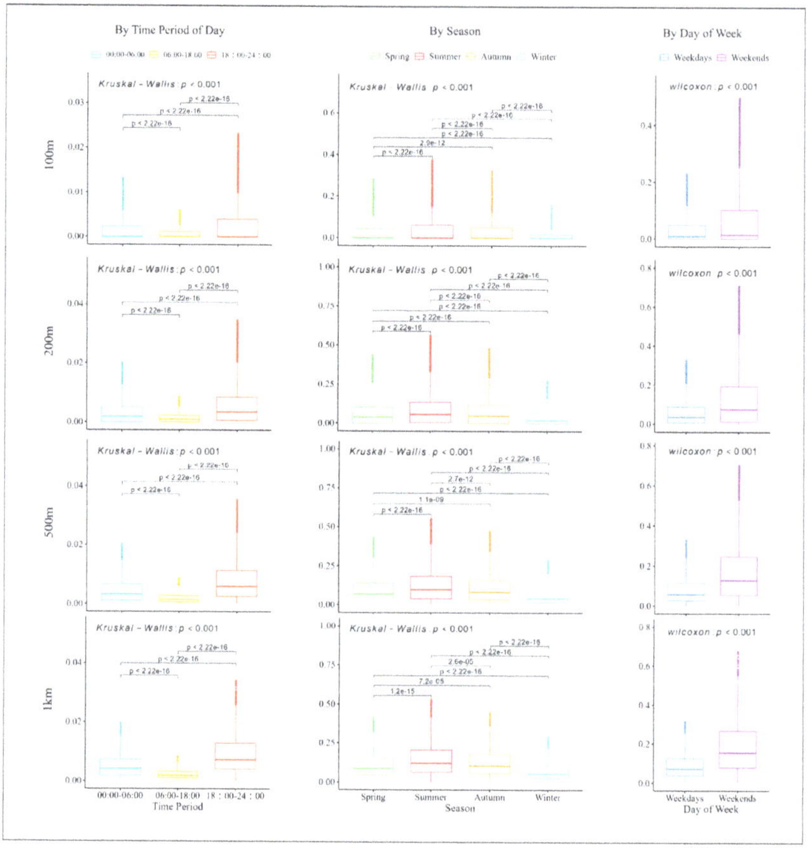

Figure 5. Distribution of NCR across different temporal divisions aggregated at four spatial scales.

Specifically, for diurnal variations, the boxplots across all spatial scales consistently showed the highest hourly NCR during nighttime and lowest during daytime. Seasonal comparisons revealed that summer months exhibited the highest daily NCR, while winter months were the lowest. The weekday–weekend comparison demonstrated significantly higher daily NCR during weekends compared to weekdays, regardless of the spatial scales used for data aggregation.

According to the boxplots of different spatial scales, the relative magnitude of the NCRs among different temporal divisions remained consistent across the four spatial scales. Nevertheless, the variations between temporal divisions with each classification scheme are more pronounced at the 1000 m scales.

3.2. Scale Effects on Urban Morphology-NCR Correlations Across Temporal Divisions

Figures 6–8 illustrate correlations between noise complaint rates (NCR) and urban morphology indicators across spatial resolutions and temporal divisions. Regardless of temporal classification, few indicators showed strong correlations at the 100 m scale, with most coefficients below 0.3. In contrast, more indicators exhibited stronger correlations at the 1000 m scale, with coefficients exceeding 0.3. This highlights the influence of grid scales on observed correlations, suggesting a larger grid scale (e.g., 1000 m) is more effective in capturing patterns. While correlation strength varies across scales, the polarity remains consistent.

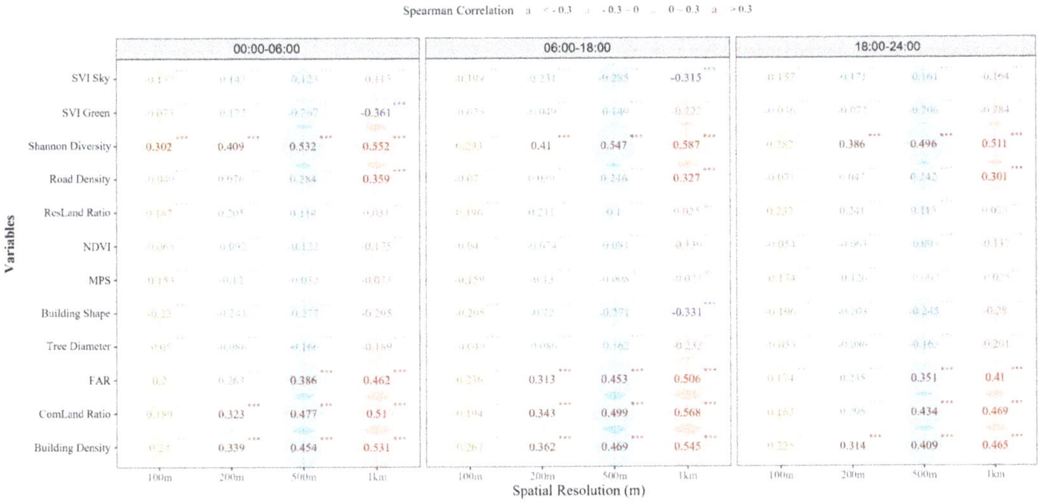

Figure 6. Correlation between urban morphology indicators and NCR at four spatial scales across different time periods of the day.

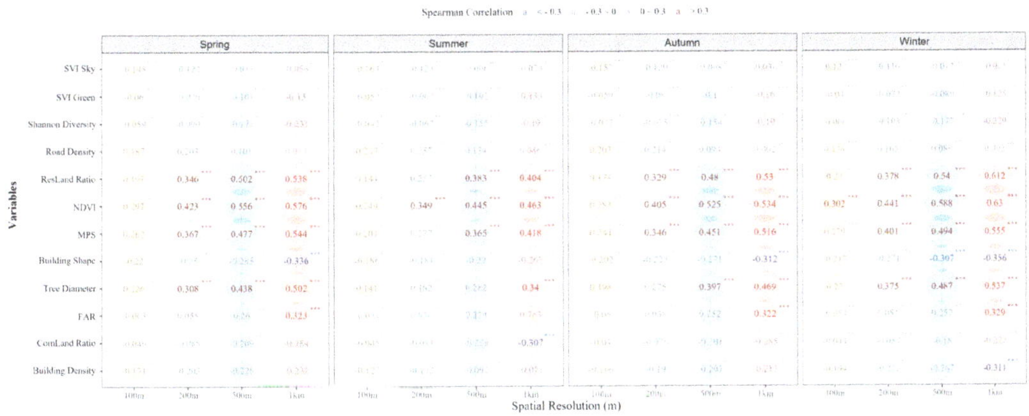

Figure 7. Correlation between urban morphology indicators and NCR at four spatial scales across different seasons.

Vegetation indicators (MPS, NDVI, tree diameter), SVI Sky, SVI Green, and building shape index negatively correlate with NCR, suggesting their potential to improve sound quality, though coefficients are generally below 0.3. Land use indicators (residential and commercial land ratios, Shannon diversity index) and building density positively correlate with NCR, indicating a potential association, though further analysis is needed to confirm causality.

Diurnal comparisons (Figure 6) show that indicators such as tree diameter, commercial land ratio, building density, and FAR have the strongest correlations with NCR during daytime (06:00–18:00). However, correlations for indicators like MPS, NDVI, tree diameter, and residential land ratio did not vary significantly across time periods.

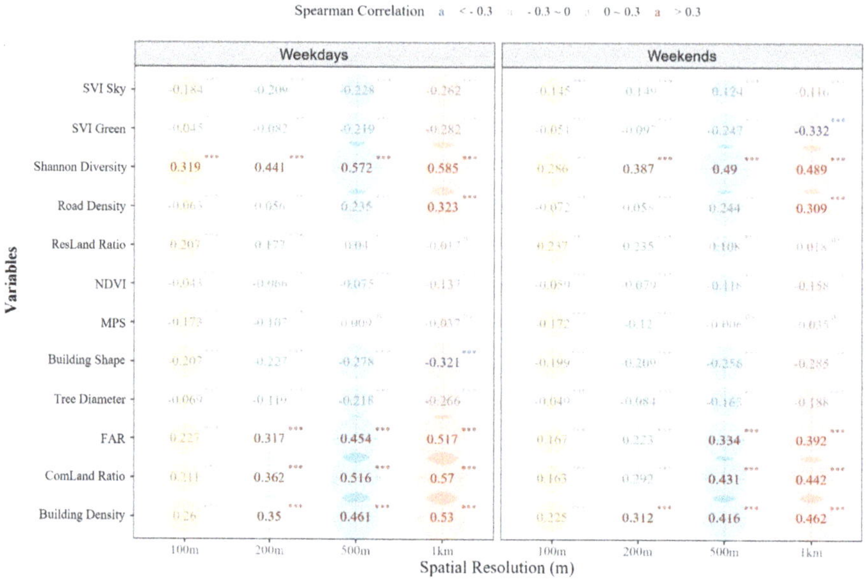

Figure 8. Correlation between urban morphology indicators and NCR at four spatial scales on weekends and weekdays.

Seasonal variations are illustrated in Figure 7. Vegetation indicators, including NDVI, MPS, and the tree diameter, maintain a relatively stable correlation with NCR across the seasons. Residential land ratio and SVI Green indicated a relatively stronger correlation during summer, while most other urban morphology indicators exhibit the strongest correlation with NCR in winter and the weakest in summer.

The workdays–weekends comparison (Figure 8) showed consistent correlations between NCR and certain urban morphology indicators, including MPS, NDVI, and road density, across both temporal divisions. However, tree diameter, commercial land use ratio, Shannon diversity index, building density, building shape index, FAR, and SVI Green exhibit stronger correlations with NCR during workdays. In contrast, residential land ratio and SVI Green show stronger correlations during weekends, suggesting that these two indicators may have a greater influence on residents' perceptions of the sound environment during non-working days.

3.3. *Scale Effects on Spatial Patterns of Noise Complaint Rates Across Temporal Divisions*

Figure 9 shows Global Moran's I values for NCR across spatial scales and temporal divisions, all indicating significant positive spatial autocorrelation (Moran's I > 0, $p < 0.001$). Values increased with larger grid scales, peaking at 1000 m, where the strongest spatial clustering was observed. This scale was selected for Local Moran's I analysis to identify spatial clusters and outliers.

Figure 10 illustrates local spatial autocorrelation patterns of NCR at the 1000 m scale. High–High clusters (high NCR values), Low–Low clusters (low NCR values), and High–Low/Low–High outliers were identified. High NCR clusters mainly occurred in Manhattan, the Bronx, Brooklyn, and Queens, while low NCR clusters were concentrated in Staten Island and parts of Queens. Weekday patterns showed high-value clusters in Manhattan and the Bronx, shifting on weekends to the Brooklyn–Queens border. Low-value clusters in Staten Island and Queens remained consistent.

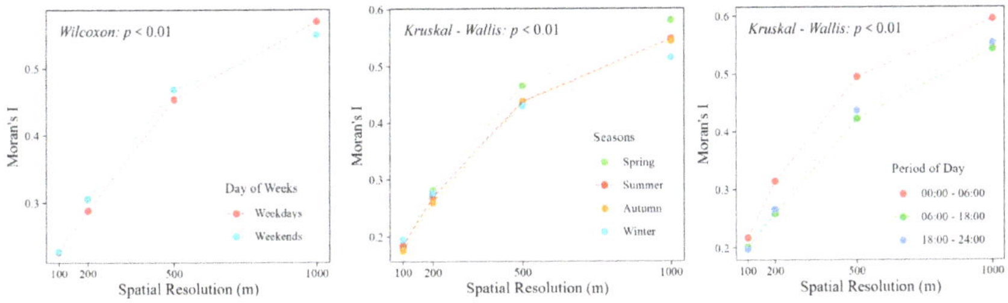

Figure 9. Global Moran's I values for NCR across the defined spatial scales and temporal divisions.

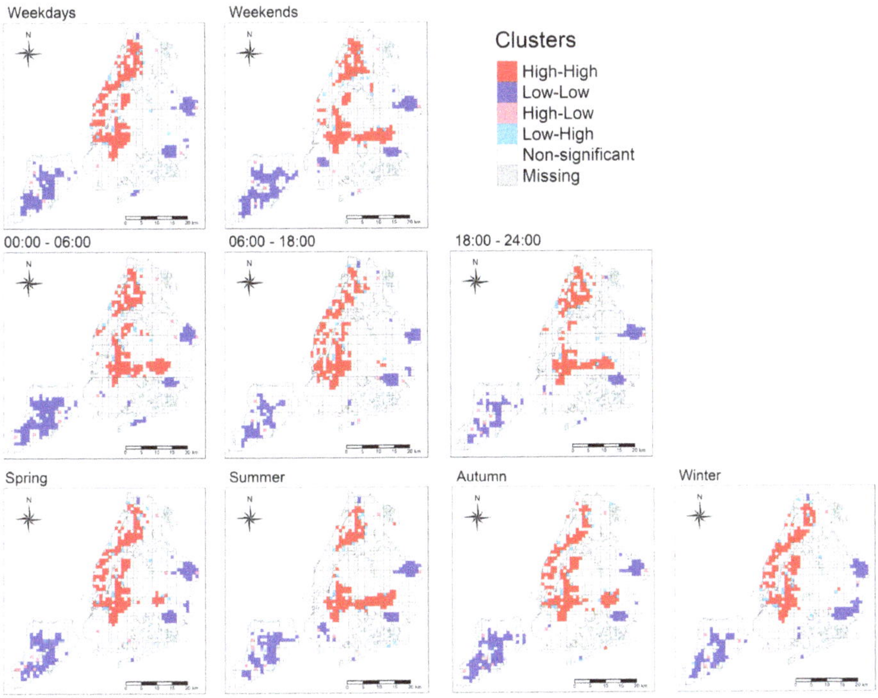

Figure 10. Local spatial autocorrelations patterns of NCRs across the defined temporal divisions at the 1000 m scale.

The diurnal analysis showed persistent low-value clusters around Kennedy Airport, Alley Pond Park in Queens, and central Staten Island, although Staten Island clusters appeared more fragmented during working hours (6:00 to 18:00) and the evening period (18:00–24:00). High-value clusters in the Bronx showed day-long persistence, while most Manhattan clusters occurred only during daytime hours (6:00 to 18:00). The Woodhaven and Richmond Hill areas of Queens exhibited a high-value cluster primarily during non-working hours (18:00–06:00).

Seasonally, high-value clusters in the Bronx and Brooklyn persisted year-round, while Manhattan clusters disappeared in summer. The Queens expanded high-value clusters in summer and saw them diminish in winter. Low-value clusters around Staten Island and Queens remained stable across seasons.

These temporal and spatial patterns reflect urban functions and morphology shaping residents' activities. To explore these relationships further, Geographically Weighted Regression (GWR) models were developed at the 1000 m scale.

3.4. Spatial Variations in Urban Morphology–NCR Relationships Across Different Temporal Contexts

Table 3 presents GWR model results for each temporal division, showing local coefficient statistics (mean and median) and model R^2 values. GWR models exhibited significantly higher explanatory power (R^2 = 0.55–0.62) compared to OLS models (R^2 = 0.27–0.45), demonstrating local spatial heterogeneity in the relationship between urban morphology and NCR.

Table 3. GWR model regression coefficient descriptive statistics.

Time	OLS-R^2	GWR-R^2	Stats	Variables											
				Svi Sky	Svi Green	RD	FAR	ME	BD	SD	CLR	RLR	TD	NDVI	MPS
00:00–06:00	0.38	0.61	Mean	−0.008	−0.011	0	−0.001	−0.002	0.007	0.002	0.003	−0.002	0	−0.002	0.001
			Median	−0.002	−0.008	0	−0.001	−0.001	0.007	0.002	0.002	−0.001	0	−0.001	0.001
06:00–18:00	0.44	0.59	Mean	−0.006	−0.002	0	0	−0.001	0	0.001	0.003	0	0	−0.001	0.001
			Median	−0.005	−0.001	0	0	−0.001	0	0.001	0.003	0	0	−0.001	0.001
18:00–24:00	0.29	0.58	Mean	−0.008	−0.005	0	0.001	−0.002	0.003	0.003	0.008	0	0	−0.002	0.005
			Median	0.002	−0.001	0	−0.001	0	0.004	0.002	0.008	0.001	0	−0.003	0.004
Spring	0.41	0.61	Mean	−0.241	−0.098	−0.001	0.012	−0.04	−0.002	0.056	0.09	0.005	−0.01	−0.041	0.026
			Median	−0.174	−0.073	−0.001	−0.006	−0.074	0.005	0.051	0.062	0.019	−0.006	−0.039	0.003
Summer	0.27	0.55	Mean	−0.119	−0.151	−0.001	−0.029	−0.127	0.076	0.067	0.072	0.023	−0.006	−0.046	0.081
			Median	0.001	−0.154	−0.001	−0.053	−0.096	0.102	0.059	0.073	0.051	−0.005	−0.042	0.05
Autumn	0.38	0.58	Mean	−0.274	−0.153	0	−0.014	−0.053	0.079	0.051	0.147	−0.022	−0.006	−0.041	0.047
			Median	−0.208	−0.156	0	−0.016	−0.027	0.071	0.047	0.153	−0.003	−0.003	−0.035	0.024
Winter	0.45	0.57	Mean	−0.192	−0.002	0	0.006	−0.029	−0.007	0.043	0.151	−0.015	−0.006	−0.024	0
			Median	−0.109	0.029	0	0.001	−0.026	−0.023	0.039	0.115	−0.011	−0.005	−0.025	−0.003
Weekdays	0.43	0.62	Mean	−0.24	−0.102	0	0.017	−0.025	−0.058	0.042	0.148	0.011	−0.01	−0.025	0.025
			Median	−0.142	−0.079	0	−0.002	−0.041	−0.047	0.041	0.142	0.015	−0.007	−0.024	0.019
Weekends	0.31	0.59	Mean	−0.15	−0.166	−0.001	−0.062	−0.127	0.177	0.059	0.103	−0.006	−0.009	−0.054	0.087
			Median	0.023	−0.158	−0.001	−0.057	−0.122	0.192	0.053	0.104	0.003	−0.005	−0.065	0.074

Note: RD is Road Density; ME is Building Shape; BD is Building Density; SD is Shannon Diversity; CLR is Commercial Land Ratio; RLR is Residential Land Ratio; TD is Tree diameter; FAR is Floor Area Ratio; NDVI is Normalized difference vegetation index; MPS is Patch Size of Vegetation.

Figures 11–13 illustrate the spatial distribution of local coefficients for three representative indicators: SVI Green, commercial land ratio and building density (the remaining indicators can be found in Appendix A). Local coefficients are categorized into six levels using equal intervals, where red generally represents positive values, blue indicates negative values, and color intensity reflects coefficient magnitude.

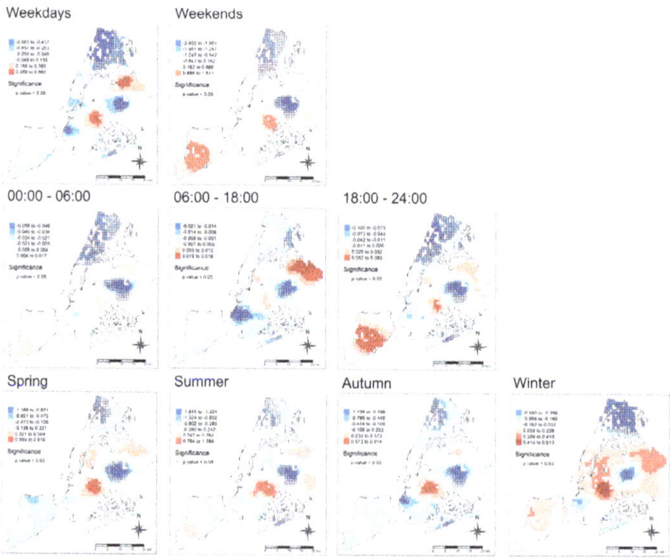

Figure 11. Spatiotemporal variations in the relationship between SVI Green and NCR.

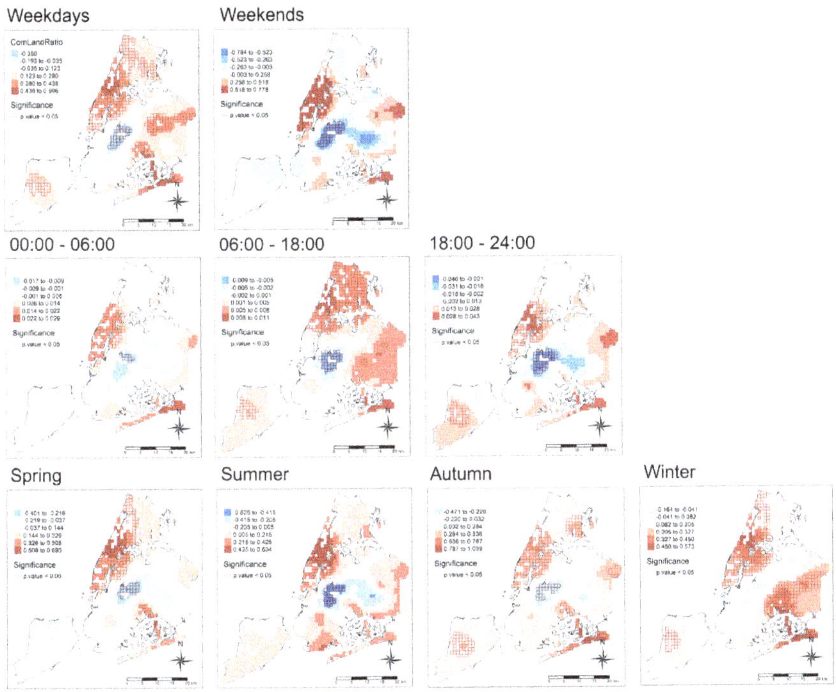

Figure 12. Spatiotemporal variations in the relationship between CLR and NCR.

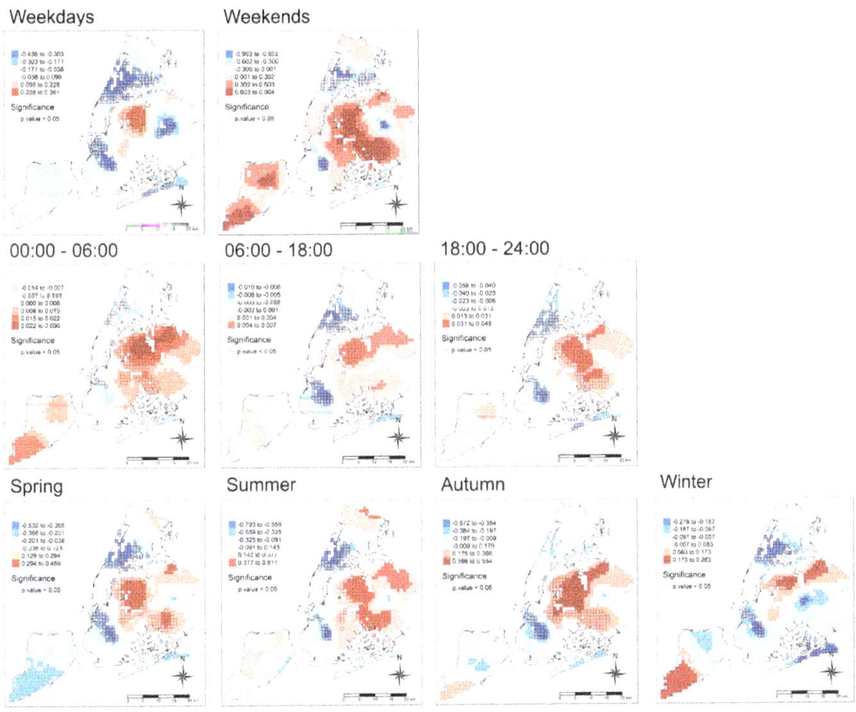

Figure 13. Spatiotemporal variations in the relationship between BD and NCR.

3.4.1. Street View Index for Green (SVI Green)

SVI Green has an overall negative association with NCR across all temporal divisions (Table 3), indicating vegetation's general noise-reduction benefits. Nevertheless, Figure 11 reveals significant spatiotemporal variations in this relationship.

The weekday–weekend comparison indicates greater spatial variations in the SVI Green—NCR relationships on weekends (local coefficients ranging from −2.655 to 1.571) compared to workdays (−0.661 to 0.562). Consistently significant negative coefficients were observed in the Bronx and Queens, while Staten Island exhibited a shift from negative coefficients on workdays to significant positive coefficients on weekends. This pattern likely emerges because green spaces, while effective at suppressing noise propagation, also serve as attractive recreational destinations. During weekends, these areas draw more visitors and recreational activities, potentially generating more noise complaints despite the vegetation's acoustic buffering properties.

Seasonal comparisons showed the smallest spatial heterogeneity in winter, with similar regional patterns in spring and autumn. This seasonal variation closely correlates with changes in vegetation coverage, as noise-reduction benefits are the greatest during periods of full foliage (summer) and diminish during seasons with reduced vegetation coverage (winter). For instance, Staten Island showed notable shifts in the SVI Green–NCR relationship from positive in winter to negative in spring. Nevertheless, the positive impact observed in Staten Island during summer can be attributed to its role as a major recreation area experiencing a seasonal tourism surge.

Diurnal comparisons revealed the largest variations at night (18:00–24:00), with Queens showing the strongest positive impact during working hours (6:00–18:00) and Staten Island during evening hours. These temporal variations indicate that the relationship between urban greenery and noise complaints is not static but varies with daily activity patterns and usage of green spaces.

3.4.2. Commercial Land Ratio (CLR)

CLR showed a positive association with NCR (Table 3), with significant spatiotemporal variations that reflect the complex interaction between land use and residents' activity patterns across different urban contexts (Figure 12).

The weekday–weekend comparison reveals more pronounced regional differences on weekends (local coefficients range from −0.784 to 0.778) than workdays (−0.350 to 0.596). In Manhattan's commercial area, positive impacts intensify on weekends, indicating increased noise complaints due to heightened commercial and entertainment activities. While northern Brooklyn, which is predominantly residential, shows stronger negative impacts on weekends, suggesting that increased leisure activities in its scattered commercial areas might be perceived more as a contributor to urban vitality than noise pollution. The area around Jamaica Station exhibits an interesting shift from positive impacts on workdays to negative on weekends, possibly because its mixed-use characteristics that commercial activities are considered more disruptive during workdays but contribute to desirable leisure opportunities on weekends.

Regarding diurnal variations, the nighttime hours (18:00–24:00) exhibit more prominent regional variations (local coefficients: −0.046 to 0.043), indicating heightened residents' sensitivity to environmental noise during these hours. South Manhattan and the Bronx demonstrate significant temporal variations: the substantial positive impact in the South Manhattan area during non-working hours (18:00–06:00) diminishes to non-significant levels (local coefficient ≈ 0, $p > 0.05$) during daytime hours (06:00–18:00), whereas the Bronx area shows the opposite pattern, reflecting different activity cycles in these distinct urban contexts.

For seasonal variations, negative coefficients are primarily found around the Brooklyn–Queens border area; however, this negative impact becomes non-significant ($p > 0.05$) in winter. High positive impacts are consistently observed in the Bronx–Manhattan border area, while the significant positive value in northern Bronx turns non-significant during summer and autumn, possibly reflecting seasonal changes in commercial activity patterns and resident sensitivity.

3.4.3. Building Density (BD)

Figure 13 demonstrates the spatiotemporal variations in the relationship between building density (BD) and NCR. High positive local impacts of BD consistently appear in Brooklyn and Queens, where dense housing clusters create more opportunities for noise transmission between residential units. In contrast, significant negative impacts persist in Manhattan and the Bronx, possibly due to better noise insulation in newer high-rise buildings or residents' different expectations of urban soundscapes. Staten Island exhibits a more unstable regional pattern, shifting between positive and negative across temporal divisions, reflecting its complex residential patterns.

The weekday–weekend comparison shows more pronounced regional differences on weekends (local coefficients range from −0.903 to 0.904) than workdays (−0.436 to 0.361). Diurnal variations show more regional differences during nighttime (18:00–24:00). These patterns suggest heightened noise sensitivity when more residents are at home and expecting quieter environments for rest and relaxation. Significant positive impacts are consistently observed in Queens throughout the day; however, the areas with significant positive influence (local coefficient > 0, $p < 0.05$) shrink during daytime hours (06:00–18:00) and expand during the dawn period (00:00–06:00). In contrast, areas with significant negative influence (local coefficient < 0, $p < 0.05$) are more widespread during the hours of 06:00 to 24:00, covering the majority of working and evening activities. This phenomenon indicates that building density's impact varies with daily activity patterns and residents' expectations of environmental noise.

Seasonal comparisons show minimal regional variations in winter, with consistent patterns across seasons, including high positive impacts around the Brooklyn–Queens border and negative impacts in northern Brooklyn and the Bronx–Manhattan border. This consistency suggests that building density's influence on noise complaints is more related to permanent structural characteristics and established residential patterns than seasonal variations in human activity.

4. Discussion

This study revealed significant scale effects and spatiotemporal heterogeneity in noise complaint patterns and their relationships with urban morphology across New York City. The identified scale dependency and spatiotemporal variations suggest a complex underlying mechanism linking urban morphology to noise complaints. In this section, we try to interpret these patterns based on existing urban theories and previous studies, aiming to understand the potential mechanisms that drive the observed scale-dependent spatial relationship between urban morphology and noise complaints, specify the implications for urban planning and noise management.

4.1. Scale Dependency in Urban Morphology–NCR Relationship

This study revealed distinct scale-dependent patterns in urban morphology-noise complaint relationships across multiple spatial scales (100 m, 200 m, 500 m, and 1000 m), reflecting the complex nature of urban noise issues at different urban planning levels.

The urban morphology–NCR relationships at different scales can be explained by both the physical and social dimensions of urban noise perception. Noise complaints result from the interaction between actual environmental noise levels (physical dimension) and residents' subjective perceptions (social dimension), while each dimension has different spatial range of influence.

From a physical perspective, noise propagation typically follows distance–decay principles, with substantial attenuation occurring within relatively short distances, the built environment's influence on noise propagation operates across multiple scales. At the immediate building surrounding scale, factors such as building materials, vegetation barriers, and building shape directly affect noise reflection and absorption. However, at the neighborhood scale, shared urban morphological characteristics—such as similar building construction standards, architectural styles, and arrangements of buildings and open spaces—can create areas with relatively uniform noise propagation patterns. These neighborhood-level morphological similarities often reflect historical development periods or zoning regulations, resulting in a broad influence on noise propagation that extends beyond individual building surroundings [53].

The social dimension of noise perception operates across multiple scales. At local scales, specific building functions and open spaces directly influence noise generation through immediate human activities. While at broader scales, previous studies have shown that urban residents' activity space typically spans 600–1200 m [53,54], and human activities within urban functional zones can be better captured at scales around 1000 m [55]. These neighborhood-level functional characteristics shape residents' expectations of urban sound environments. For instance, loud music may be less likely to be perceived as noise in commercial zones than in residential areas. The dual role of human activities, both as noise sources and factors shaping noise perception, helps explain the observed multi-scale relationships in noise complaints.

The results at different scales reveal distinct patterns of urban morphology-noise relationships. While the strongest spatial autocorrelation was observed at the 1000 m scale, suggesting coherent neighborhood-level patterns, the smaller scales (100–200 m) capture more localized relationships that may be more relevant for specific noise interventions.

4.2. The Temporal Variations in Urban Morphology-NCR Relationship

The results revealed distinct temporal patterns in noise complaint rates (NCR), and their relationships with urban morphology indicators appear to align closely with residents' activity patterns across diurnal, weekday–weekend, and seasonal divisions.

Diurnal patterns showed stronger correlations between urban morphology indicators and NCR during daytime hours (06:00–18:00), when human activities are typically more intense. This likely reflects that daytime noise, largely from street-level activities and traffic, is more influenced by urban morphology. In contrast, nighttime noise sources are more localized within buildings, reducing urban morphology's impact.

GWR results reflected these temporal variations, showing stronger spatial heterogeneity in morphology–NCR relationships during non-working periods, such as weekends or nighttime hours. This indicates that residents may focus more on their living environment during leisure periods, affecting their perception and reporting of noise issues.

These findings underscore the importance of considering temporal patterns in evaluating urban morphology's impact on noise complaints, guiding urban design and noise management strategies.

4.3. The Spatiotemporal Dynamics in Urban Morphology-NCR Relationships

The multi-temporal GWR analysis reveals that urban morphology's influence on noise complaints varies significantly across spatial and temporal context, challenging the effectiveness of uniform noise management strategies. The results have identified several key patterns in how urban morphology influences noise complaints and highlighted several important implications for urban planning and noise management.

For example, vegetation's effectiveness in noise mitigation shows significant seasonal and locational variations. While vegetation generally reduces noise complaints, its varying effectiveness across seasons and its dual role as both noise buffer and activity attractor suggest that noise mitigation strategies need to be regionally and seasonally adaptive. In residential areas like the Bronx and Queens where vegetation consistently reduces noise complaints, planners might prioritize dense vegetation barriers and street trees. However, in recreational areas like Staten Island, green space design should incorporate buffer zones between active recreation areas and nearby residential zones, and seasonal tourism management strategies should be integrated with noise control measures.

The spatial arrangement of land uses is also a critical factor. The complex relationship between commercial activities and noise complaints across different urban contexts highlights the importance of temporal considerations in land use planning. The contrasting patterns observed in areas like Jamaica Station area demonstrate how the same urban form can have opposing effects on noise complaints depending on the time period. This suggests that future mixed-use development plans should consider not just the spatial arrangement of different land uses, but also their temporal activity patterns.

Building density's relationship with noise complaints reveals strong temporal variations, particularly in residential areas. The intensified relationship during non-working periods highlights how urban morphology interacts with human activity patterns to influence noise perception. The contrasting patterns between dense housing clusters and newer high-rise residential buildings suggest that building typology and insulation techniques might play crucial roles in mediating the relationship between density and noise complaints. This implies that density-related noise issues might be better addressed through improved building standards and design guidelines rather than density restrictions alone.

These findings highlight the need for context-specific approaches to urban noise management that consider both spatial and temporal variations. Simple correlations between urban morphology and noise complaints may mask important local variations and temporal dynamics that are crucial for effective noise control strategies. This understanding suggests that urban noise management strategies should be flexible and adaptable, varying across different urban contexts and time periods rather than applying uniform solutions across entire cities [56].

5. Conclusions

This study investigated the scale-dependent and spatiotemporal relationships between urban morphology and noise complaint rates (NCR) in New York City using multiple statistical approaches, including Spearman correlation, Moran's I test, and multi-temporal Geographically Weighted Regression (GWR).

Key findings revealed three main aspects of urban morphology–NCR relationships:

(1) Scale Dependency: Analysis across four spatial scales (100 m, 200 m, 500 m, 1000 m) reveals distinct patterns of urban morphology–noise relationships. While the strongest spatial autocorrelation was observed at the 1000 m scale, suggesting coherent neighborhood-level patterns, the smaller scales capture more localized relationships that may be more relevant for specific local noise interventions.

(2) Temporal Variations: The relationship between urban morphology and NCR demonstrated distinct temporal variations that align with human activity patterns. Particularly, urban morphology indicators showed stronger correlations with NCR during daytime hours (06:00–18:00) when street-level activities dominate noise generation. During non-working periods (nighttime and weekends), the relationships exhibited greater spatial heterogeneity, suggesting that residents' changing activity patterns and heightened environmental awareness during leisure periods influence how urban form affects noise perception and reporting.
(3) Spatiotemporal Variations: Multi-temporal GWR models revealed complex interactions between urban morphology and noise complaints across different temporal contexts. Street vegetation demonstrates varying effectiveness between seasons and dual functionality as both noise buffer and activity attractor. Commercial areas show contrasting impacts between workdays and weekends, particularly in mixed land use areas. Building density's influence is most pronounced during non-working periods in residential areas, with distinct patterns between dense housing clusters and high-rise residential buildings, reflecting the interaction between urban morphology and human activity cycles.

These findings demonstrate that urban morphology's relationship with the sound environment is inherently scale-dependent and temporally dynamic, suggesting that universal approaches to urban noise management may be insufficient. Instead, effective noise mitigation strategies should consider appropriate spatial scale for analyzing and implementing interventions, temporal variations in both noise patterns and the effectiveness of morphological features, and local contextual factors that influence how urban morphological characteristics affect noise perception.

Nevertheless, several main limitations should be noted: First, while our study uses noise complaints as a proxy for residents' noise perception, reflecting a complex interaction between objective noise levels and subjective responses, the lack of corresponding objective noise measurements means making it difficult for us to fully disentangle the relationship between actual noise levels and perceived disturbance.

Second, while NCR data captured temporal variations, urban morphology indicators remained static due to data availability constraints, limiting our understanding of temporal dynamics. Future studies could extend this research by investigating the urban morphology–NCR relationships over longer time periods (years to decades) to understand how urban development and morphological changes influence noise patterns over time.

Third, while analysis across scales (100–1000 m) showed increasing correlations at larger scales, the potential peak correlation scale remains unknown beyond our maximum analysis unit of 1000 m. Additionally, the predetermined scales, though aligned with administrative units for practical application, may not capture all optimal scales at which urban morphology influences noise perception in different urban contexts.

Last, although we identified spatial variations in urban morphology–NCR relationships, our interpretations of these local patterns were based on inferences rather than direct evidence of local contexts, necessitating further investigation to verify these contextual effects.

Author Contributions: Conceptualization, S.C. and P.H.; methodology, S.C., P.H. and B.Y.; software, G.S., Y.C. and Y.W.; validation, S.C. and P.H.; formal analysis, S.C., B.Y. and P.H.; investigation, G.S. and Y.C.; resources, S.C. and P.H.; data curation, S.C. and P.H.; writing—original draft preparation, P.H.; writing—review and editing, S.C. and B.Y.; visualization, S.C. and P.H.; supervision, S.C. and B.Y.; project administration, S.C.; funding acquisition, S.C. and Y.W. All authors have read and agreed to the published version of the manuscript.

Funding: This research was funded by the Natural Science Foundation of Sichuan, China (Grant No. 24NSFSC0923) and the National Natural Science Foundation of China (Grant No. 52208035). The APC was funded by S.C.

Data Availability Statement: The raw data supporting the conclusions of this article will be made available by the authors on request.

Acknowledgments: The authors would like to thank all the organizations that provided data sources for this manuscript.

Conflicts of Interest: The authors declare that they have no known competing financial interests or personal relationships that could have appeared to influence the work reported in this paper.

Appendix A. Spatiotemporal Variations in the Relationship Between the Remaining Urban Morphology Indicators and NCR

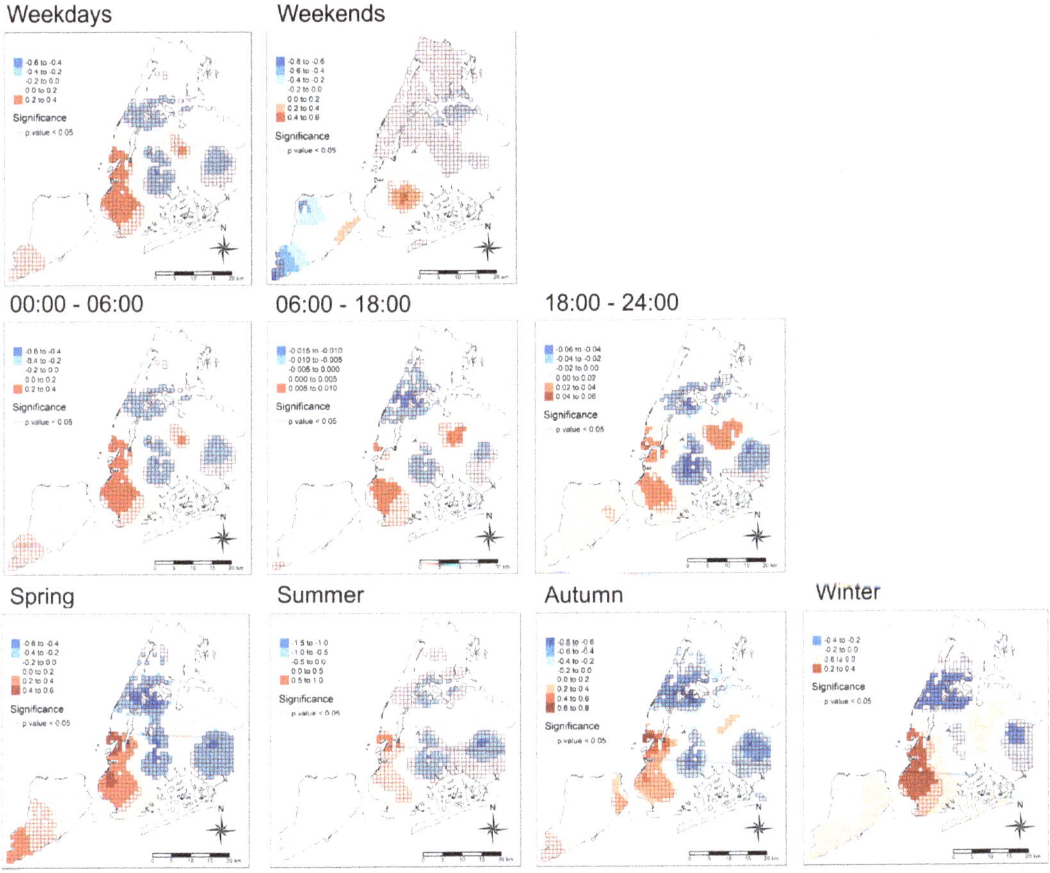

Figure A1. Spatiotemporal variations in the relationship between BD and NCR.

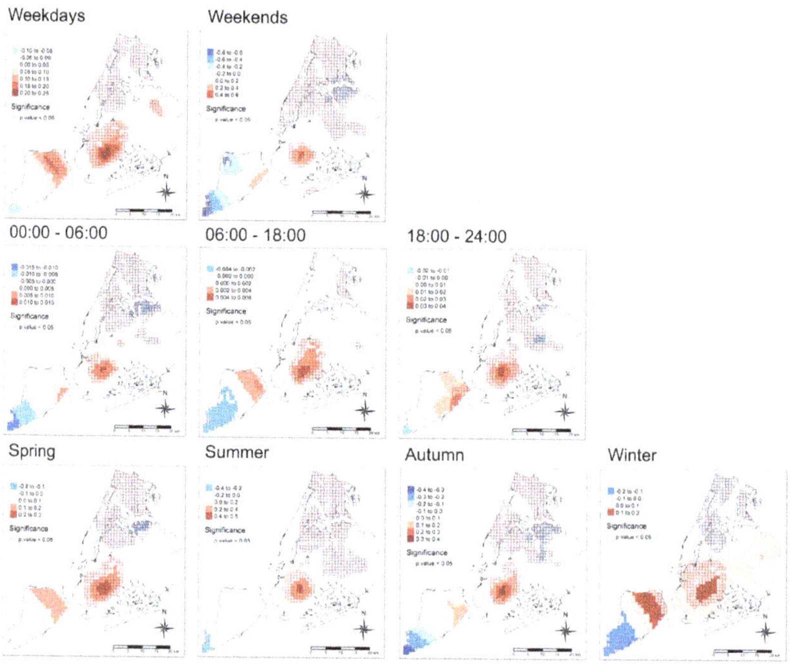

Figure A2. Spatiotemporal variations in the relationship between FAR and NCR.

Figure A3. Spatiotemporal variations in the relationship between MPS and NCR.

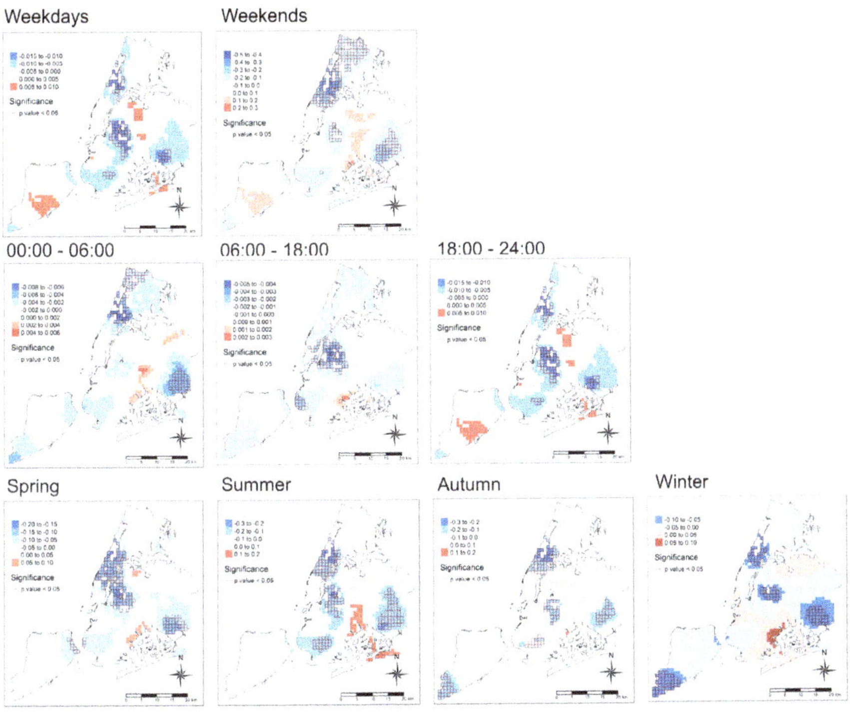

Figure A4. Spatiotemporal variations in the relationship between NDVI and NCR.

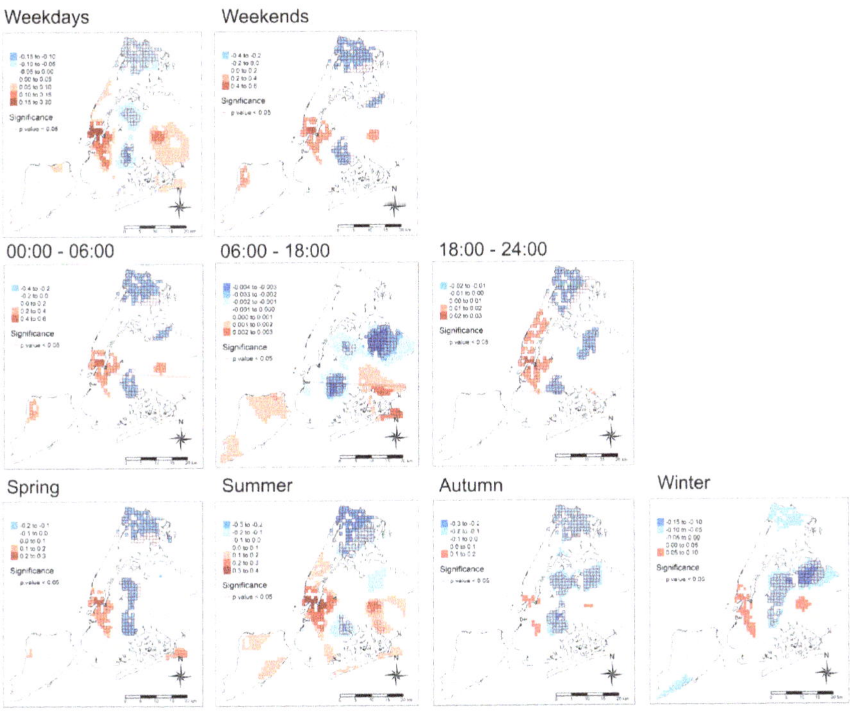

Figure A5. Spatiotemporal variations in the relationship between RLR and NCR.

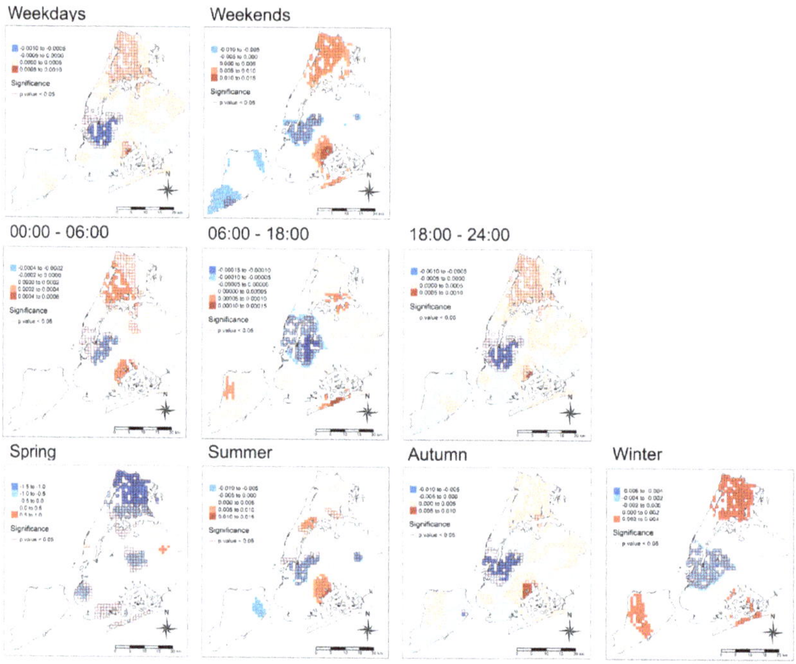

Figure A6. Spatiotemporal variations in the relationship between road density and NCR.

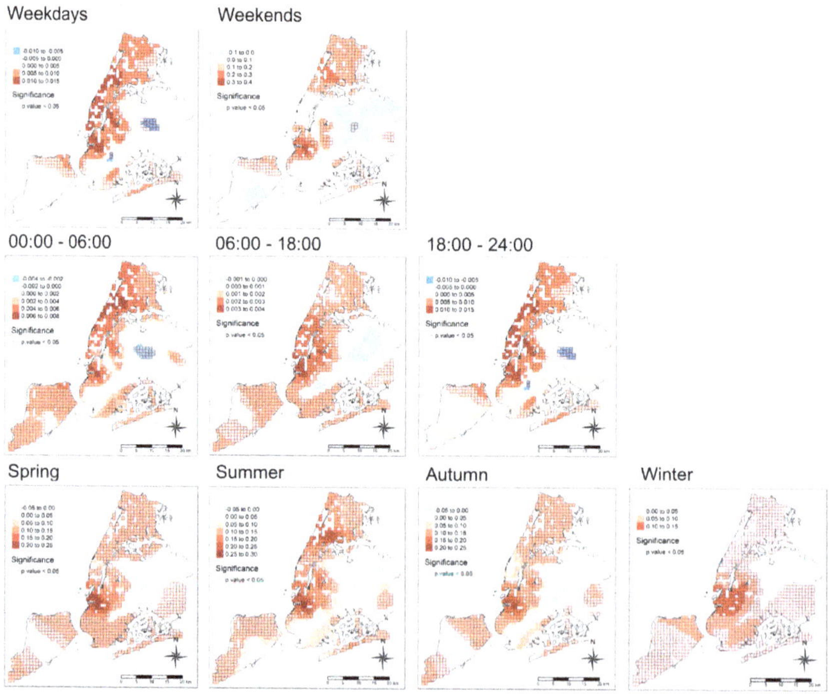

Figure A7. Spatiotemporal variations in the relationship between Shannon diversity and NCR.

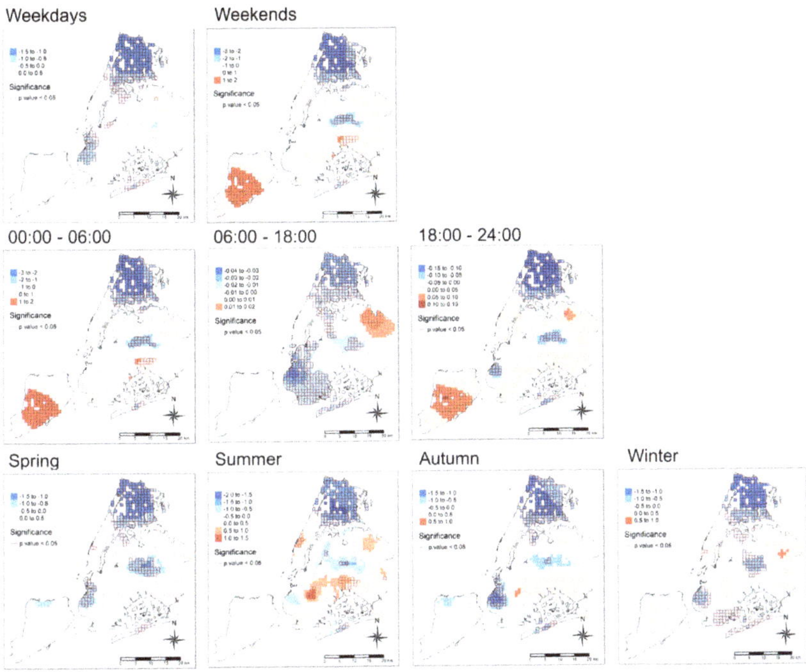

Figure A8. Spatiotemporal variations in the relationship between GVI Sky and NCR.

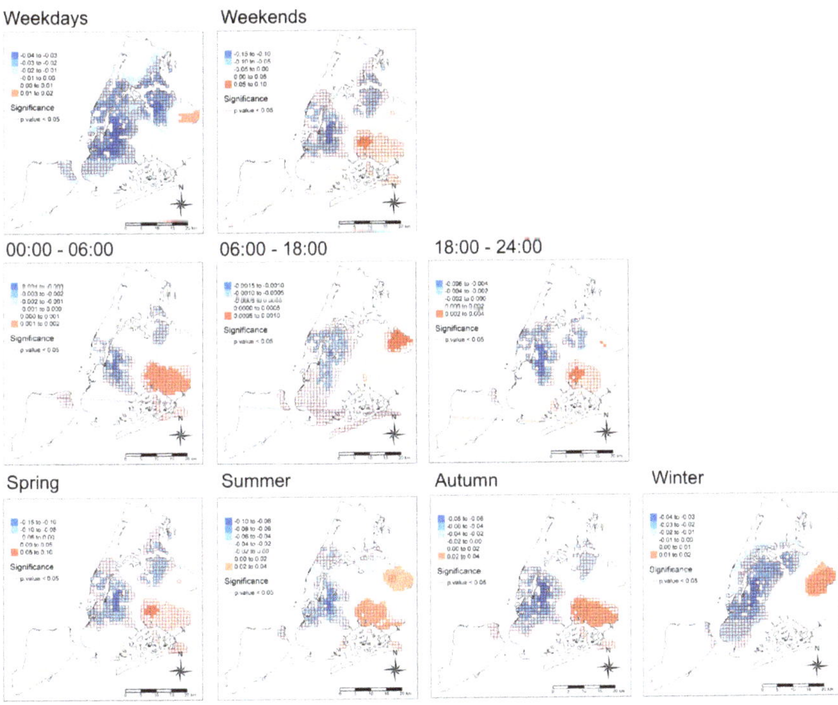

Figure A9. Spatiotemporal variations in the relationship between Tree diameter and NCR.

References

1. Basner, M.; Babisch, W.; Davis, A.; Brink, M.; Clark, C.; Janssen, S.; Stansfeld, S. Auditory and Non-Auditory Effects of Noise on Health. *Lancet* **2014**, *383*, 1325–1332. [CrossRef] [PubMed]
2. Bouzir, T.A.K.; Zemmouri, N. Effect of Urban Morphology on Road Noise Distribution. *Energy Procedia* **2017**, *119*, 376–385. [CrossRef]
3. Chen, S. The Challenge of Noise Pollution in High-Density Urban Areas: Relationship between 2D/3D Urban Morphology and Noise Perception. *Build. Environ.* **2024**, *253*, 111313. [CrossRef]
4. Han, X.; Huang, X.; Liang, H.; Ma, S.; Gong, J. Analysis of the Relationships between Environmental Noise and Urban Morphology. *Environ. Pollut.* **2018**, *233*, 755–763. [CrossRef]
5. Fotheringham, S.; Wong, D.W.S. The Modifiable Areal Unit Problem in Multivariate Statistical Analysis. *Environ. Plan. A* **1991**, *23*, 1025–1044. [CrossRef]
6. Margaritis, E.; Kang, J. Relationship between Urban Green Spaces and Other Features of Urban Morphology with Traffic Noise Distribution. *Urban For. Urban Green.* **2016**, *15*, 174–185. [CrossRef]
7. Qu, F.; Kang, J. Effects of Built Environment Morphology on Wind Turbine Noise Exposure at Building Façades. *Renew. Energy* **2017**, *107*, 629–638. [CrossRef]
8. Hong, J.Y.; Jeon, J.Y. Relationship between Spatiotemporal Variability of Soundscape and Urban Morphology in a Multifunctional Urban Area: A Case Study in Seoul, Korea. *Build. Environ.* **2017**, *126*, 382–395. [CrossRef]
9. Lan, Z.; He, C.; Cai, M. Urban Road Traffic Noise Spatiotemporal Distribution Mapping Using Multisource Data. *Transp. Res. Part D Transp. Environ.* **2020**, *82*. [CrossRef]
10. Guedes, I.C.M.; Bertoli, S.R.; Zannin, P.H.T. Influence of Urban Shapes on Environmental Noise: A Case Study in Aracaju—Brazil. *Sci. Total Environ.* **2011**, *412–413*, 66–76. [CrossRef]
11. Yildirim, Y.; Arefi, M. Noise Complaints and Transportation Inequality Assessment. *Transp. Res. Part D Transp. Environ.* **2021**, *99*, 103021. [CrossRef]
12. Tang, U.W.; Wang, Z.S. Influences of Urban Forms on Traffic-Induced Noise and Air Pollution: Results from a Modelling System. *Environ. Model. Softw.* **2007**, *22*, 1750–1764. [CrossRef]
13. Fleischmann, M.; Romice, O.; Porta, S. Measuring Urban Form: Overcoming Terminological Inconsistencies for a Quantitative and Comprehensive Morphologic Analysis of Cities. *Environ. Plan. B Urban Anal. City Sci.* **2021**, *48*, 2133–2150. [CrossRef]
14. Guo, F.; Schlink, U.; Wu, W.; Hu, D.; Sun, J. Scale-Dependent and Season-Dependent Impacts of 2D/3D Building Morphology on Land Surface Temperature. *Sustain. Cities Soc.* **2023**, *97*, 104788. [CrossRef]
15. Li, Z.; Hu, D. Exploring the Relationship between the 2D/3D Architectural Morphology and Urban Land Surface Temperature Based on a Boosted Regression Tree: A Case Study of Beijing, China. *Sustain. Cities Soc.* **2022**, *78*, 103392. [CrossRef]
16. Han, D.; An, H.; Wang, F.; Xu, X.; Qiao, Z.; Wang, M.; Sui, X.; Liang, S.; Hou, X.; Cai, H.; et al. Understanding Seasonal Contributions of Urban Morphology to Thermal Environment Based on Boosted Regression Tree Approach. *Build. Environ.* **2022**, *226*, 109770. [CrossRef]
17. Cao, Q.; Luan, Q.; Liu, Y.; Wang, R. The Effects of 2D and 3D Building Morphology on Urban Environments: A Multi-Scale Analysis in the Beijing Metropolitan Region. *Build. Environ.* **2021**, *192*, 107635. [CrossRef]
18. Dong, Q.; Huang, Z.; Zhou, X.; Guo, Y.; Scheuer, B.; Liu, Y. How Building and Street Morphology Affect CO_2 Emissions: Evidence from a Spatially Varying Relationship Analysis in Beijing. *Build. Environ.* **2023**, *236*, 110258. [CrossRef]
19. Liu, J.; Huang, Y.-J.; Chen, Z.; Hong, X.-C. Multi-Scale Effects of Landscape Pattern on Soundscape Perception in Residential Green Spaces. *Forests* **2023**, *14*, 2323. [CrossRef]
20. Song, J.; Meng, Q.; Kang, J.; Yang, D.; Li, M. Effects of Planning Variables on Urban Traffic Noise at Different Scales. *Sustain. Cities Soc.* **2024**, *100*, 105006. [CrossRef]
21. Rahman, M.M.; Najaf, P.; Fields, M.G.; Thill, J.-C. Traffic Congestion and Its Urban Scale Factors: Empirical Evidence from American Urban Areas. *Int. J. Sustain. Transp.* **2022**, *16*, 406–421. [CrossRef]
22. Wu, C.; Ren, F.; Hu, W.; Du, Q. Multiscale Geographically and Temporally Weighted Regression: Exploring the Spatiotemporal Determinants of Housing Prices. *Int. J. Geogr. Inf. Sci.* **2019**, *33*, 489–511. [CrossRef]
23. Liu, J.; Kang, J.; Luo, T.; Behm, H.; Coppack, T. Spatiotemporal Variability of Soundscapes in a Multiple Functional Urban Area. *Landsc. Urban Plan.* **2013**, *115*, 1–9. [CrossRef]
24. Guo, L.-H.; Cheng, S.; Liu, J.; Wang, Y.; Cai, Y.; Hong, X.-C. Does Social Perception Data Express the Spatio-Temporal Pattern of Perceived Urban Noise? A Case Study Based on 3,137 Noise Complaints in Fuzhou, China. *Appl. Acoust.* **2022**, *201*, 109129. [CrossRef]
25. Wang, S.; Liu, Y.; Corcoran, J. A Season for Complaints: How Does Weather Affect Noise Complaints between Neighbors? *Weather. Clim. Soc.* **2021**, *13*, 753–768. [CrossRef]
26. Zhou, Z. Analysis of Traffic Noise Spatial Distribution Characteristics and Influencing Factors in High-Density Cities. *Appl. Acoust.* **2024**, *217*, 109838. [CrossRef]

27. Zhao, Y.; Cheng, J.; Wang, S.; Qin, L.; Zhang, X. Spatiotemporal Analytics of Environmental Sounds and Influencing Factors Based on Urban Sensor Network Data. *IJGI* **2023**, *12*, 386. [CrossRef]
28. Karapostoli, A.; Votsi, N.-E. Urban Soundscapes in the Historic Centre of Thessaloniki: Sonic Architecture and Sonic Identity. *Sound Stud.* **2018**, *4*, 162–177. [CrossRef]
29. He, P.; Yu, B.; Ma, J.; Luo, K.; Chen, S.; Shen, Z. Exploring the Non-Linear Relationship and Synergistic Effect between Urban Built Environment and Public Sentiment Integrating Macro- and Micro-Level Perspective: A Case Study in San Francisco. *Front. Psychol.* **2024**, *15*, 1276923. [CrossRef]
30. Ariza-Villaverde, A.B.; Jiménez-Hornero, F.J.; Gutiérrez De Ravé, E. Influence of Urban Morphology on Total Noise Pollution: Multifractal Description. *Sci. Total Environ.* **2014**, *472*, 1–8. [CrossRef]
31. Zhang, Y.; Kwan, M.-P.; Ma, H. Sensing Noise Exposure and Its Inequality Based on Noise Complaint Data through Vision-Language Hybrid Method. *Appl. Geogr.* **2024**, *171*, 103369. [CrossRef]
32. Marry, S.; Defrance, J. Analysis of the Perception and Representation of Sonic Public Spaces through on Site Survey, Acoustic Indicators and in-Depth Interviews. *Appl. Acoust.* **2013**, *74*, 282–292. [CrossRef]
33. Benavides, J.; Rowland, S.T.; Do, V.; Goldsmith, J.; Kioumourtzoglou, M.-A. Unintended Impacts of the Open Streets Program on Noise Complaints in New York City. *Environ. Res.* **2023**, *224*, 115501. [CrossRef] [PubMed]
34. Yuan, M.; Yin, C.; Sun, Y.; Chen, W. Examining the Associations between Urban Built Environment and Noise Pollution in High-Density High-Rise Urban Areas: A Case Study in Wuhan, China. *Sustain. Cities Soc.* **2019**, *50*, 101678. [CrossRef]
35. Zhang, P.; Ghosh, D.; Park, S. Spatial Measures and Methods in Sustainable Urban Morphology: A Systematic Review. *Landsc. Urban Plan.* **2023**, *237*, 104776. [CrossRef]
36. Yang, L.; Lu, Y.; Cao, M.; Wang, R.; Chen, J. Assessing Accessibility to Peri-Urban Parks Considering Supply, Demand, and Traffic Conditions. *Landsc. Urban Plan.* **2025**, *257*, 105313. [CrossRef]
37. Yang, L.; Yang, H.; Yu, B.; Lu, Y.; Cui, J.; Lin, D. Exploring Non-Linear and Synergistic Effects of Green Spaces on Active Travel Using Crowdsourced Data and Interpretable Machine Learning. *Travel Behav. Soc.* **2024**, *34*, 100673. [CrossRef]
38. Tong, H.; Kang, J. Characteristics of Noise Complaints and the Associations with Urban Morphology: A Comparison across Densities. *Environ. Res.* **2021**, *197*, 111045. [CrossRef]
39. Luo, P.; Yu, B.; Li, P.; Liang, P.; Liang, Y.; Yang, L. How 2D and 3D Built Environments Impact Urban Surface Temperature under Extreme Heat: A Study in Chengdu, China. *Build. Environ.* **2023**, *231*, 110035. [CrossRef]
40. Yang, J.; Shi, B.; Shi, Y.; Marvin, S.; Zheng, Y.; Xia, G. Air Pollution Dispersal in High Density Urban Areas: Research on the Triadic Relation of Wind, Air Pollution, and Urban Form. *Sustain. Cities Soc.* **2020**, *54*, 101941. [CrossRef]
41. Longley, P.A.; Adnan, M. Geo-Temporal Twitter Demographics. *Int. J. Geogr. Inf. Sci.* **2016**, *30*, 369–389. [CrossRef]
42. Hu, J.; Fan, T.; Tang, X.; Yang, Z.; Ren, Y. Nonlinear Relations of Urban Morphology to Thermal Anomalies: A Cross-Time Comparative Study Based on Grad-CAM and SHAP. *Ecol. Indic.* **2024**, *162*, 112024. [CrossRef]
43. Zhang, S.; Yuan, C.; Chen, T.; Ma, B.; Liu, N. A Cross-Scale Indicator Framework for the Study of Annual Stability of Land Surface Temperature in Different Land Uses. *Sustain. Cities Soc.* **2024**, *116*, 105936. [CrossRef]
44. Chen, H.-C.; Han, Q.; De Vries, B. Urban Morphology Indicator Analyzes for Urban Energy Modeling. *Sustain. Cities Soc.* **2020**, *52*, 101863. [CrossRef]
45. Li, N. Spatiotemporal Scale-Dependent Effects of Urban Morphology on Meteorology: A Case Study in Beijing Using Observations and Simulations. *Build. Environ.* **2023**, *240*, 110383. [CrossRef]
46. United States Census Bureau. New York. Available online: https://www.census.gov/geographies/reference-files/2010/geo/state-local-geo-guides-2010/new-york.html (accessed on 21 November 2024).
47. Fleischmann, M. Momepy: Urban Morphology Measuring Toolkit. *JOSS* **2019**, *4*, 1807. [CrossRef]
48. Yang, L.; Yu, B.; Liang, Y.; Lu, Y.; Li, W. Time-Varying and Non-Linear Associations between Metro Ridership and the Built Environment. *Tunn. Undergr. Space Technol.* **2023**, *132*, 104931. [CrossRef]
49. Yang, L.; Ao, Y.; Ke, J.; Lu, Y.; Liang, Y. To Walk or Not to Walk? Examining Non-Linear Effects of Streetscape Greenery on Walking Propensity of Older Adults. *J. Transp. Geogr.* **2021**, *94*, 103099. [CrossRef]
50. Chen, Y. An Analytical Process of Spatial Autocorrelation Functions Based on Moran's Index. *PLoS ONE* **2021**, *16*, e0249589. [CrossRef]
51. Lu, B.; Brunsdon, C.; Charlton, M.; Harris, P. Geographically Weighted Regression with Parameter-Specific Distance Metrics. *Int. J. Geogr. Inf. Sci.* **2017**, *31*, 982–998. [CrossRef]
52. Feuillet, T.; Commenges, H.; Menai, M.; Salze, P.; Perchoux, C.; Reuillon, R.; Kesse-Guyot, E.; Enaux, C.; Nazare, J.-A.; Hercberg, S.; et al. A Massive Geographically Weighted Regression Model of Walking-Environment Relationships. *J. Transp. Geogr.* **2018**, *68*, 118–129. [CrossRef]
53. Millward, H.; Spinney, J.; Scott, D. Active-Transport Walking Behavior: Destinations, Durations, Distances. *J. Transp. Geogr.* **2013**, *28*, 101–110. [CrossRef]

54. Scheiner, J. Interrelations between Travel Mode Choice and Trip Distance: Trends in Germany 1976–2002. *J. Transp. Geogr.* **2010**, *18*, 75–84. [CrossRef]
55. Asad, K.M.B.; Yuan, Y. The Impact of Scale on Extracting Individual Mobility Patterns from Location-Based Social Media. *Sensors* **2024**, *24*, 3796. [CrossRef]
56. Bian, Q.; Wang, C.; Sun, Z.; Yin, L.; Jiang, S.; Cheng, H.; Zhao, Y. Research on Spatiotemporal Variation Characteristics of Soundscapes in a Newly Established Suburban Forest Park. *Urban For. Urban Green.* **2022**, *78*, 127766. [CrossRef]

Disclaimer/Publisher's Note: The statements, opinions and data contained in all publications are solely those of the individual author(s) and contributor(s) and not of MDPI and/or the editor(s). MDPI and/or the editor(s) disclaim responsibility for any injury to people or property resulting from any ideas, methods, instructions or products referred to in the content.

land

Article

Spatial Patterns of Land Take in a Mediterranean City: An Assessment of the SDG Indicator 11.3.1 in the Peri-Urban Area of Thessaloniki

Athena Yiannakou * and George Zografos

School of Spatial Planning and Development, Aristotle University of Thessaloniki, 541 24 Thessaloniki, Greece; georgezografos07@gmail.com
* Correspondence: adgianna@plandevel.auth.gr

Abstract: Urban sprawl, characterized by continuous or discontinuous spatial patterns of artificial surface expansion, has been a common trend in most cities, even in those with longstanding compact features, such as cities in the northern Mediterranean region. This paper assesses the land take patterns in the peri-urban area of a typical compact city that experienced significant sprawl trends after the mid-1990s, which are closely linked to the specificities of planning regulations regarding the development in peri-urban settlements as well as outside planned areas. Using the rapidly suburbanized southern peri-urban area of Thessaloniki, Greece, as a case study, the paper analyzes the factors influencing the land cover change in the middle-class-led peri-urbanization during the period 2000–2018 and provides an estimate of the SDG indicator 11.3.1 "ratio of land consumption rate to population growth rate", a suitable indicator for monitoring spatial changes. The main conclusions of the study indicate that, during the period examined, the peri-urban zone investigated in the case study exhibited a higher rate of population growth compared to that of artificial surfaces, with the latter showing a higher change during 2006–2012. However, the spatial pattern of urban expansion displays a fragmented yet linear form, creating fragmented enclaves of agricultural land.

Keywords: land take; land consumption; urban sprawl; SDG indicator 11.3.1; urban infill indicator; peri-urbanization; Mediterranean city; Thessaloniki; Greece

1. Introduction

The rapid expansion of urban areas and their associated infrastructure presents a global challenge with significant environmental, economic, and social implications. Environmental degradation and increased carbon dioxide emissions are linked to the urbanization and urban sprawl processes, which result in the loss of agricultural, forest, and natural land to urban and essentially artificial land. This latter refers to the areas sealed off by construction and urban infrastructure, green spaces, and sports and leisure facilities [1–6]. Urban sprawl is generally characterized by a decrease in the urban density, the decentralization of urban functions, and the transformation of a compact urban form into an irregular, discontinuous, and scattered pattern. In addition to population growth that drives land consumption, built-up areas are expanding faster than the population even where the population is declining [4,5,7].

Urban sprawl is often associated with unplanned urban development characterized by a mix of land uses—including residential and typically large-scale commercial, office, service, and leisure functions—as well as low-density building in peri-urban areas [4], thereby indicating limited urban planning control, particularly in relation to land subdivision. Sprawling cities, in contrast to compact or cohesive cities, are characterized by fragmented and empty spaces, which demonstrate inadequacy in controlling urban growth and highlight the consequences of unplanned development [4]. Urban sprawl is caused by various factors. Economic geographers consider urban sprawl to result from population growth, income growth, and reduced travel costs [8]. The literature acknowledges the significant variations in this theory, noting the increasing complexity of the factors contributing to urban sprawl. Among the factors contributing to urban sprawl, the implementation of land use policies and urban planning is particularly significant, especially concerning the access to land, housing ownership, building processes, and policies influencing urban and residential development, resulting in substantial variations in the urban sprawl across countries [9].

The conversion of undeveloped, usually natural land, into residential urban areas, industrial sites, or transportation infrastructure is closely linked to low-density and/or dispersed urban development [2,10]. "Land take" is the official term used in the EU for this conversion phenomenon after the introduction of the "no net land take by 2050" target, an ambitious, in many respects, target of reaching land take neutrality by 2050 [2]. Land take is recognized as a sustainability problem. The substantial loss of soil functions and ecosystem services associated with land take is regarded as a major environmental challenge both in the European Union and globally, particularly where quantitative estimations regarding land-cover changes show that urban area expansion is strongly negatively correlated with changes in forest, cropland, and grassland. It is also considered as a factor that affects socio-economic parameters [11–13]. It is worth noting that land take does not coincide with urban sprawl. Despite the wide range of definitions regarding urban sprawl, there is a broad consensus that this phenomenon occurs on the urban fringe in rapidly growing areas and that it designates a form of urban development that consumes large amounts of land, usually taking the form of low-density or dispersed development [2,3].

The wide variety of the terms used in official reports and the literature, such as land take, land consumption, soil sealing, artificialization, and their specificity has been thoroughly studied by Marquard et al. [10]. Here, "land take" is regarded as a process that converts, and thereby diminishes, natural, semi-natural, forest, or agricultural land. A "land take indicator" therefore addresses the change in the area of agricultural, forest, and other semi-natural and natural areas that are converted to artificial surfaces and sealed by construction and urban infrastructure as well as sports and leisure facilities and urban greening [10,12]. Artificial surfaces as defined by [10] to refer to land that is assigned to one of the following classes: urban fabric (continuous and discontinuous); industrial, commercial, and transport units; mine, dump, and construction sites; artificial, non-agricultural vegetated areas (green urban areas, sport and leisure facilities). The conversion of natural, semi-natural, forest, or agricultural land to artificial surfaces impairs the land's ecological functions and reduces the ecosystem resilience [12], thereby undermining the urban resilience and sustainability. The related and additionally used term "land consumption" includes three aspects, the expansion of built-up area which can be directly measured, the total land area used for agriculture, forestry, or other economic activities, and the over-intensive exploitation of land used for agriculture and forestry [10]. Additionally, land consumption can be defined as the rate at which land is annually consumed by cities for urban purposes, including open spaces [14].

These different definitions highlight the complexity of the activities in space and the challenge of observing and interpreting the spatial changes, with implications for spatial planning [15]. It also highlights the need for a clear, coherent, multi-scalar, and legally anchored framework to elaborate on indicators, taking into consideration the concepts of credibility, relevance, and legitimacy used at the science–policy interface [16]. It is also worth noting here that in various European countries such as Romania, Greece, Italy, Portugal, and Estonia, legally or otherwise binding definitions for the term "land take" do not seem to exist [17].

Urban development patterns vary from region to region, and consequently, urban land cover changes also vary due to the socioeconomic, cultural, historical, and environmental disparities [13]. Regarding the urbanization and urban sprawl trends in the Mediterranean region, in particular, it has been argued that they are characterized by informality, deregulated planning, and weak public policies [18]. Since the 1980s, several Mediterranean cities—and, in some cases, since the 1990s—have undergone a rapid transition from the traditional "compact" model to more dispersed forms of suburban and other types of urban sprawl in their outskirts, as well as tourism-driven urbanization along the coastal areas. Studies have revealed considerable differences in the urban growth trends across Mediterranean cities regarding the different urban forms, the degree of compactness and dispersal, and the fragmentation patterns and hence regarding the land take patterns [19,20]. Most of the relevant studies concerning Mediterranean cities, particularly those focusing on Italian and Greek urban contexts, primarily address the urban sprawl processes, highlighting the significant transformations in the traditional compact urban form that characterized these cities in the past [18–21]. Other important studies place particular emphasis on developing methods to estimate the land take within the context of Mediterranean urbanization [22,23]. However, there remains a considerable scope for further research, especially regarding how the critical drivers of urban sprawl—such as specific features of the development process and regulatory frameworks—are interrelated with the land take patterns.

In Greece, there were three main drivers of land take since the 1980s. The first was urban sprawl with all the forms of development (housing, commercial, service, leisure, and a widespread road network needed to connect all these areas), characterized by continuous, suburban, and dispersed patterns in the outskirts of both larger and smaller cities. These processes were more intense up until 2010 and the outbreak of the financial crisis, followed by the subsequent recession that halted the construction activity in urban centers for almost the entire 2010s. Second, tourist and secondary residence development with dispersed patterns, accompanied by a widespread road network needed to connect all these areas, occurred throughout the coastal areas. The third was the construction of transport infrastructure, especially large motorways [21,22].

The main goal of this paper is to advance the understanding of land take by examining it from the perspective of a city in the northern Mediterranean region with a long-standing tradition of compact urban form. In particular, the paper delves more deeply into areas that have experienced middle-class-led peri-urbanization since the mid-1990s, a process drastically facilitated by national legislation concerning the development in small, formerly rural settlements, and areas outside the official urban plan. These provisions have, in turn, facilitated certain types of development while undermining the formal spatial planning. In this context, the paper estimates Sustainable Development Goal (SDG) indicator 11.3.1, which measures the ratio of land consumption rate to population growth rate, in order to highlight the importance of using indicators to quantify the land cover and land use changes over time and to analyze the specific spatial patterns of urban sprawl and land take.

2. Materials and Methods

2.1. Methodological Framework and Calculation Steps for SDG 11.3.1

From a methodological point of view, a mixed-methods approach comprising both qualitative and quantitative analyses was applied. This approach is structured as presented in Figure 1 and is further examined in the following paragraphs.

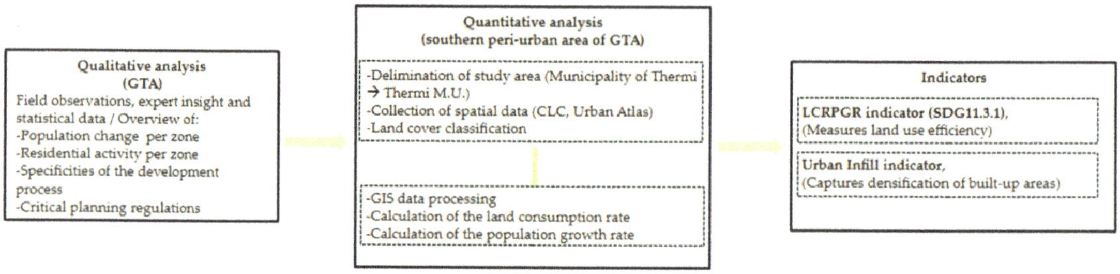

Figure 1. Methodological framework of the case study.

First, a primarily qualitative analysis provides an overview of the spatial patterns of the urban development and sprawl trends in Thessaloniki. This analysis focuses on the Greater Thessaloniki Area (GTA), encompassing its compact urban core and a broader peri-urban zone, which includes several suburban settlements experiencing significant development beyond the planned areas. It incorporates population census data, yearly statistical surveys on the legal building activity, and observations from multiple on-site surveys conducted during the preparation of General Urban Plans, as well as other previous research in Thessaloniki by one of the authors. Here, the authors' expert opinion, derived from long-term professional planning experience, contributes essentially to this analysis.

Second, the study elaborates a quantitative estimation of the land take in the southern peri-urban zone of Thessaloniki that exhibited the most intense urban sprawl trends after 1990.

In terms of measurement, the United Nations has established SDG 11, which focuses on creating inclusive, safe, resilient, and sustainable cities and human settlements. One of its key sub-targets, 11.3, aims by 2030, to improve inclusive and sustainable urbanization for all, as well as enhance the capacities for the participatory, integrated, and sustainable planning and management of human settlements in all countries. The proposed indicator for this target, 11.3.1, is defined as the "the ratio of land consumption rate to population growth rate" [14].

To calculate this indicator, two parameters are required: population growth and the rate of land consumption. While the population growth rate is relatively simple to calculate, widely available, and easier to interpret, determining the rate of land consumption necessitates the adoption of innovative methodologies and the establishment of a well-defined analytical framework. Our calculation is based on the UN-approved metadata [24].

In addition to the SDG 11.3.1 (*LCRPGR*) indicator, we also included the Urban Infill indicator, a sub-indicator suggested by [24]. This indicator captures the complementary aspects of spatial development dynamics. Whereas *LCR* measures the expansion of artificial surfaces relative to the *PGR*, the Urban Infill indicator estimates the internal densification within the study area. This facilitates a more thorough understanding of the land take patterns and spatial fragmentation in peri-urban areas.

The calculation of SDG indicator 11.3.1 involved six key steps:

1. Determination of the analysis period: The years 2001, 2006, 2012, and 2018 were selected as the reference points for the analysis.
2. Delimitation of the study area: The spatial extent of the study area was defined to ensure the consistency and relevance to the research objectives.
3. Spatial analysis and calculation of the land consumption rate (*LCR*): The extent of land consumption over the selected time periods was quantified through spatial data analysis. The *LCR* is calculated using the following formula:

$$LCR = \frac{ln\left(\frac{Urb_{t+n}}{Urb_t}\right)}{\gamma} \quad (1)$$

4. Spatial analysis and calculation of the population growth rate (*PGR*): Population growth rates were calculated for the same periods using demographic data. The *PGR* is calculated using the following formula:

$$PGR = \frac{ln\left(\frac{Pop_{t+n}}{Pop_t}\right)}{\gamma} \quad (2)$$

5. Calculation of the *LCRPGR* ratio: The ratio of land consumption rate to population growth rate was computed as the primary indicator (SDG 11.3.1). The *LCRPGR* indicator is calculated using the following formula:

$$LCRPGR = \text{Land Consumption Rate/Population Growth Rate} \quad (3)$$

6. Calculation of the Urban Infill indicator: A secondary indicator, the Urban Infill indicator, was calculated to provide a deeper insight into the spatial patterns of the land take within the case study area. The Urban Infill indicator is calculated using the following formula:

$$Total\ change\ built\text{-}up\ area(\%) = \frac{(UrBU_{t+n} - UrBU_t)}{UrBU_t} \quad (4)$$

The methodology for calculating each rate and indicator is analytically presented in the Section 3, where each computational step is explained in detail. It is worth commenting here that from a research perspective, the assessment of the 11.3.1 indicator remains relatively limited. Notable studies utilizing this indicator to analyze urban sprawl include a study that focuses on the North Rhine-Westphalia region [25], and one that examines urbanization in China [26].

2.2. Case Study Area and the Spatial Data Sources

This part of the study was conducted in the Municipality of Thermi in the peri-urban zone of Thessaloniki, with a specific focus on the Thermi Municipal Unit (M.U.), an area characterized by rapid population growth and significant residential expansion within as well as outside the planned areas. These trends have shaped a distinctive spatial pattern of suburban development, encompassing a large suburban settlement, Thermi, that today functions as a small urban center, a number of smaller suburban settlements, and extensive unplanned developments scattered throughout the entire territory of Thermi M.U. Therefore, Thermi M.U. serves as a representative example of middle-class-led peri-urbanization and provides substantial material for examining how the specific features of the development process and regulatory frameworks are interrelated with the land take patterns.

Figure 2 presents the key spatial zones of the GTA using color-coded boundary lines: black delineates the extent of the GTA; yellow outlines Poleodomiko Sigrotima Thessalonikis (PSTh); orange marks the administrative boundary of the Municipality of Thermi; and red highlights the Municipal Unit of Thermi (Thermi M.U.), which constitutes the study area. The imagery layer integrates high-resolution satellite and aerial photography, with most of the content captured within the past three to five years. It provides spatial coverage at resolutions of one meter or finer for the majority of the Earth's land surface [27].

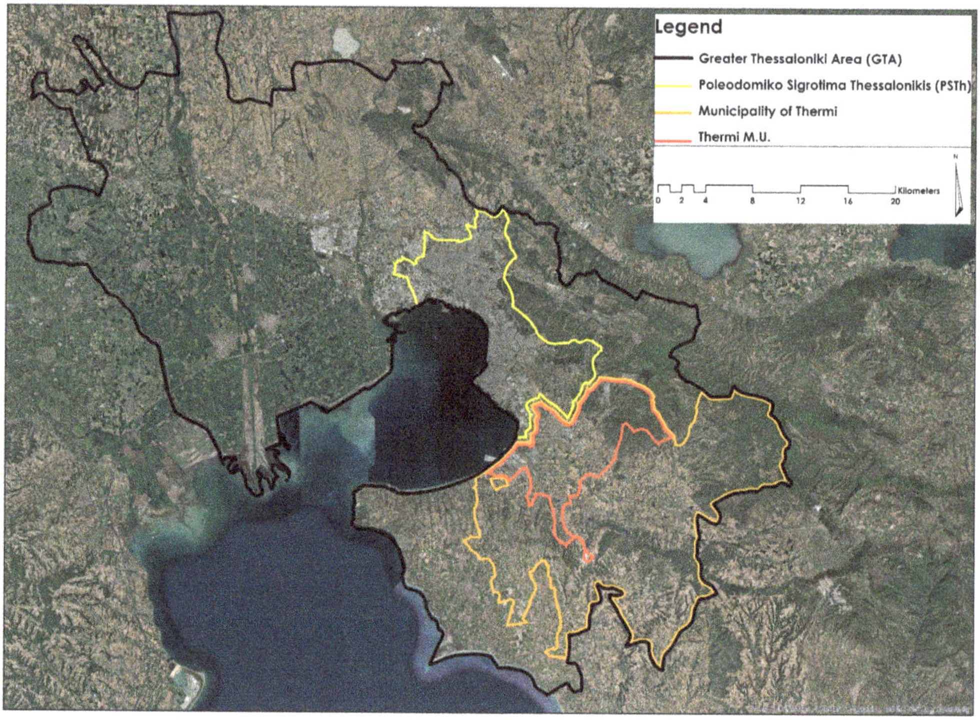

Figure 2. Spatial zones of Greater Thessaloniki Area and the Municipal Unit of Thermi.

To analyze SDG indicator 11.3.1, the study investigated the artificial surface changes over time for the years 2001, 2006, 2012, and 2018. The case study area was delineated and mapped within a GIS environment, with the primary objective of collecting satellite imagery and spatial information, including the administrative boundaries, population data, transport infrastructure, and land cover changes over time. This mapping process also established precise geospatial coordinates reflecting the current conditions, facilitating the subsequent analysis.

The spatial data collected, processed and analyzed, were drawn from the European Copernicus Earth Observation Programme. Specifically, the study utilized data from the Copernicus database on European Land Cover, which adheres to a standardized methodology to support environmental policy development. The land cover categories and classes were determined based on the standards and guidelines outlined in the technical reports of the CORINE Land Cover (CLC) and the European Urban Atlas databases. It should be mentioned that the CLC, which first used 1990 as the reference year for its initial census, was updated in 2000 and has adhered to a six-year update cycle since. It serves as a baseline, utilizing ortho-corrected satellite imagery of high spatial resolution. To ensure the comparability across inventories, its technical parameters remain consistent, including the description of land cover types through a three-level composite classification system, a minimum mapping unit of 25 hectares, and a minimum mapping width of 100 m. The raw data are provided in a vector format with polygonal topology, achieving thematic accuracy that exceeds the specified minimum threshold of 85%. To analyze the land cover changes in the Municipality of Thermi, cartographic visualizations of the CLC data were produced to capture the temporal changes accurately and offer a detailed understanding of the causes of these changes and the evolution of the typology of the land cover.

Furthermore, data for the years 2006, 2012, and 2018 were obtained from the Urban Atlas, a land-monitoring service under the European Copernicus Programme. This resource provides reliable, high-resolution land use maps for major urban zones and their peripheries, addressing the gaps in the knowledge about land use evolution and urban dynamics. The Urban Atlas was chosen for this part of the study due to its ability to track the developments and temporal changes in land use, including low-density urban fabric, at a resolution 100 times higher than that of the CLC. This part of the study was conducted at the scale of Thermi M.U., which was deemed optimal due to its large cartographic scale. This scale effectively captures the geometry of spatial elements and highlights the fragmentation of urban and non-urban landscapes. For this part of the study, primary land use data for 2001 were sourced from the archives of the Geochoros SA Consultancy.

3. Results

3.1. *Spatial Patterns of the Urban Sprawl in Thessaloniki and the Planning Provisions Shaping Them*

3.1.1. General Trends in the Spatial Patterns of Urban Sprawl

In the urban development pattern of Greek cities, significant changes have been observed since the 1980s in Athens and the 1990s in Thessaloniki and other major cities. The traditional cohesive and high-density city model has given way to suburbanization and urban diffusion throughout the peri-urban space. This has led to the process of "peri-urbanization", that is, along with the further intensification of compact zones, the big cities experienced a wide expansion with the development of the suburban housing within and outside the peri-urban settlements, as well as the development of all the other urban activities and functions—such as commercial centers, leisure facilities, and transport infrastructure—in their peri-urban zones [28–31]. These urban development trends essentially occurred over a 15- to 20-year period, from the 1990s to the end of the 2000s, coinciding with the country's fast growth prior to the onset of the 2009 financial crisis. Residential construction, tourism, commerce, and services were the main economic activities related to urban sprawl. Agricultural land, along with other natural areas in close proximity to cities, was considerably restricted in the quickly suburbanized peri-urban zones. In the 2010s, the financial crisis in the country and the longstanding recession resulted in a drastic decline in the housing construction sector, while also significantly halting the urban sprawl trends during this period.

In Thessaloniki, in particular, the clear trend of population migration from the so-called "Poleodomiko Sigrotima Thessalonikis" (PSTh), i.e., from the compact urban area to the surrounding peri-urban zone, is a relatively recent phenomenon. Before 2009, urban sprawl was a phenomenon that occurred over a period of just 15 years, from the mid-1990s to the late 2000s, although the first signs had emerged by the late 1980s with the sporadic out-of-plan construction of luxury homes or large detached houses, primarily by affluent households. However, it was in the second half of the 1990s—when the for-profit construction sector began producing housing in peri-urban areas—that the suburbanization in Thessaloniki expanded significantly. During this period, the urban sprawl throughout the peri-urban area became consolidated, not only for residential purposes but also for a variety of other urban activities, including large department stores, exhibitions, leisure facilities, and services of all kinds [21,30].

During the 20-year period from 1991 to 2011, the proportion of the population of the GTA living in non-PSTh areas relative to the entire GTA increased from 13.6% in 1991 to 17% in 2001, and to 21.8% in 2011. Practically speaking, by 2001, the peri-urban zone had regained the population percentage that it held in 1961; however, its social geography had radically transformed, with the previously rural areas now functioning as purely suburban ones.

In the 1990s, the population of the GTA as a whole grew by 1% annually, at a higher rate than the national average. PSTh, the compact urban area, grew by 0.6% annually, the inner peri-urban zone grew by 4.7% annually, and the outer peri-urban zone grew at an annual rate of 2.5%. The first time in the city's contemporary history that the compact zone experienced a population decline was during the period from 2001 to 2011. The inflow of economic migrants from Balkan and former Soviet Union countries had halted the compact zone's population loss in the 1990s, whereas it kept the rate of decline in the 2000s relatively low. During the 2010s, these trends changed considerably. This was the first decade in 70 years that Thessaloniki lost population, although at a lower rate than the national average. At the same time, PSTh experienced a smaller population decline than the GTA average, whereas the inner peri-urban zone increased, though at a lower rate than in previous decades. Of special interest is the fact that this increase primarily affected the two main settlements in the inner suburban zone: Thermi M.U. in the southern inner suburban zone (the case study area) and Oreokastro in the northern inner suburban zone, both of which are in close proximity to PSTh. Table 1 presents the population change in Thessaloniki by zone for the 30-year period 1991–2021.

Table 1. Population change in Thessaloniki by zone, 1991–2021. Source: elaboration of data from EL.STAT Population Census, 1991–2021.

Spatial Zones	1991	2001	2011	2021	Average Annual Rate of Change		
					1991–2001	2001–2011	2011–2021
PSTh	780,948	830,355	793,583	788,957	0.62%	0.45%	0.06%
Inner peri-urban zone	42,368	67,042	96,176	102,707	4.70%	3.67%	0.66%
Outer peri-urban zone	81,186	103,627	125,031	114,934	2.47%	1.90%	0.84%
GTA	906,493	1,003,025	1,016,801	1,008,619	1.02%	0.14%	0.08%
Regional Unit of Thessa-loniki	973,100	1,084,001	1,110,551	1,092,919	1.09%	0.24%	0.16%

From the perspective of housing construction activity, during the period 1995–2008, 34% of the total housing production occurred in areas outside PSTh. Based on the statistical data on housing construction presented in Table 2, the share of new dwellings built in the peri-urban zone, relative to the entire GTA, was 27.7% during the five-year period of 1995–1999, increased to 38.6% during 2000–2004, and then decreased slightly to 36% during 2005–2008. It is worth noting that during the period 2009–2019, a period of drastic decline in housing production in the entire GTA—reaching a situation of almost no production—the share of housing production in the peri-urban zone was 48.8%, the largest share of which concerned housing construction in Thermi M.U. Figure 3 presents statistical data on the annual production of new dwellings in the GTA by zone during the period 1994–2019.

Table 2. Number of new dwellings constructed in the period 1994–2019. Source: elaboration of data from EL.STAT. Surveys of Legal Building Activity, 1995–2019.

Spatial Zones	1995–1999	2000–2004	2005–2008	1995–2008	2009–2012	2013–2015	2016–2019	2009–2019
PSTh	29,879	32,317	33,548	95,744	4269	487	1021	5777
Inner peri-urban zone	4830	7987	8085	20,902	2128	351	357	2836
Outer peri-urban zone	6603	12,285	9937	28,825	2050	350	266	2666
GTA	41,312	52,589	51,570	145,471	8447	1188	1644	11,279
Peri-urban zone/GTA	27.67%	38.55%	34.95%	34.18%	49.46%	59.01%	37.90%	48.78%

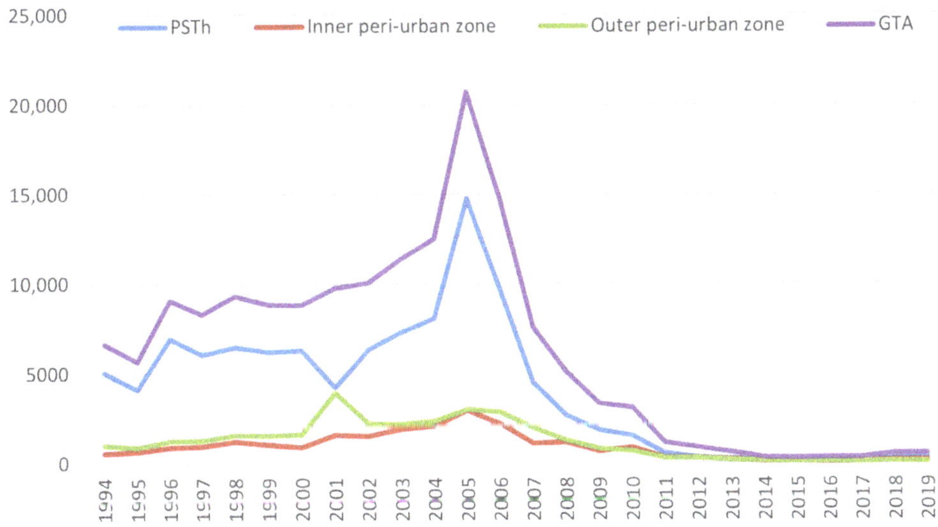

Figure 3. Annual production of new dwellings (legal building activity) 1995–2019. Source: elaboration of data from EL.STAT., Surveys of Legal Building Activity, 1995–2019.

3.1.2. Development Patterns

Based on Yiannakou [21,32] the following development patterns can be identified in Thessaloniki during the period 1991–2011:

- Major reconstruction inside and on the fringes of PSTh: This includes undeveloped plots in the inner compact urban area but primarily areas on its fringe that were incorporated into the town plan from the 1990s to the early 2000s. Since the late 1990s, all newly constructed residential buildings in planned areas have included a considerable percentage of unofficial apartment floor space (i.e., not appearing in the planning permit), with the unauthorized conversion of various elements, such as semi-outdoor spaces, balconies, and garages, into main apartment floor space.

Analytical surveys indicate that in most residential buildings, nearly 40% of the main apartment floor space was constructed without an official permit. This practice effectively increased, unofficially, the official building coefficients by almost 40% and, therefore, intensified the official densities in these areas.

- Major reconstruction within the statutory boundaries of peri-urban settlements: Initially, reconstruction occurred in the settlements with available land within the town plan area and sufficient official building coefficients to attract speculative housing development. Notable examples include Perea, Neoi Epivates, and Agia Triada in the outer peri-urban coastal zone, as well as Thermi and Oreokastro in the inner urban zone. Later, house-building activity expanded to all the settlements in the peri-urban zone. Since the mid-1980s, all these settlements have had official settlement boundaries (not necessarily planned areas) and building coefficients defined by general law (see below). These legal provisions were sufficient to attract speculative housing development and, consequently, middle-class households seeking suburban housing. As in PSTh, all the elements capable of artificially increasing the building coefficients—including, in this case, the conversion of basements into main house floor area—were fully integrated into the main house floor space.
- Out-of-plan residential development: This process followed a chronological progression, beginning with the construction of luxury detached houses, followed by double and triple houses for large families, and culminating in the development of small speculative housing complexes. The speculative house-building sector's expansion into out-of-plan residential development was largely driven by the informal increase in the officially permitted maximum building area on estates outside the official plan, effectively creating building coefficients nearly equal to those within the planned areas. This was achieved by incorporating various elements that expanded the main floor space. This type of out-of-plan construction was also a common practice in second-home areas. Analytical surveys indicate that in most house complexes of this type, nearly 75% of the main floor space was constructed informally, i.e., without appearing in the official permit. A large number of high-income and middle-income maisonettes in the peri-urban zone, many of which feature outstanding architectural design, have been built with this level of informality.
- Out-of-plan development of all types of other uses/facilities: this category encompasses all the other types of off-plan construction, facilitated by building variances that permitted deviations from the standard regulations.

From the perspective of spatial structure, the key feature throughout the period 1991-2011 was the transformation of the city from a typically linear, coherent urban form—shaped by natural barriers that influenced the outline of the urban fabric—into a sprawled city characterized by local concentrations of relatively dense construction within peri-urban settlements and widespread building across the entire peri-urban zone.

3.1.3. The Main Regulatory Framework

From an institutional perspective, the supply of planned areas in the peri-urban zone remained limited to small statutory expansions in certain settlements, while a few much larger expansions were never completed. Even the expansion in Thermi M.U., the largest ever in the peri-urban area, was completed only 25 years after its launch and thus entered the housing market only in the late 2000s. Two crucial pieces of legislation formed the regulatory framework for the peri-urban model of development:

- Presidential Decree 24.4/3.5.1985: This decree established regulatory settlement limits for settlements with populations of 2000 inhabitants or fewer in the 1981 census year. In most cases, these limits were defined over a larger area than the actual built-up area.

Within these extended official settlement boundaries, building coefficients were set at nearly the same levels as those of fully planned areas on the fringe of PSTh.

- PD 31.5.1985: This decree governs building provisions and restrictions for areas outside the city plan. In Greece, one of the main causes of dispersed development is the potential for the so-called "ektos schediou domisi", i.e., "out-of-plan" development, which refers to construction outside planned urban zones, as allowed by Greek statutory planning. According to this provision, every property outside the official settlement plan with a minimum size of 4000 m^2 can be developed with a maximum built-up area of 200 m^2 for residential use. Similar regulations also allow "out-of-plan" development for other purposes, including industrial, commercial, and service uses. These development processes were intensified either by unauthorized construction, defined as building beyond the area permitted by official regulations, or, in the case of commercial, industrial, and service development, by exploiting deviations provided for in the relevant legal regulations [33,34].

The above institutional framework, originally intended to regulate the building policies in rural settlements until they acquire an official plan or to control construction on the periphery of cities and towns, ultimately shaped a peculiar stock of land for development. This stock featured official building coefficients that could reach values as high as those in planned areas and was combined with permanent informal construction that far exceeded these limits. As a result, the groundwork was laid for the significant expansion of speculative construction in peri-urban areas. A large portion of these suburbanized settlements lack proper urban planning in parts or even entirely. These areas exhibit deficiencies in the spaces for collective use, lack comprehensive street planning, and feature a network of public spaces limited to roads serving individual properties, often created solely for accessibility needs.

During the period under review, new planned areas that were proposed to be released for development by the General Urban Plans enacted in the 2000s were at scales unprecedented for the traditionally restrictive release of planned areas of previous decades. This created yet another peculiar "in-formation" stock of land intended for future development, in which prices increased sharply, approaching those of the land inside the plan. The extremely slow pace of preparing the necessary plans for these areas created a form of "virtual stock", as the actual release of planned land for development has remained largely unchanged to this day. What changed during the 2010s, however, was the drastic slowdown of the peri-urbanization process in broad sprawl patterns.

3.2. *An Assessment of the SDG Indicator 11.3.1 in Thermi M.U.*

3.2.1. Land Use/Land Cover Change During the Study Period, Municipality of Thermi and Thermi M.U.

Throughout the Municipality of Thermi, in the years 1990, 2000, 2006, 2012, and 2018, the presence of extensive artificial surfaces was evident. Specifically, urban areas consist of scattered buildings that extend into and fragment vegetation zones and natural terrain, creating a discontinuous spatial pattern. Additionally, industrial and commercial zones, as well as transport networks, are well-developed across the study area. These include the existing road network and the city's airport infrastructure, located in the northwest near the coastal border. Areas of mining and quarrying activities are also identified, with a primary focus on the open-pit extraction of industrial and other minerals. Moreover, there are artificial, non-agricultural green zones, which include structured urban green spaces, sports complexes, and recreational facilities.

The majority of the land in The Municipality of Thermi is dominated by agricultural areas, with the primary subcategory being non-irrigated arable land. These areas support a variety of crops, including flowers, orchards, nurseries, and medicinal, aromatic, and culinary plants. Smaller portions of the land are dedicated to permanent irrigated areas, where the crops are sustained using fixed infrastructure such as drainage networks and irrigation canals. Permanent crops, such as vineyards, occupy the land for extended periods and provide repeated harvests. Grassland areas, characterized by dense herbaceous vegetation, also constitute a smaller percentage of the total land use. Figure 4 presents the land cover in the Municipality of Thermi.

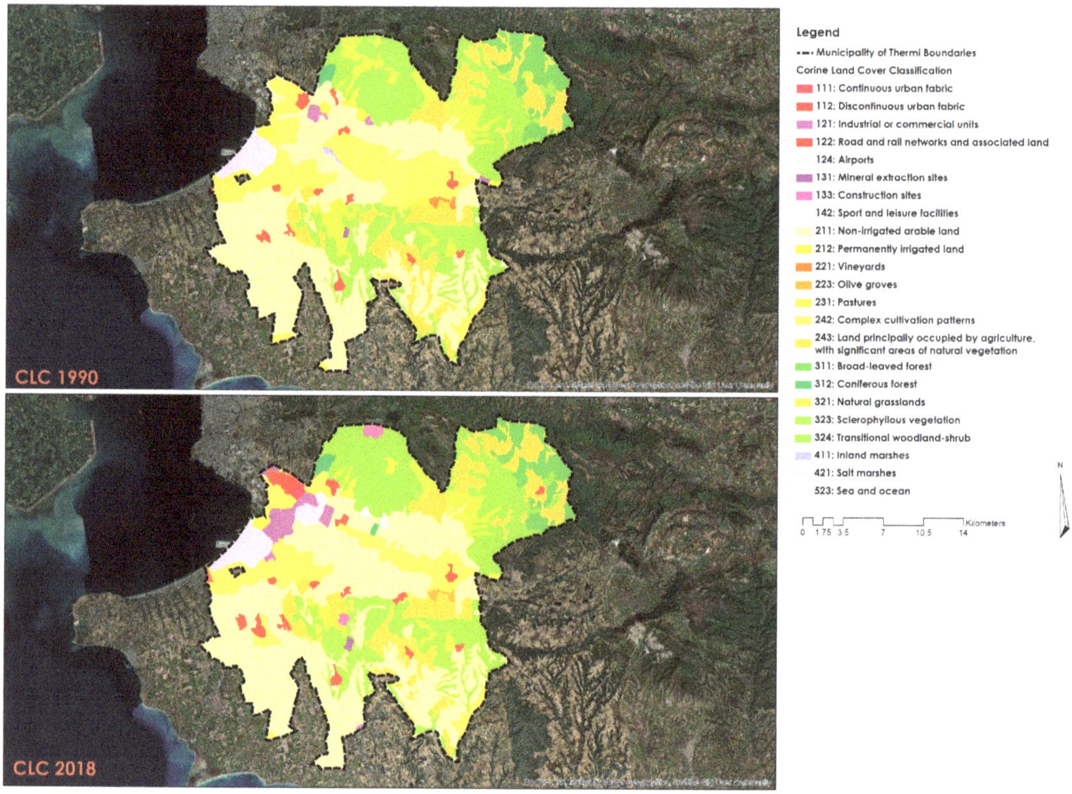

Figure 4. Land cover in the Municipality of Thermi, (**top**) Corine land cover 1990; (**bottom**) Corine land cover 2018.

Figure 5 illustrates the temporal evolution of various land use and land cover categories within Thermi M.U. These categories were classified based on the standards of the Urban Atlas and are broadly distinguished into the following: continuous and discontinuous urban fabric; industrial and commercial zones; transportation infrastructure, including roads, railways, airports, and port areas; mining and mineral extraction and dump sites; construction sites; land without a current use; green urban areas; sports and leisure facilities; arable land; permanent crops; pastures; complex and mixed cultivation patterns; orchards; forests; herbaceous vegetation associations; open spaces with little or no vegetation; wetlands and water bodies.

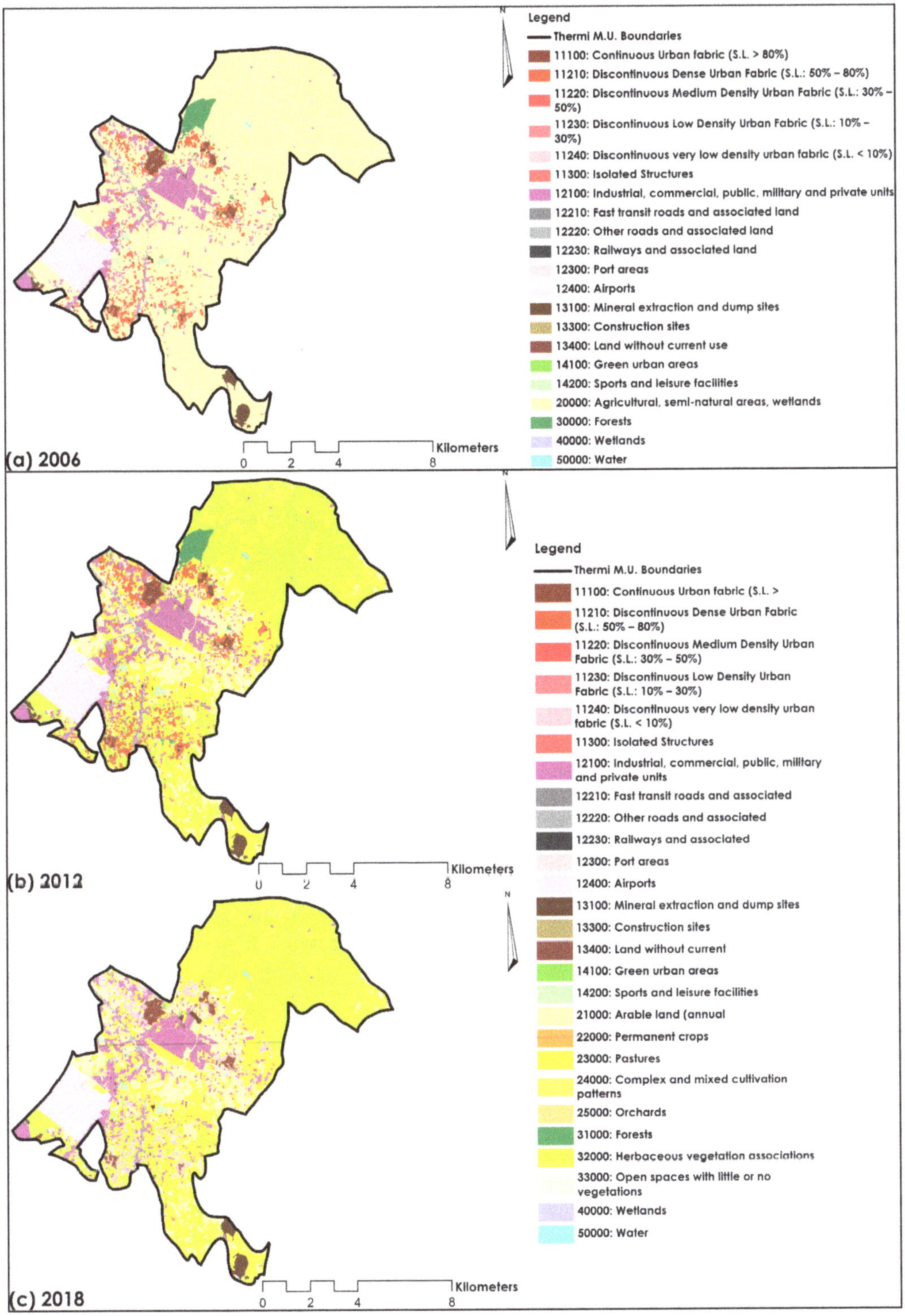

Figure 5. Changes over time in the different categories of land use/land cover in Thermi M.U.

The full classification scheme for Land Use/Land Cover (LU/LC) in UA2006, UA2012, and UA 2018, across all hierarchical levels, is detailed in the official documentation [35]. In 2012, the original 2006 nomenclature was expanded to include a distinct wetland category and more detailed subdivisions within agricultural and natural areas. Notably, the 2012 and 2018 classification system introduces finer granularity in categories 2 and 3 by breaking them into additional subclasses and formally distinguishing wetlands as a separate class. It should be clarified that the classes "open spaces with little or no vegetation" (320 00) and "combinations of herbaceous vegetation" (330 00) from the Urban Atlas 2012 and 2018 database (category 3) are included within the "rural and semi-natural space" category (category 2) in the Urban Atlas 2006 database. In 2006, agricultural and semi-natural areas dominated, covering 74.3% of Thermi M.U, followed by industrial and commercial infrastructure at 8%. Smaller percentages of land coverage were occupied by airport facilities (5.1%), discontinuous urban fabric (4%), road axes (2.7%), and continuous urban fabric (1.2%). All other land cover categories accounted for less than 1% of the total area.

The 2012 land cover analysis reveals notable trends of landscape change. These changes stem both from modifications in the structure and methodology of the Urban Atlas tool but mostly from the socio-economic pressures that drove the transformations in the case study area's landscape. Herbaceous vegetation combinations (36.9%), pasture (20.1%), and arable land (15.4%) remained the dominant land cover. However, incremental increases in industrial and commercial areas (8.9%) and discontinuous urban fabric (4.7%, primarily for residential use), as well as expansions in individual structures, construction sites, and mining sites, indicate a diffuse pattern of development and fragmentation of the rural landscape.

By 2018, a slight decrease in arable land and a parallel increase in scattered urban infrastructure were observed compared to 2012. Notable changes that require attention include shifts in agricultural land use, primarily driven by pressures to abandon marginal lands and the urbanization of farmland—all of which contribute to land take. The results also underscore the conversion of land into agricultural use at the expense of natural areas, along with changes in forested areas, which have had considerable impacts on the region's ecosystem.

Through the extraction and processing of data, 17 land cover classes were identified and subsequently reclassified based on their characteristics to create four new land cover classes, as detailed in Table 3 and visualized in Figure 6. This grouping was deemed necessary for an in-depth analysis of the land consumption, aligning with the literature reviewed and the definition of SDG indicator 11.3.1 [24]. Moreover, this approach facilitates the comparative analysis of both qualitative and quantitative aspects across different time periods for the study area. It also enables an assessment of the rate and nature of spatial transformations over time.

The apparent reclassification of artificial surfaces to agricultural land between 2001 and 2006 does not necessarily reflect the actual land use reversals. Rather, these changes can be attributed to the methodological discrepancies between the datasets used between the Urban Atlas classifications and the historical land cover data obtained from the records of the consultancy firm, Geochoros SA. From the reclassification and creation of the four land cover classes, the total artificial surface for each year of analysis was calculated. These calculations are presented in Figure 7, which depicts the area of artificial surfaces for each year.

Table 3. Reclassification of existing land cover categories into new classes.

Urban Atlas Code	Land Cover Class	Final Classification
111 00	Continuous urban fabric	Artificial surfaces
112 10	Discontinuous dense urban fabric	
113 00	Isolated Structures	
121 00	Industrial, commercial, public, military, and private units	
122 10	Fast transit roads and associated land	
124 00	Airports	
131 00	Mineral extraction and dump sites	
132 00	Construction sites	
133 00	Land without a current use	
141 00	Green urban areas	
142 00	Sports and leisure facilities	
210 00	Arable land	Agricultural and semi natural areas
230 00	Pastures	
310 00	Forests	Forests
320 00	Herbaceous vegetation associations	
330 00	Open spaces with little or no vegetation	
500 00	Water	Water bodies and wetlands

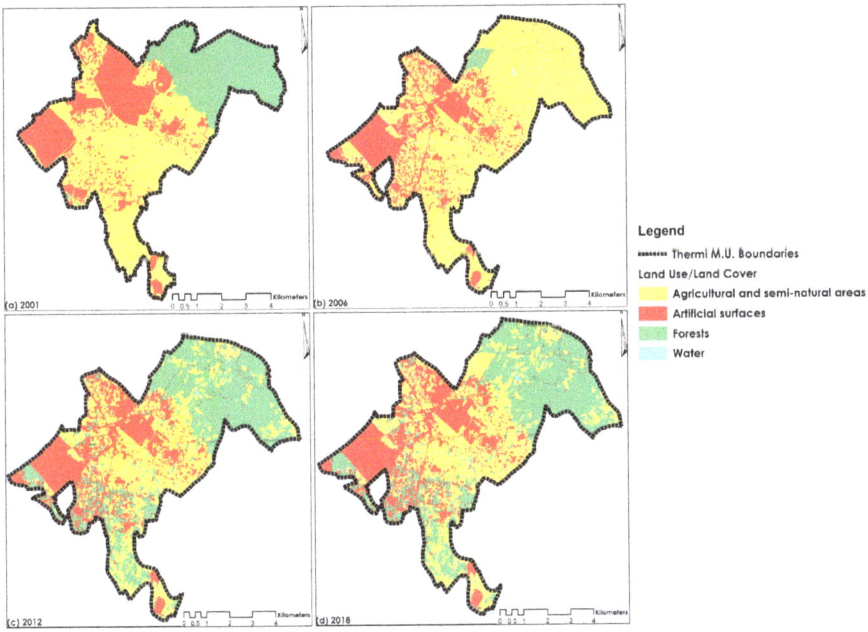

Figure 6. Visualization of the four land cover classes for the years 2001, 2006, 2012, and 2018.

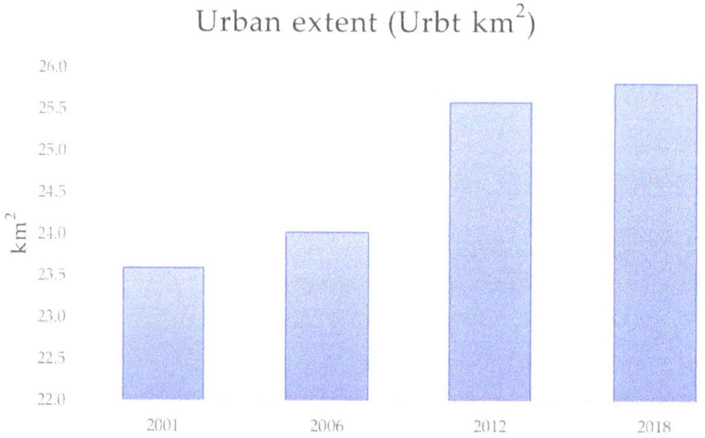

Figure 7. Area of artificial surfaces for each year of analysis.

3.2.2. Computation of the Land Consumption Rate (*LCR*) and Population Growth Rates (*PGRs*)

To calculate the land consumption rate (*LCR*) in Thermi M.U., based on these values of Urban Extent we determined the area of artificial surfaces (Urbtx) for the years under analysis (2001, 2006, 2012, and 2018). The *LCR* was then calculated using the following Equation (5):

$$LCR = \frac{ln\left(\frac{Urb_{t+n}}{Urb_t}\right)}{\gamma} \quad (5)$$

where:

- Urb_t: the total built-up area in the initial year (km^2);
- Urb_{t+n}: the total built-up area in the previous year (km^2);
- y: the number of years between $V_{present}$ and V_{past} (the considered period in years).

The results in Table 4 indicate that, from 2001 to 2018, the artificial surfaces in the study area appropriated land from other uses at an annual rate of 0.5%. The highest rate of land consumption occurred during the period 2006–2012, with a rate of 1.05%. The smaller percentage increase observed during the years 2001–2006 can be attributed to the fact that new artificial surfaces were developed as extensions of the existing residential cores. In contrast, during the years 2006–2012, development occurred more peripherally and in a diffuse form.

Table 4. LCR values.

Period	UBRt2	UBRt1	LCR	LCR (%)
2001–2006	24.032049	23.595843	0.0037	0.37%
2006–2012	25.592702	24.032049	0.0105	1.05%
2012–2018	25.817089	25.592702	0.0015	0.15%
2001–2018	25.817089	23.595843	0.0053	0.53%

The population growth rate (*PGR*) is calculated using the total population within the study area during the analysis period, following Equation (6) below, and the results are presented in Table 5.

$$PGR = \frac{ln\left(\frac{Pop_{t+n}}{Pop_t}\right)}{\gamma} \tag{6}$$

where:

- ln the value of the natural logarithm;
- Pop_t: the total population within the study area in the initial year;
- Pop_{t+n}: the total population within the study area in the final year;
- Y: the number of years between the two measurement periods.

Table 5. PGR values.

Period	Pop2	Pop1	PGR	PGR (%)
2001–2006	20,442	16,546	0.0423	4.23%
2006–2012	25,574	20,442	0.0373	3.73%
2012–2018	27,573	25,574	0.0125	1.25%
2001–2018	27,573	16,546	0.0300	3.00%

To estimate the population evolution for the years 2006, 2012, and 2018, population data from the EL.STAT censuses for 2001, 2011, and 2021 were utilized. Reductions were applied as follows: for 2006 and 2012, the average annual growth rate (AGR) for the period 2001–2011 was used. For 2018, a significantly reduced MERM was employed, considering the population growth rate of the entire Municipality of Thermi during 2011–2021. For the population estimation of Thermi M.U., this rate was adjusted upward by a percentage. The adjustment was based on construction activity data, which indicated that Thermi M.U. exhibited relatively significant construction activity during the 2010s compared to other M.U.s within the municipality and the wider Thessaloniki area.

3.2.3. LCRPGR Indicator Analysis

The final index, *LCRPGR* (Land Consumption Rate and Population Growth Rate), is summarized in the following Equation (7) and the results are presented in Table 6.

$$LCRPGR = Land\ Consumption\ Rate/Population\ Growth\ Rate$$

$$LCRPGR = \frac{\left(\frac{ln\left(\frac{Urb_{t+n}}{Urb_t}\right)}{\gamma}\right)}{\left(\frac{ln\left(\frac{Pop_{t+n}}{Pop_t}\right)}{\gamma}\right)} \tag{7}$$

Table 6. LCRPGR values (SDG indicator 11.3.1).

Period	LCR	PGR	LCRPGR = LCR/PGR
2001–2006	0.0037	0.0423	0.087
2006–2012	0.0105	0.0373	0.281
2012–2018	0.0015	0.0125	0.116
2001–2018	0.0053	0.0300	0.176

Values approaching and slightly below unity indicate a compact growth model characterized by efficient spatial organization, proximity among activities and services, and congruent population growth relative to new developments. Conversely, values significantly below unity often reflect urban areas experiencing challenges such as urban sprawl, high infrastructure costs, reliance on motorized transport, and widespread environmental degradation.

In the case of Thermi M.U., the results reveal a population growth rate that surpasses the growth rate of artificial surfaces. This trend is attributed to the suburbanization observed in previous years, driven by a demand for housing outside the urban core and a preference for improved living conditions. The value of *LCRPPGR* (SDG indicator 11.3.1) for all the analyzed years indicates an inefficient use of the available land. However, urban development is shaped by a multitude of factors that cannot be fully captured or generalized through a single index.

3.2.4. Urban Infill Indicator Analysis

To provide deeper insights, it is essential to incorporate secondary measures of SDG indicator 11.3.1, such as the total change in the built-up area, which is a measure of the total increase in artificial areas within the urban area over time. Additionally, the rate of change in "urban voids" (urban infill) serves as an important metric for interpreting and monitoring the urban development patterns. This indicator evaluates the densification of the built environment over time, capturing the emergence of new developments.

The urban infill is calculated using the same inputs as the land consumption rate for the different analysis years, based on the below Equation (8):

$$Total\ change\ built\text{-}up\ area(\%) = \frac{(UrBU_{t+n} - UrBU_t)}{UrBU_t} \tag{8}$$

where:

- $UrBU_{t+n}$ is the total built-up area in the urban area/city over time for the current/final year
- $UrBU_t$ is the total built-up area in the urban area/city over time for the past/initial year

Table 7 presents the results of the Urban Infill indicator. The most significant change occurred between 2006 and 2012, with a 6.5% increase. In comparison, the change during 2001–2006 was relatively modest at 1.8%, while it declined further to 0.9% between 2012 and 2018, coinciding with the economic recession in Greece. Overall, between 2001 and 2018, urban areas experienced a cumulative change of 9.4%.

Table 7. Urban infill values per period.

Period	Urban Infill (%)
2001–2006	1.8%
2006–2012	6.5%
2012–2018	0.9%
2001–2018	9.4%

Finally, Figure 8 illustrates the spatial patterns of urban sprawl and land take over time along with their relationship to the designated settlement area within the formal plan.

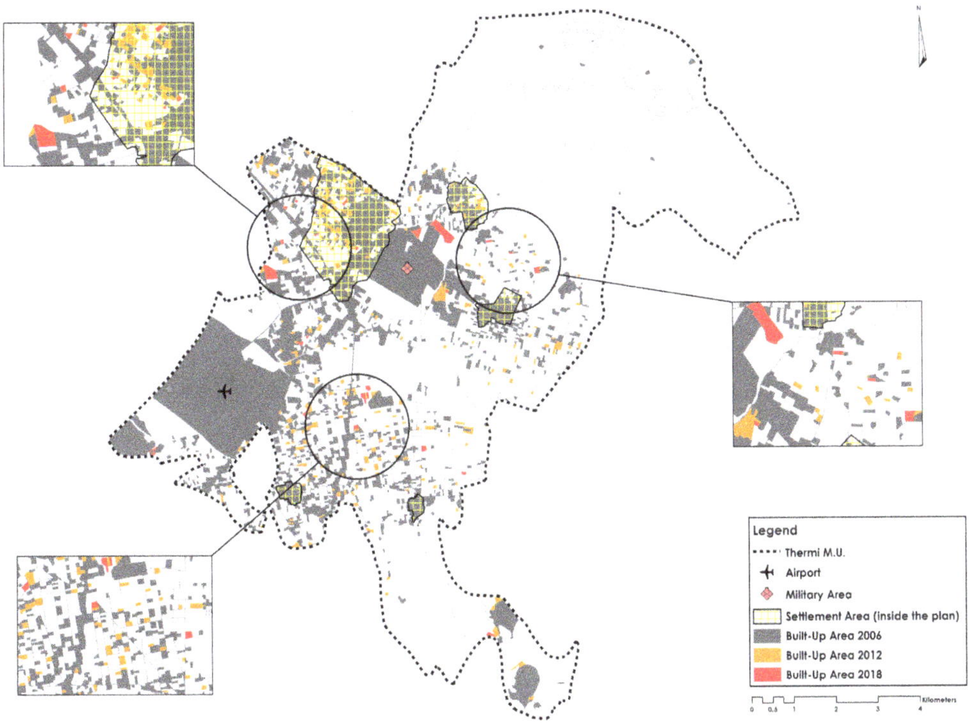

Figure 8. Spatial patterns of urban sprawl and land take in Thermi M.U. over time.

4. Discussion and Conclusions

This paper attempts to enhance the study of land take from the perspective of a Mediterranean city and investigates the phenomenon of land take in the city of Thessaloniki, which had typical compact features until 30 years ago. It provides, on the one hand, an overview of the spatial patterns of urban sprawl—the main driver of land take—and, on the other, a quantitative perspective through the calculation of SDG indicator 11.3.1, which measures the ratio of the rate of land consumption to the rate of population growth.

The first part of our study provided an overview of the urban sprawl in Thessaloniki and highlighted that the urban sprawl in Thessaloniki is a relatively recent phenomenon that began after the mid-1990s, as was the case in other Mediterranean cities [19], and continuing until the late 2000s. As the literature points out, the form of dispersal and the fragmentation patterns, and consequently land take, vary from region to region [13,19,20]. The detailed analysis of land use and land cover changes in the peri-urban areas of Thessaloniki that experienced intense development in the period under review provides ample evidence of these variations, along with valuable insights into the patterns of urban sprawl and land take. These findings reveal significant trends in the spatial and landscape transformation.

One of the crucial aspects of urban growth and change in the twenty years preceding the outbreak of the economic crisis was the "artificial" and largely unrestricted supply of land for residential development. This was enabled by formal regulations of general laws, whose implementation ultimately weakened the spatial planning system, as well as due to the widespread informality. All these processes led to a great dispersion of urban land uses throughout the peri-urban areas and consequently led to a continuous loss of control over urban sprawl, with critical economic, political, social, and environmental impacts.

The second part of our study focused on the land take, defined as the conversion of primarily natural land into artificial surfaces, which compromises the land's ecological functions. Its aim was to estimate the land take in Thessaloniki using simple metrics such as the SDG 11.3.1 indicator and the complementary Urban Infill indicator, and employing as a case study a rapidly growing peri-urban area rather than the entire urban region. This approach enables a more nuanced understanding of the relationship between population growth and land consumption within the sprawled structure of a metropolitan area. Such insights help fill the gap regarding how the critical drivers of urban sprawl, such as the specific features of the development process and regulatory framework, are interrelated with the land take patterns. In addition, they are essential for informing policies, including the EU's objective of achieving land take neutrality.

In the study area, the Municipality of Thermi, specifically Thermi M.U., a typical example of what we called in this paper middle-class-led peri-urbanization, the population growth rate exceeds the rate of artificial surface expansion. While this finding may seem to contradict the existing literature and policy reports [2–5], it was not unexpected, as urban development patterns vary significantly between regions due to their socio-economic and geo-spatial differences. Nonetheless, the study highlights the alarming reduction and fragmentation of large, homogeneous open spaces, primarily caused by the expansion of artificial surfaces at the expense of agricultural and natural areas. This process contributes to agricultural abandonment, degrades natural processes, and disrupts ecosystem services.

The urban sprawl in the study area followed a patchy, to a large extent linear pattern along the major road axes, with clear tendencies toward discontinuity, particularly during the 2006–2012 period. This linear development was concentrated along the main road corridors connecting Thessaloniki with the airport and the resort areas of Halkidiki. Additionally, scattered land cover changes fragmented the agricultural and natural landscapes throughout the study area, particularly between the main settlement, Thermi, and the two nearest smaller settlements. The expansion of artificial surfaces was also evident around the airport, originating from the developments established before 2001, which served as focal points for the further peri-urban sprawl.

During the 2012–2018 period, the urban sprawl was markedly less dynamic, primarily due to the economic recession and its impact on the building activity and suburban growth. During this time, only 0.3 km^2 of new artificial surfaces were added, with a more spatially limited and homogeneous pattern of expansion. This growth was, in fact, concentrated mainly within the Thermi planned area, the officially designated zone for urban development that was enacted and released for development by the end of the 2000s. This decade also marked a shift in the subsequent housing demand, either toward PSTh or the two largest and largely self-sufficient settlements in the peri-urban zone, Thermi and Oreokastro, as indicated by the 2021 census data.

The above findings provide valuable insights for the regulation and management of urban sprawl. As cities grow, the land take also increases, but it occurs in diverse and complex forms. More compact forms of urban growth and expansion are generally considered more compatible with natural land ecosystems, although such forms are not without environmental implications. This presupposes that planning should accommodate a wide range of spatial forms, including sustainable suburban configurations that meet the need for housing and urban functions, while simultaneously adopting more environmentally responsible physical structures. In any case, urban planning must also take into account the multiple factors that influence urban development and contribute to land take rather than rely on linear interpretations of indicators. The main limitations of this study outline the future research directions. It is important to note that, given the scale of mapping and the inherent limitations of such spatial analyses, the findings are not intended to provide precise descriptions of the spatial changes. Instead, they offer general conclusions that serve as a basis for further investigation. It is also important that such a study encompasses the entire urban area, including its distinct urban zones, in order to more effectively highlight the overall patterns of land take across the urban system.

Moreover, a significant limitation relates to the nature of the SDG 11.3.1 indicator and its interpretation. The factors influencing the urban development patterns are multifaceted, making it challenging to derive comprehensive conclusions from a single indicator. Incorporating secondary indicators, derived from the same datasets, offers deeper insights and a more nuanced understanding of the results. For instance, comparing the rates of land consumption and population change across areas with similar characteristics provides a more robust analytical framework. No dataset, however, can perfectly capture the detailed dynamics of land cover changes. Developing localized methods for quantitative and automated spatial analysis can improve the objectivity and accuracy of assessments and predictions. These efforts can be strengthened by leveraging extensive, calibrated, and validated datasets.

Nonetheless, the availability of core datasets with consistent specifications for cross-domain applications remains limited. Thus, the creation of up-to-date and regularly updated land cover and land use maps is essential. Such resources would support the decision making at the local, regional, and national levels by facilitating the sustainable management of natural environments and urban areas. The study of land take can guide the spatial planning processes, inform the coordination of policies, and ensure the preservation of natural reserves. Moreover, these efforts enable the timely monitoring of the changes resulting from anthropogenic activities (e.g., tourism, holiday home construction, unplanned development) and natural disasters, while providing critical information for assessing the environmental impacts of the socio-economic processes. This, in turn, supports the formulation of effective spatial planning supporting development and at the same time protecting natural landscapes and preserving their ecosystem services.

Author Contributions: Conceptualization, A.Y. and G.Z.; methodology, A.Y. and G.Z.; software, G.Z.; validation, A.Y. and G.Z.; formal analysis, G.Z.; investigation, A.Y. and G.Z.; writing—original draft preparation, A.Y. and G.Z.; writing—review and editing, A.Y. and G.Z.; visualization, G.Z.; supervision, A.Y. All authors have read and agreed to the published version of the manuscript.

Funding: This research received no external funding.

Data Availability Statement: Data are contained within the article.

Conflicts of Interest: The authors declare no conflicts of interest.

References

1. Johnson, M.P. Environmental Impacts of Urban Sprawl: A Survey of the Literature and Proposed Research Agenda. *Environ. Plan. A: Econ. Space* **2001**, *33*, 717–735. [CrossRef]
2. Colsaet, A.; Laurans, Y.; Levrel, H. What drives land take and urban land expansion? A systematic review. *Land Use Policy* **2018**, *79*, 339–349. [CrossRef]
3. Couch, C.; Leontidou, L.; Petschel-Held, G. *Urban Sprawl in Europe: Landscapes, Land-Use Change & Policy*; Blackwell Publishing: Oxford, UK, 2007. [CrossRef]
4. European Environment Agency. Urban Sprawl in Europe, The Ignored Challenge. EEA Report No 10/2006. Available online: https://www.eea.europa.eu/publications/eea_report_2006_10/eea_report_10_2006.pdf/view (accessed on 1 October 2024).
5. European Environment Agency. Urban Sprawl in Europe Joint EEA-FOEN Report. EEA Report 11/2016. Available online: https://www.eea.europa.eu/en/analysis/publications/urban-sprawl-in-europe (accessed on 1 October 2024).
6. European Environment Agency. Land Take and Land Degradation in Functional Areas. EEA Report 17/2021. Available online: https://www.eea.europa.eu/en/analysis/publications/land-take-and-land-degradation (accessed on 10 October 2024).
7. Siedentop, S.; Fina, S. Monitoring urban sprawl in Germany: Towards a GIS-based measurement and assessment approach. *J. Land Use Sci.* **2010**, *5*, 73–104. [CrossRef]

8. Bruechner, J.K. Urban Sprawl: Diagnosis and Remedies. *Int. Reg. Sci. Rev.* **2000**, *23*, 160–171. [CrossRef]
9. Ehrlich, M.V.; Hilber, C.A.L.; Schöni, O. Institutional settings and urban sprawl: Evidence from Europe. *J. Hous. Econ.* **2018**, *42*, 4–16. [CrossRef]
10. Marquard, E.; Bartke, S.; Gifreu i Font, J.; Humer, A.; Jonkman, A.; Jürgenson, E.; Marot, N.; Poelmans, L.; Repe, B.; Rybski, R.; et al. Land Consumption and Land Take: Enhancing Conceptual Clarity for Evaluating Spatial Governance in the EU Context. *Sustainability* **2020**, *12*, 8269. [CrossRef]
11. European Commission. The Communication from the Commission to the European Parliament, the Council, the European Economic and Social Committee and the Committee of the Regions: "Roadmap to a Resource Efficient Europe". COM/2011/0571 Final 2011. Available online: https://eur-lex.europa.eu/legal-content/EN/TXT/PDF/?uri=CELEX:52011DC0571 (accessed on 10 October 2024).
12. European Environment Agency. Net Land Take in Cities and Commuting Zones in Europe. Indicator Assessment 2023. Available online: https://www.eea.europa.eu/data-and-maps/indicators/land-take-3/assessment (accessed on 10 October 2024).
13. Bagan, H.; Yamagata, Y. Land-cover change analysis in 50 global cities by using a combination of Landsat data and analysis of grid cells. *Environ. Res. Lett.* **2014**, *9*, 064015. [CrossRef]
14. UN-Habitat. *SDG Indicator 11.3.1 Training Module: Land Use Efficiency*; United Nations Human Settlement Program (UN-Habitat): Nairobi, Kenya, 2018.
15. Eichhorn, S.; Adam, B.; Schürholt, K.; Jansen, H.; Kötter, T.; Terfrüchte, T.; Eichfuss, S.; Rohde, N.; Wilberz, J.; Stielike, J.M. No net land take policy in practice: Applications and potentials of planning instruments in municipalities. Results of an online survey in North Rhine-Westphalia. *Raumforsch. Und Raumordn.* **2024**, *82*, 68–84. [CrossRef]
16. Decoville, A.; Feltgen, V. Towards a credible, relevant and legitimate implementation framework for the EU no net land take objective. *J. Environ. Plan. Manag.* **2025**, 1–21. [CrossRef]
17. Bovet, J.; Marquard, E.; Schröter-Schlaack, C. *International Expert Workshop on Land Take, 4–5 April 2019 in Berlin, Germany. Workshop Report*; Helmholtz Centre for Environmental Research (UFZ): Leipzig, Germany, 2019.
18. Salvati, L.; Carlucci, M. Patterns of Sprawl: The Socioeconomic and Territorial Profile of Dispersed Urban Areas in Italy. *Reg. Stud.* **2015**, *50*, 1346–1359. [CrossRef]
19. Salvati, L.; Morelli, V.G. Unveiling Urban Sprawl in the Mediterranean Region: Towards a Latent Urban Transformation? *Int. J. Urban Reg. Res.* **2014**, *38*, 1935–1953. [CrossRef]
20. Lagarias, A.; Sayas, J. Urban sprawl in the Mediterranean: Evidence from coastal medium-sized cities. *Reg. Sci. Inq.* **2018**, *10*, 15–32.
21. Yiannakou, A. Patterns of suburban growth and sprawl in Thessaloniki: Spatial aspects of a multifaceted crisis. In Proceedings of the 10th Regular Scientific Conference of the Greek Section of ERSA: Economic Crisis and Development and Cohesion Policies, Thessaloniki, Greece, 1–2 June 2012. (In Greek).
22. Lagarias, A. Impervious Land Expansion as a Control Parameter for Climate-Resilient Planning on the Mediterranean Coast: Evidence from Greece. *Land* **2023**, *12*, 1844. [CrossRef]
23. Lagarias, A.; Stathakis, D. Towards a more realistic estimation of urban land take by combining cadastral parcels and building footprints. *Environ. Plan.* **2024**, *0*. [CrossRef]
24. UN-Habitat. *SDG Indicator 11.3.1: Ratio of Land Consumption Rate to Population Growth Rate—Metadata Document*; United Nations Human Settlements Programme (UN-Habitat): Nairobi, Kenya, 2021. Available online: https://sdg.data.gov/11-3-1/ (accessed on 1 April 2025).
25. Ghazaryan, G.; Rienow, A.; Oldenburg, C.; Thonfeld, F.; Trampnau, B.; Sticksel, S.; Jürgens, C. Monitoring of Urban Sprawl and Densification Processes in Western Germany in the Light of SDG Indicator 11.3.1 Based on an Automated Retrospective Classification Approach. *Remote Sens.* **2021**, *13*, 1694. [CrossRef]
26. Wang, Y.; Huang, C.; Feng, Y.; Zhao, M.; Gu, J. Using Earth Observation for Monitoring SDG 11.3.1-Ratio of Land Consumption Rate to Population Growth Rate in Mainland China. *Remote Sens.* **2020**, *12*, 357. [CrossRef]
27. Esri, Maxar, Earthstar Geographics, and the GIS User Community. Available online: https://www.arcgis.com/home/item.html?id=10df2279f9684e4a9f6a7f08febac2a9 (accessed on 1 April 2025).
28. Lagarias, A. Exploring land use policy scenarios with the use of a cellular automata-based model: Urban sprawl containment and sustainable development in Thessaloniki. *Geocarto Int.* **2015**, *30*, 1033–1051. [CrossRef]
29. Yiannakou, A. Urban Plans for Thessaloniki: Ideology and practice during the 20th century. In *Thessaloniki at the Verge: The City as a Transformation Process*; Kafkalas, G., Lambrianidis, L., Papamichos, N., Eds.; Kritiki: Athens, Greece, 2008; pp. 447–487. (In Greek)
30. Salvati, L. Neither ordinary nor global: A reflection on the 'extra-ordinary' expansion of Athens. *Urban Plan. Transp. Res.* **2014**, *2*, 49–56. [CrossRef]
31. Karadimitriou, N.; Maloutas, T.; Arapoglou, V.P. Multiple Deprivation and Urban Development in Athens, Greece: Spatial Trends and the Role of Access to Housing. *Land* **2021**, *10*, 290. [CrossRef]

32. Yiannakou, A. From sprawl to smog and the roles of spatial planning: Aspects of the "Greek crisis". In Proceedings of the Joint AESOP/ACSP Congress, Dublin, Ireland, 15–19 July 2013.
33. Delladetsima, P.M. Sustainable Development and Spatial Planning: Some considerations arising from the Greek case. *Eur. J. Spat. Dev.* **2012**, *12*, 46.
34. Economou, D. The planning system and rural land use control in Greece: A European perspective. *Eur. Plan. Stud.* **1997**, *5*, 461–476. [CrossRef]
35. Copernicus. *Mapping Guide for an European Urban Atlas*; European Union: Brussels, Belgium, 2016.

Disclaimer/Publisher's Note: The statements, opinions and data contained in all publications are solely those of the individual author(s) and contributor(s) and not of MDPI and/or the editor(s). MDPI and/or the editor(s) disclaim responsibility for any injury to people or property resulting from any ideas, methods, instructions or products referred to in the content.

land

MDPI

Article

Impact of Urbanization on Surface Temperature in Morocco: A Multi-City Comparative Study

Mohamed Amine Lachkham [1], Lahouari Bounoua [2,*], Noura Ed-dahmany [1] and Mohammed Yacoubi Khebiza [1]

[1] Laboratory of Water Sciences, Microbial Biotechnologies, and Natural Resources Sustainability (AQUABIOTECH), Faculty of Sciences Semlalia, Cadi Ayyad University, Marrakesh 40000, Morocco; m.lachkham.ced@uca.ac.ma (M.A.L.); n.eddahmany.ced@uca.ac.ma (N.E.-d.); yacoubi@uca.ac.ma (M.Y.K.)

[2] Biospheric Sciences Laboratory, National Aeronautics and Space Administration, Goddard Space Flight Center, Greenbelt, MD 20771, USA

* Correspondence: lahouari.bounoua@nasa.gov

Abstract: Morocco, like many nations undergoing significant economic and social transformation, is experiencing rapid urbanization alongside an ongoing rural exodus. This, coupled with the country's diverse climate and heterogeneous geography, warrants a detailed exploration of urbanization's effect on surface climate. Utilizing the Simple Biosphere (SiB2) model's simulated surface temperature, this study analyses summer's urban heat structure of seven Moroccan urban areas and their surroundings, assessing the urban impact on surface temperature at the city center, and the intensity and spatial distribution of the urban heat island (UHI) effect at different spatial resolutions. Results show wide-ranging dissimilarities in urban thermal profiles, with the maximum UHI intensity recorded at 8.7 °C in the Dakhla peninsula. Urban heat sink (UHS) effects were observed in six of the seven studied cities, with Marrakech being the exception, only exhibiting UHI effects. A more detailed examination of the thermal profile in Rabat's metropole at a finer scale, using Landsat-observed land surface temperature (LST), yields additional insights into UHI characteristics, and the findings are contrasted with the existing literature to provide broader insights. The implications of this study strongly resonate within the Moroccan context and its neighboring regions with similar environmental and socio-economic features and should aid in the development of sustainable regional urban planning.

Keywords: urban climate; urban heat island; urban heat sink; land surface temperature; urbanization; impervious surface area; MODIS; Landsat; Morocco

1. Introduction

The current state of academic knowledge regarding the urban heat island (UHI) effect reflects a continuous recognition of its global significance, while acknowledging its nuances across biomes and world regions. Research has elucidated the mechanisms driving the UHI's formation, and advances in remote sensing and modeling techniques have enabled the quantification and characterization of its intensity with greater precision. The UHI phenomenon manifests when urban temperatures are higher than those of the surrounding vegetated, or otherwise rural, areas, as a result of the natural alterations caused by urbanization upon land surface properties and energy balance. Consequently, urban areas often exhibit elevated temperatures, particularly during periods of calm wind and high insolation [1]. Furthermore, the urban heat sink (UHS), an inverted UHI, manifests when urban areas are cooler than their rural surroundings due to factors such as proximity to water bodies providing a cooling effect, or the presence of resource-augmented urban vegetation within cities in arid climate zones [2,3]. This study investigates surface temperature differences derived from modeled data; thus, the phenomena examined correspond to a surface urban heat island (SUHI) and a surface urban heat sink (SUHS), although for simplicity, the terms urban heat island (UHI) and urban heat sink (UHS) are used throughout.

Four primary components influence the UHI effect: urban fabric characteristics, anthropogenic activities, urban morphology, and global temperature patterns. Indeed, the UHI effect can occur naturally, even without human activities, due to the higher capacities of urban materials to absorb and retain solar energy compared to the rural surroundings. Building materials in urban areas, like concrete and asphalt, possess low albedo, high thermal conductivity, and a significant capacity to absorb and retain solar heat which results in elevated surface temperatures [4]. Secondly, human activities such as industrial processes, transportation, and energy consumption generate substantial heat emissions in urban areas, further contributing to temperature increases [5,6]. Thirdly, numerous scholars have highlighted the impact of urban morphological parameters—including building size, shape, compactness, vegetation cover types, and population densities—on the UHI effect [7–10]. Lastly, rising regional and global temperatures driven by increased greenhouse gas emissions contribute to baseline temperature increases, further intensifying UHI effects [11–13].

While the basic principles of the UHI effect seem evident, several studies have highlighted the influence of regional climate conditions, urban planning policies, and socio-economic factors on its intensity and spatial distribution [14–16]. For instance, its intensity tends to be more pronounced in densely populated metropolitan areas with extensive impervious surfaces and limited vegetation cover, exacerbating heat-related health risks and energy demands [17]. Nonetheless, a UHI's manifestation can vary, as subtropical cities experience distinct thermal profiles compared to temperate or tropical counterparts [18,19].

The UHI effect is not a static phenomenon but a dynamic process influenced by urban typology and the condition of surrounding ecosystems. It is essential to distinguish between the occurrence of the UHI and changes in its amplitude. The UHI effect, defined as the temperature difference between an urban core and its surrounding rural area, is shaped by both regional and global warming, and its intensity primarily varies based on differences in land cover. Since different land cover types respond differently to rising temperatures, the impact of warming on the UHI will vary nonlinearly across urban areas, depending on the city's characteristics and the health of the nearby vegetation. This suggests that as urban temperatures rise with global warming, the response of the surrounding rural areas will largely depend on the physiological state of their vegetation. For instance, if the vegetation is heat tolerant and unstressed, it will continue to photosynthesize and transpire, cooling the rural area and amplifying the urban–rural temperature contrast, consequently intensifying the UHI effect. Conversely, if the vegetation is stressed due to either temperature or drought, it will downregulate its stomata and reduce transpiration, releasing most absorbed solar energy as sensible heat, which leads to increased rural temperatures, potentially surpassing urban ones and resulting in a UHS effect.

Across the African continent, the UHI effect manifests as a multifaceted phenomenon, influenced by diverse geographical, climatic, socio-economic, and cultural factors. From the metropolises of Yaoundé (Cameroon) and Nairobi (Kenya) to the urban centers of Addis Ababa (Ethiopia) and Lagos (Nigeria), it has been shown that urban areas across the African continent experience elevated temperatures compared to their surrounding rural vicinities [20–22].

Morocco, located in northwestern Africa, offers a compelling case for studying urban thermal profiles due to its unique geographical and socio-environmental characteristics.

Bordered by the vast Sahara Desert to the south, the Atlantic Ocean to the west, and the Mediterranean Sea to the north, the country experiences a pronounced environmental gradient that shapes diverse thermal profiles. The coexistence of densely populated urban centers with contrasting natural landscapes creates microclimatic variations, which adds to the country's strong meridional temperature and precipitation gradients, leading to significant land cover differences. Additionally, most of the population is concentrated in dispersed agglomerations along the northern and western coasts, where the pressure of urbanization and the growing rural exodus [23] heighten the relevance of studying the UHI effect. Beyond its environmental implications, understanding the urban thermal structure in Morocco is crucial for understanding related socio-economic challenges across sectors such as public health, infrastructure, and energy demand. Comprehensive insights into the country's UHI dynamics should support the development of effective, region-specific adaptation strategies.

By analyzing specific examples from Morocco—including Casablanca, Marrakech, Agadir, Oujda, Errachidia, Dakhla, and the metropolitan area of Rabat, which encompasses the capital city Rabat, along with neighboring cities Salé and Temara—and drawing on findings from previous studies, this contribution aims to provide deeper insights into factors influencing urban thermal structures and the change in their intensities in the context of global and regional climate change. The findings are intended to inform urban heat mitigation strategies and climate adaptation efforts in Morocco, as well as in neighboring regions with comparable environmental, and socio-economic characteristics.

2. Study Area, Data, and Model

2.1. Study Area

Seven Moroccan cities from different administrative regions were selected to carry out this study. Figure 1 shows the location of these urban areas, selected for their varied urban attributes, such as different climate zones, altitudes, urban morphology, and land use patterns. Studying the UHI effect in the Rabat metropolitan area is particularly important due to its unique geographic position, flanked by the commuter urban conurbations of Salé to the north and Temara to the south (Figure 1), forming a metropolis for which the UHI effect has not been previously studied. Additionally, the urban thermal profiles of Errachidia and Dakhla offer novel findings with a perspective of cities within arid and Saharan climates.

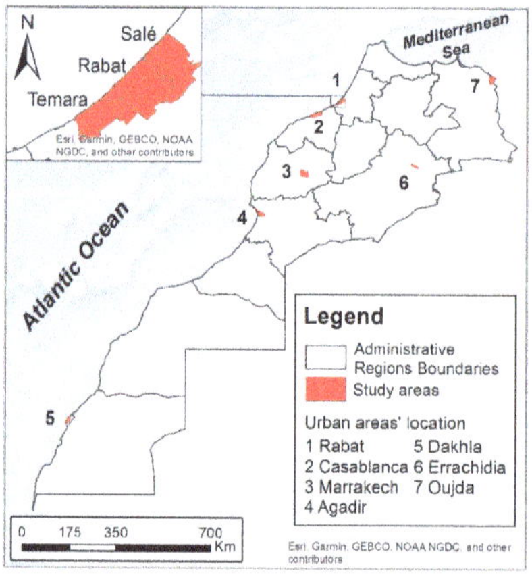

Figure 1. Geographic location of the study areas, including the cities of Rabat, Casablanca, Marrakech, Agadir, Dakhla, Errachidia, and Oujda.

2.2. Data

The detailed methodology of the data fusion is described in [24], and a summary is presented here for reference. Landsat-based Impervious Surface Area (ISA) from the Global Land Survey (GLS) [25] was used to characterize urban areas at 30 m spatial resolution, and MODIS-based land cover products [26] at 500 m spatial resolution were used to describe the land cover vegetation within a global climate modeling grid (CMG) of $0.10^{\circ} \times 0.10^{\circ}$ of latitude–longitude (approximately 10×10 km).

The data fusion followed three distinct steps: (1) we first aggregated the different land cover types from MODIS 500 m and calculated fractions of each land cover type within the CMG; (2) we co-registered and aggregated the ISA from 30 m to the same CMG; and (3) we replaced into the CMG the build-up fraction from MODIS with the Landsat-based ISA. In doing so, we recognize that the accuracy of Landsat is superior to MODIS in characterizing the urban area. Since these two fractions are not equal, MODIS buildup being larger than Landsat ISA at the CMG level, these differences were redistributed proportionally over other non-urban classes within the CMG using their fractions as weights. This led to a global 0.1×0.1 degrees CMG in which each pixel may have up to 13 distinct land cover classes. For each CMG land cover type, biophysical parameters were generated every 16 days using MODIS data.

2.3. Model

The land cover map along with the biophysical parameters were used as boundary and lateral conditions for the Simple Biosphere model (SiB2) [27] to simulate the impact of urbanization as a form of land use on the surface energy, water, and carbon cycles at a global level for the year 2010, selected due to the availability of comprehensive biophysical and climatological data, and as a reliable global ISA product which allowed accurate urban characterization.

SiB2 is a biophysically based land surface model (LSM) that calculates the exchanges of energy, carbon, water, and momentum, explicitly considering 13 land cover classes [27], and includes an urban canopy model [28]. The model accounts for the hydraulic and thermal properties of various soil types. The prognostic variables include the canopy, ground, and deep soil temperatures, the water content in three soil layers, the canopy and ground interception stores, as well as canopy stomatal conductance. SiB2 also incorporates an integrated photosynthesis-conductance sub-model that simultaneously calculates the carbon assimilation and water vapor exchanges between the canopy and the canopy air space [29]. Alongside the biophysical fields that describe the vegetation characteristics and phenology, SiB2 is forced by short- and longwave radiation, large-scale and convective precipitation, canopy-level temperature, wind speed, specific humidity, and surface pressure. It outputs emitted and reflected radiation fluxes, soil and canopy latent and sensible heat fluxes, and the carbon assimilation rate, among other diagnostics [28]. The SiB2 surface temperature outputs have been validated in many studies [30–33], demonstrating a high degree of similarity to observations made by satellite instruments, such as Landsat and MODIS, highlighting the model's reliability and accuracy in estimating surface temperature and thus the UHI effect.

In this study, the model was used in a stand-alone mode forced by meteorological data obtained from MERRA analysis [34]. This approach helps to isolate the effects of land cover change on local surface temperature independent of atmospheric circulation. This also allowed us to run the model globally at 0.10×0.10-degree spatial resolution. The model was spun up for 5 years, forced by hourly climate data. It was integrated independently for each land cover class and outputs from different integrations are stratified by land cover type, as well as a weighted average obtained as a response from each land cover weighted by its fraction within the CMG. After spin-up, each simulation was integrated forward another 3 years using the same atmospheric drivers so that model responses are exclusively attributable to changes in land cover class, a method proven to capture the observed changes in land surface. In this analysis, we use the model surface temperature outputs averaged over the last three years of integration over Morocco, for the summer season (JJA), which is characterized by conditions that typically amplify urban–rural thermal contrasts, offering a meaningful lens through which to evaluate land cover impacts on surface climate.

2.4. Data Organization

Surface (skin) temperatures obtained from SiB2 were time-averaged over the months of June, July, and August using daily average temperatures to create a seasonal summer mean temperature. A buffer zone was created at 0.1-degree intervals from the representative center of each urban area and expanding outward towards the peripheries (Figure 2). The city center was defined by the center of the CMG having the highest ISA fraction. A series of axes were then drawn, connecting the city center to the buffer edge, with a spacing of 5 degrees between each axis, making a total of 72 axes to best capture the temperature structure distinguishing the urban core from its surrounding landscape. The original gridded model temperatures were interpolated to the radial axes to obtain temperatures around a city. This process took in consideration elevation and distances between cities. Along each axis, temperatures were extracted every 0.1 degree (radial spacing), and were subsequently averaged across axes within defined quadrants to derive temperature values representing key directional transects adapted for each urban area. This approach facilitated the visualization of the temperature structure within cities, and their peri-urban and rural areas, although having some limitation near large topographic features and coastal areas.

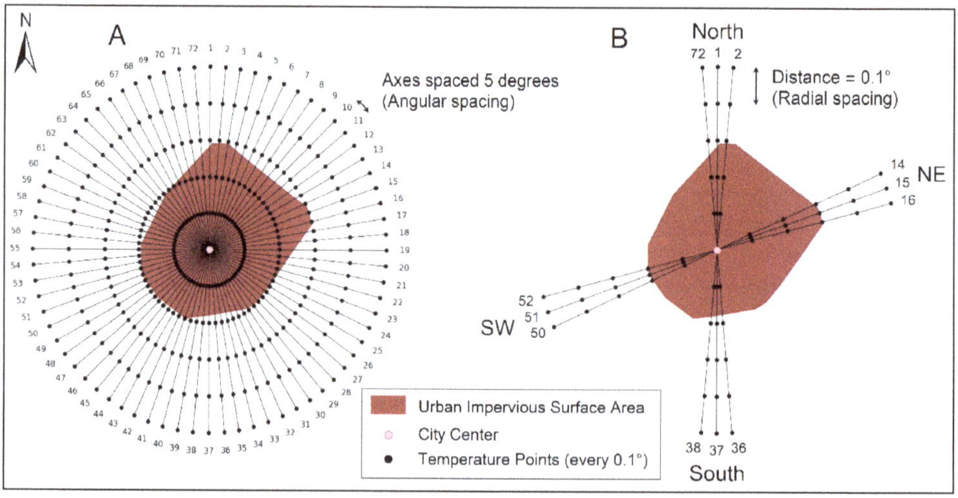

Figure 2. Illustration of axes and buffer zone for LST analysis: (**A**) Distribution of axes radiating outward from the urban center to delineate directions and the buffer zone. (**B**) Selected quadrants with their respective axes defining north–south and northeast–southwest directional transects.

To analyze the UHI effect, two representative cross-sections of each city were selected based on geographical characteristics and land cover composition. For instance, in the case of coastal cities, cross-sections were designed to capture temperature profiles both perpendicular and parallel to the coastline, whereas, for inland cities, cross-sections extended in both north–south and east–west directions, adjusted based on land cover features, taking into consideration any natural or anthropogenic obstacles that could influence the temperature structure.

3. Results and Discussion

Table 1 summarizes the results for the seven urban areas, presenting the impact of urbanization on surface temperature in the city center and the corresponding temperature differences between urban and surrounding rural areas. These values indicate the presence of either a UHI or a UHS, depending on the direction of the transect. The urban impact on surface temperature is calculated as the difference between the actual temperature—defined as the average temperature of all land cover classes (including urban) coexisting in each CMG weighted by their fractions—and the average temperature of vegetation classes only (excluding urban) weighted by their fractions. UHI and UHS values are obtained as differences between the actual temperature at the city center and that of the surrounding rural area along each transect, with positive values indicating a UHI and negative values indicating a UHS.

Table 1. Urban impact on surface temperature and UHI/UHS intensity across the selected Moroccan cities, positive values represent UHI and negative values represent UHS.

	City	Urban Impact on LST (°C)	Transect Direction	Urban Center–Rural Difference (UHI/UHS)	
Coastal Cities	Rabat	0.2	Perpendicular	NW (10 km) 4.8 °C	SE (10 km) −0.5 °C
			Parallel	NE (20 km) −0.2 °C	SW (20 km) 0.0 °C
	Casablanca	0.2	Perpendicular	NW (10 km) 5.1 °C	SE (10 km) 0.0 °C
			Parallel	NE (20 km) 0.0 °C	SW (10 km) 0.4 °C
	Agadir	0.2	Perpendicular	W (10 km) 6.1 °C	E (10 km) −0.6 °C
			Parallel	N (10 km) −0.1 °C	S (20 km) 0.5 °C
	Dakhla	0.2	NE-SW	NE (10 km) 8.7 °C	SW (10 km) 8.7 °C
			NW-SE	NW (10 km) 8.7 °C	SE (10 km) 8.7 °C
Inland Cities	Errachidia	0.1	N-S	N (10 km) 0.7 °C	S (10 km) −0.5 °C
			W-E	W (10 km) 1.1 °C	E (10 km) −0.7 °C
	Marrakech	0.2	N-S	N (10 km) 0.3 °C	S (10 km) 0.1 °C
			W-E	W (10 km) 0.2 °C	E (10 km) 0.6 °C
	Oujda	0.4	N-S	N (10 km) −0.2 °C	S (10 km) 0.6 °C
			W-E	W (10 km) 0.3 °C	E (10 km) 0.2 °C

The urban impact on surface temperature at the city center varied across urban areas. Errachidia experienced the lowest impact, with an increase of only 0.1 °C. This was expected as this city, located in central eastern Morocco, has a hot desert climate characterized by high temperatures, peaking in summer, with very little rainfall throughout the year. The region's landscape is dominated by desert and rocky terrain while the vegetation cover is sparse, typical of a semi-arid to arid environment, with drought-resistant plants such as shrubs, grasses, and scattered palm groves, making the natural vegetation limited outside a few irrigated zones. Within this arid desert environment, urbanization and the increase in impervious surfaces have had a minimal impact on surface temperature. In contrast, Rabat, Casablanca, Agadir, and Dakhla, all coastal cities, have had a similar urban-induced increase in temperature of about 0.2 °C, despite their differences in size and climate. This consistency underlines how factors beyond city size, such as proximity to water and urban vegetation, play a role in moderating this difference. In Marrakech, an inland city, urbanization has had an impact of 0.2 °C on surface temperature, which is likely due to local vegetation being stressed during the summer season. Indeed, in this semi-arid region, daytime surface temperature can get as high as 40 °C, stressing the local vegetation which shunts all of the absorbed solar radiation into sensible heating, thus increasing its temperature and decreasing the difference from the urban areas. The city of Oujda, located in the northeastern part of the country, and not too far from the Mediterranean shores, has experienced the largest urban-induced temperature of 0.4 °C. There, the summer is moderately warm, and the vegetation is less stressed than in Marrakech. Furthermore, most of the urban vegetation is irrigated during summertime, making the urban area warmer than the vegetated landscape.

A detailed investigation is addressed in the following sections through a closer examination at each city's landscape and how it influences the ambient thermal structure.

3.1. Case of Coastal Cities

3.1.1. Temperature Structure in the Metropolitan Area of Rabat

The city of Rabat is characterized by a humid to sub-humid Mediterranean climate and is distinguished by a blend of modern and traditional architectural styles. Rabat benefits from the maritime influence of the Atlantic Ocean modulating its temperature, particularly during the warm and dry summer months (JJA). The most relevant temperature cross-sections in Rabat are those oriented perpendicular and parallel to its coastline (Figures 3 and 4).

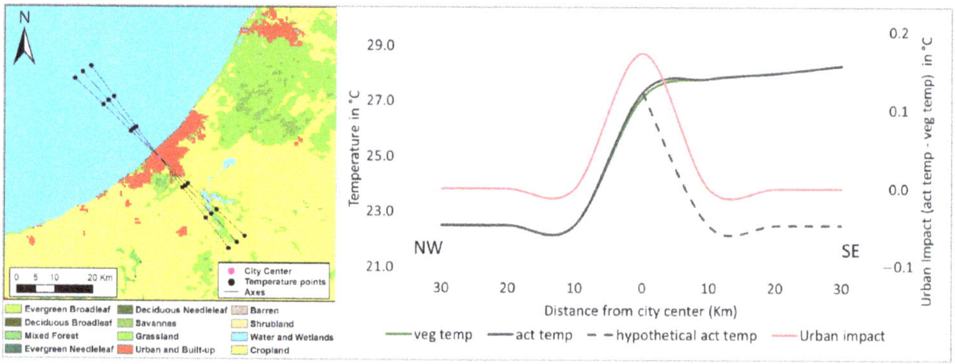

Figure 3. Summertime (JJA) simulated average surface temperature cross-section along a northwest NW–southeast SE transect perpendicular to the coastline and crossing the city of Rabat. The left panel shows the land cover for the year 2010, and the selected axes for the transect. The right panel displays the surface temperature profile along the transect: The left y-axis shows the model-simulated temperatures in degrees Celsius, representing the actual temperature ('act temp') and the vegetation-only temperature ('veg temp'). The dashed curve shows the hypothetical temperature profile if the city were surrounded by water. The right y-axis shows the difference between 'act temp' and 'veg temp' representing the impact of urbanization on surface temperature.

The quadrant perpendicular to the coast originating from the city center towards the northwest and the southeast allowed for the construction of the cross-section of average surface temperature represented in Figure 3.

From the northwest versant of the city, temperature increases from 22.5 °C over the ocean to 27.3 °C in the city center, creating a UHI with an amplitude of 4.8 °C. Along the same transect, and at 10 km towards the southeastern periphery, the landscape becomes rural, and the temperature increases by 0.5 °C, indicating the presence of a UHS. Further south (approx. 30 km), low-albedo fallow lands reach 28.3 °C, indicating a pronounced UHS of 1.0 °C. During dry summers, irrigated urban green spaces facilitate significant evapotranspiration and evaporation. This leads to a high latent heat flux that moderates urban temperatures through evaporative cooling. Unlike the surrounding dry rural lands which dissipate solar radiation mainly as sensible heat, the urban environment leverages evaporative cooling, creating a strong thermal gradient. Additionally, coastal sea breezes cool the urban area more effectively than the hot, moisture-deficient rural periphery due to factors like urban thermal mass and ventilation.

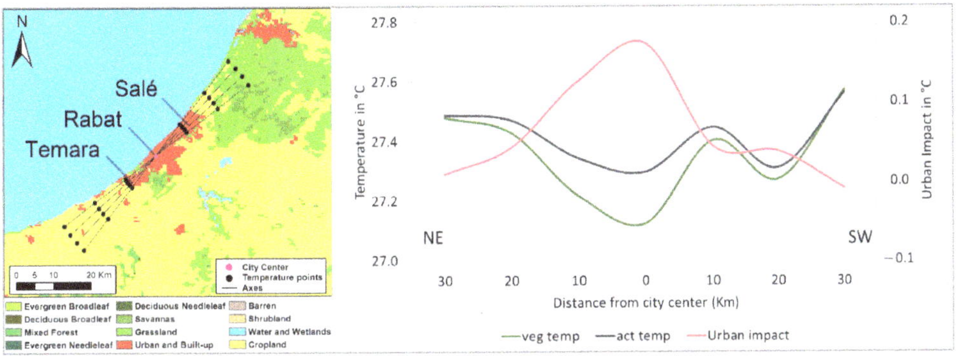

Figure 4. Same as Figure 3, except for simulated surface temperature cross-section along a northeast–southwest transect parallel to the coastline and crossing the metropole of Rabat–Salé–Temara.

This underscores that the intensity and extent of UHIs are not uniformly distributed around a city, challenging the conventional notion of symmetrical temperature patterns which would likely manifest in a manner similar to the symmetrical 'hypothetical act temp' curve presented in Figure 3. Instead, the magnitude of the UHI effect is closely tied to the distance from the city center, influenced by the city's geographic location and the prevailing climate zone. Factors such as land surface characteristics and urban morphology further shape the asymmetric nature of UHIs as seen in Rabat's urban area, and their recognition is essential to accurately assess urban temperature dynamics. Even though Rabat is known for its well-maintained urban forests and irrigated lawns and gardens, its impervious surfaces resulted in this moderate increase of 0.2 °C added to its baseline surface temperature. This relatively restrained urban impact on surface temperature can be attributed to the city's strong oceanic influence and its well-ventilated environment. The proximity to the Atlantic Ocean enhances air circulation, mitigating the heat retention typically associated with urban surfaces. In the city center, however, the simulated impervious surfaces' temperature is 27.8 °C, or 0.5 °C higher than the actual temperature, suggesting that Rabat's greenery plays a significant role in mitigating the city's heating, regulating its climate through evapotranspiration cooling. In contrast, at the 10 km mark in the southeastern outskirts dominated by croplands that lie fallow during the dry summer season, the model-simulated vegetation and actual temperatures are virtually the same, indicating the absence of impervious surfaces.

To better assess the oceanic effect on the surface temperature structure in the metropolitan area of Rabat, we construct a cross-section along a transect parallel to the coastline (Figure 4). Since the effect of the ocean is practically homogeneous along this transect, the difference between the actual temperature, including the urban impervious areas' temperature, and the vegetation-only temperature is directly attributable to the land cover characteristics. Similarly, differences between temperature in the city center and its surroundings are solely due to land surface characteristics.

Along this transect, Rabat exhibits unique land use characteristic as its surroundings remain predominantly impervious at approximately 10 km in both directions. This is due to the presence of the neighboring commuter cities of Salé to the NE and Temara to the SW. The actual temperature profile indicates that both of these commuter cities are warmer than Rabat's center, suggesting a UHS at 10 km of less than 0.1 °C from the NE, and a UHS of approximately 0.1 °C from the SW. At the 30 km marks, the UHS is more significant with an amplitude of 0.2 °C to the NE and 0.3 °C from the SW. The vegetation-only temperature is slightly cooler than the actual temperature all along the transect with a maximum difference recorded in Rabat's center, where the simulated impact of the ISA increased the baseline temperature by about 0.2 °C. The complexity of the temperature structure between the three cities of Temara, Rabat, and Salé and their rural immediate vicinities is further analyzed at a higher spatial resolution.

Neighborhood Level LST Analysis Across Rabat.

Land surface temperature (LST) data from a Landsat 5 scene captured on 16 July 2010 were used for analysis due to minimal cloud cover. Landsat 5 LSTs are originally captured at 120 m spatial resolution for the thermal band and commonly resampled to 30 m for analysis [35]. A 25 km southwest–northeast transect, crossing Rabat, Salé, and Témara, was selected to track the region's architectural diversity. The transect provided 24 observations at 1 km intervals. A supervised maximum likelihood classification of land cover using a Landsat 7 image from the same day identified six classes: urban and built-up, water and wetland, mixed forest, bare land, cropland, and grassland. Accuracy was validated using historical Google Earth imagery, achieving an overall classification accuracy of 94%. Figure 5 shows the LST profile spanning from Rabat's city center along a southwest–northeast transect crossing the cities of Salé and Temara.

Average LSTs were computed using data points from the three cities, Temara (points 1–6), Rabat (points 7–17), and Salé (points 18–24) (Figure 5). The results, consistent with the trends simulated by the SiB2 model (Figure 4), indicate that Rabat exhibited a lower average LST of 37.1 °C, while both Temara and Salé recorded similar average LSTs of 38.5 °C. Notably, the LST values presented in the cross-sections in Figures 3 and 4 represent modeled seasonal mean (day and night) surface temperatures, while the Landsat LSTs presented in Figure 5C reflect observed mid-morning acquisitions, which accounts for their higher readings compared to the seasonal means. Additionally, UHS phenomena were identified towards both the NE and SW edges of the transect, with LST differences of −4.0 °C and −3.3 °C, respectively, compared to Rabat city center's LST. These findings indicate that, despite the moderating influence of the Atlantic Ocean on the LSTs for all the three cities, Rabat experienced lower urban heat. A key factor in this difference is the extensive amount of vegetation cover found in Rabat, including urban forests such as Dar Essalam and Ibn Sina, as well as numerous parks and irrigated lawns. In contrast, neighboring Salé and Temara have limited green infrastructure and higher population densities, driven by more affordable housing. Consequently, they experience elevated LSTs, which underscores the role of urban vegetation in moderating heat through shade provision, enhanced evapotranspiration, and improved microclimate regulation.

To better understand the effect caused by the different urban morphologies and land use patterns of different neighborhoods on LST changes, we analyze the latter over the transect covering the city of Rabat using data points 7–17 (Figure 5). The minimum temperature reads 35.7 °C and is recorded at the northernmost point number 17. This measurement point is in the Mellah neighborhood, predominantly residential, and characterized by a traditional architectural tissue, comprised of low-rise individual housing units, called Ryads which feature an open-roof area in the center of the house, hosting a vegetative cover, usually trees, playing the role of a traditional air conditioning system. Adding to that, the Mellah neighborhood is flanked by the Bouregreg River and the Atlantic corniche, two large water bodies whose proximity contributes to urban heat mitigation, resulting in the recorded minimum LST of 35.7 °C. In contrast, the maximum observed LST of 39.0 °C was recorded at the most southern periphery of the city (point 7), in barren lands separating Rabat and Temara.

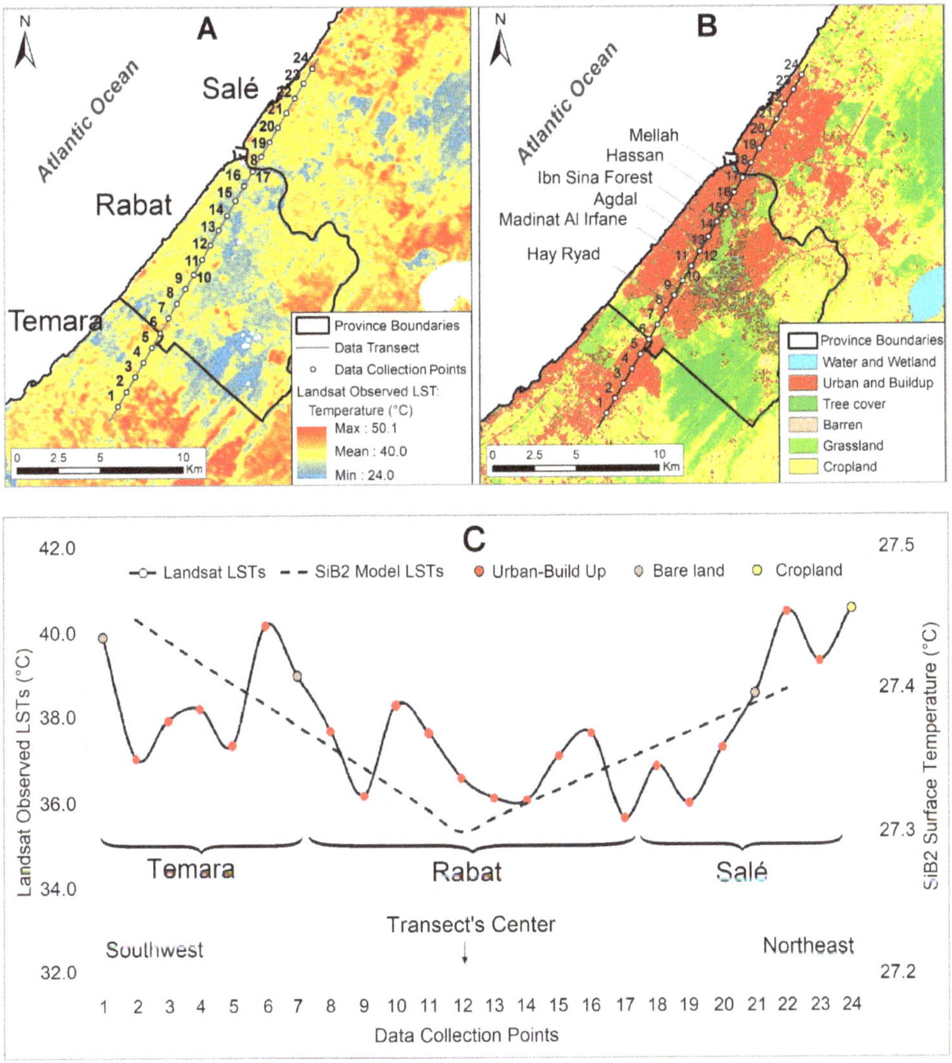

Figure 5. Spatial analysis of LST and land cover over Rabat's metropolitan area. (**A**) Landsat 5-observed LST scene (30 m resolution) with the SW-NE transect crossing Temara, Rabat, and Salé over 24 data collection points spaced 1 km apart. (**B**) Study area land cover map with the same transect overlaid, showing the three cities' urban expansion. (**C**) LST variation along the transect, annotated with the corresponding land cover type of each data collection point.

Within Rabat, Hay Ryad neighborhood (points 8–10) recorded some of the highest observed LSTs. This neighborhood is formed by a diverse urban tissue, in the form of high-rise institutional and commercial buildings, administrative settings, and corporate headquarters. This concentration of anthropogenic activity and urban density explains why this district, serving as the vital economic hub of the capital, experiences elevated LSTs compared to the rest of the neighborhoods. Further north along the transect, Madinat

Al Irfane neighborhood (points 11–12) showed a decrease in LSTs compared to Hay Ryad. As its name suggests, translating literally to "City of knowledge", this district hosts the majority of Rabat's academic institutions, with over ten public universities, and features some of the capital's biggest medical centers. These types of buildings feature substantial green cover and exhibit lower urban density given their horizontal rather than vertical expansion, which justifies the moderate LSTs recorded in this neighborhood. Further north, Agdal neighborhood (points 13–15) recorded some of the lowest LSTs in Rabat, owing to the urban morphology which is mainly characterized by high-rise buildings acting as urban canyons whose shading contributes to mitigating accumulated heat. Agdal's lower LSTs are further supported by the evapotranspiration cooling from the large adjacent Ibn Sina Urban Forest (Figure 5B). Further north is located the Hassan neighborhood (point 16). Reminiscent of the old colonial architectural influence, this district is known for its high-rise buildings and a high amount of irrigation-maintained green areas, and like Agdal, the urban canyon's shading combined with the vegetation's cooling play an essential role in maintaining a relatively moderate LST of 27.3 °C (Figure 5). Overall, this temperature profile indicated that neighborhoods with the highest green cover recorded lower LSTs, while busy commercial and economic centers hosting higher ISA densities and anthropogenic activity experienced warmer LSTs in Rabat. Additionally, this profile confirms that Salé and Temara are warmer than Rabat, highlighting the need for tailored urban policies to address the heat disparities observed across these three cities forming a metropolitan area.

3.1.2. Temperature Structure in Agadir

Located in west-central Morocco along the Atlantic Ocean, the city of Agadir features a semi-arid climate characterized by warm summers and mild winters. To assess its thermal environment, we conducted two cross-sectional profiles of LST: one oriented perpendicular to the Atlantic shore and another extending inland through the city, aligned parallel to the coastline.

The perpendicular cross-section extends westwards towards the Atlantic Ocean and eastwards through grasslands and croplands (Figure 6). The seafront temperature of 22.5 °C is significantly cooler than that of Agadir's city center simulated at 28.7 °C, resulting in a pronounced UHI with an intensity of 6.1 °C. In contrast, a UHS of −0.6 °C is noted to the east, where grasslands and fallow croplands dominate the landscape. The city's irrigated green cover contributes to localized cooling over the urban area, while the simulated surface temperature continues to gradually increase eastwards as one gets further away from the coastline, to reach 30.8 °C at the transect's edge at around 30 km.

The cross-section parallel to the coastline spans 30 km from Agadir's city center, extending northwards through grasslands and shrublands and southwards through the commuter city of Ait Melloul, before continuing through grasslands (Figure 6). Agadir's city center recorded a surface temperature of 28.7 °C, marking a UHS of −0.1 °C from the northern rural vicinity around 10 km, and a UHI of 0.5 °C to the southern rural lands, at around 20 km.

Similar to the coastal city of Rabat, the urban expansion continues encroaching towards the south of Agadir. The urban impact on surface temperature is clearly marked over the center of the commuter city of Ait Melloul with a peak of approximately 0.3 °C, higher than that of 0.2 °C simulated in Agadir's city center (Figure 6). Ait Melloul is noted for its comparatively lower housing costs, which translates to a higher density of impervious surfaces at the expense of urban greenery. The interplay of these factors, confirmed in Figure 7, leads to the higher urban impact on surface temperature recorded in Ait Melloul.

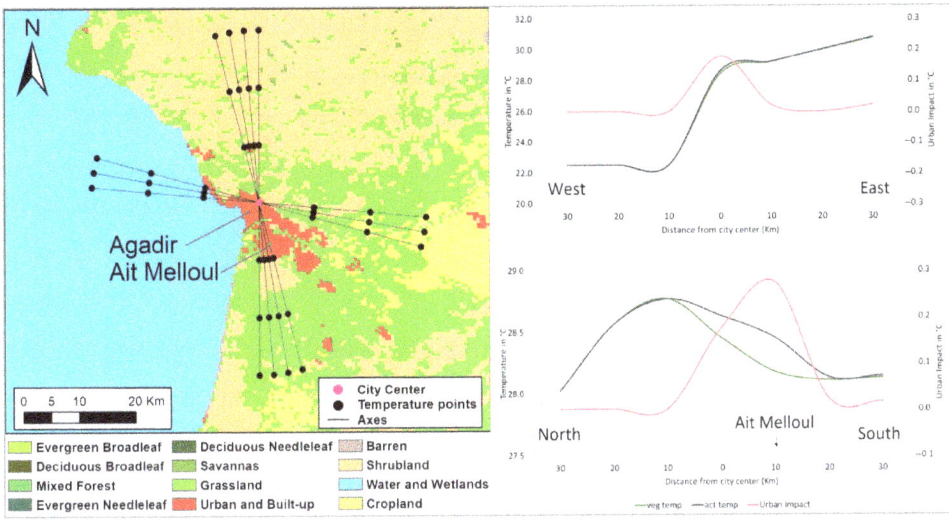

Figure 6. Same as Figure 3, except for simulated surface temperature cross-sections along two transects: a west–east transect perpendicular to the shore, and a north–south transect parallel to it, crossing the city of Agadir.

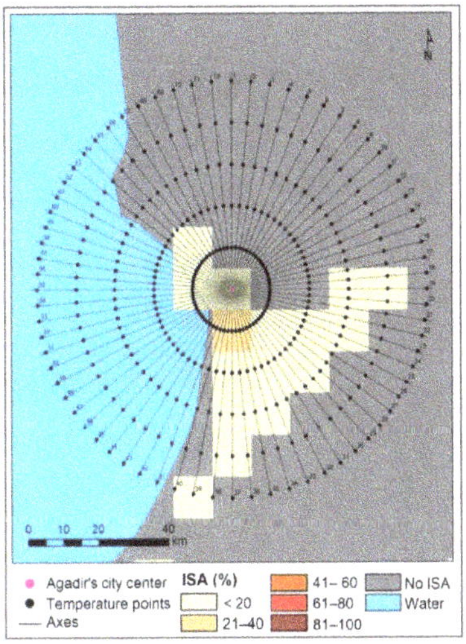

Figure 7. Impervious surface area distribution over the cities of Agadir and Ait Melloul.

In their study of the UHI effect across several Moroccan cities during the vegetative growing season, Fathi et al. (2019) [36] found a daytime UHI amplitude of 1.1 °C and a nighttime UHI amplitude of 0.8 °C in Agadir. For Ait Melloul, the study found a daytime UHS amplitude of −0.3 °C and a nighttime UHI amplitude of 0.5 °C. These findings show how inter-seasonal variability can affect the UHI and UHS, as during the growing season in this region, usually occurring during the months from November to March, the surrounding rural areas benefit from moderate insolation and mild precipitations, optimizing the vegetation cover's density. This means that the presence of higher evapotranspiration cooling correlates with the UHI effects noted by [36]. In contrast, during the dry summer months (JJA), the lack of precipitation and the absence of irrigation contribute to elevated rural area temperatures, thus leading to the formation of the UHS observed in our results. Further contributing to the observed UHS in coastal cities are the coastal oceanic breezes. These thermally driven circulations occur as warmer land heats up during the day, drawing cooler, more humid air from the ocean inland, directly lowering urban air temperatures through advection. Indeed, while sea breezes provide some cooling to the surrounding rural areas, the effect is often more pronounced and sustained within the urban environment due to the urban area's greater thermal mass (slower heating/cooling of materials) and the channeling effect of urban canyons which can enhance turbulent mixing and ventilation, particularly in the immediate coastal zone [36]. This differential cooling, where the urban area benefits disproportionately from the incoming cool air compared to the intensely heated, dry rural surroundings, facilitates the observed UHS phenomena in coastal environments like those of Agadir, Rabat, and Casablanca.

3.1.3. Temperature Structure in Casablanca

The city of Casablanca is the largest and most populated urban center acting as the economic capital of Morocco [37]. The city is located along the Atlantic Ocean approximately 100 km south of Rabat (Figure 1). They share a similar Mediterranean climate, influenced by the ocean's proximity. Two cross-section analyses were conducted: one traversing the city center in a northwest–southeast direction, perpendicular to the coastline, and a second quasi-parallel to the shore (Figure 8).

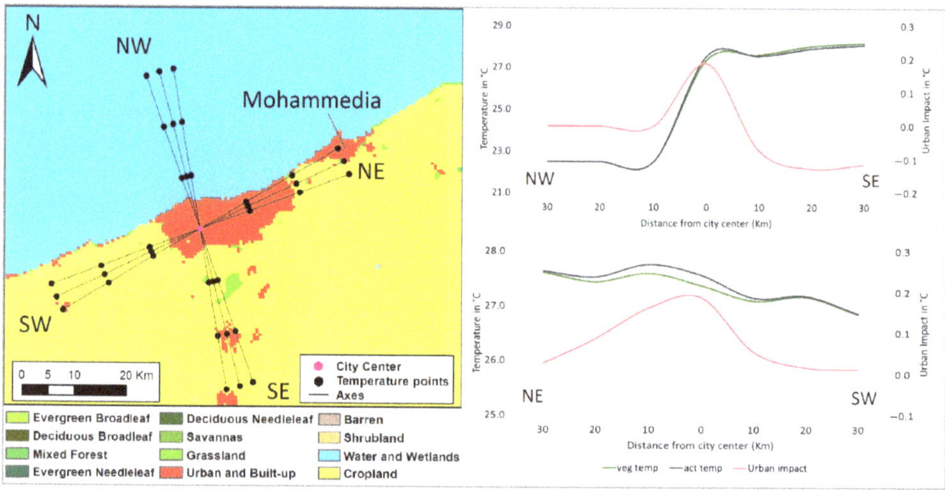

Figure 8. Same as Figure 3, except for simulated surface temperature cross-sections along two transects: a NW-SE transect perpendicular to the shore and a NE-SW transect parallel to it, crossing the city of Casablanca.

Similar to the perpendicular cross-sections of Rabat and Agadir, Casablanca's one (Figure 8) clearly captures the difference in temperatures between the city and the Atlantic Ocean, as the seafront temperature of 22.5 °C sharply contrasts the 27.6 °C of the city center. Overland and at around 10 km southeast of Casablanca, the simulated surface temperature in the rural vicinity stayed relatively constant at around 27.6 °C, before initiating a slow increase along the transect to reach the value of 28.1 °C around 30 km southeast of the city. Our findings corroborate those of [38], who analyzed observed data and identified comparable UHI and UHS effects in the city of Casablanca. Nevertheless, the amplitudes in our study diverge from theirs, a discrepancy largely attributable to our use of simulated seasonal averages rather than specific daily temperature measurements as employed by [39].

The second temperature profile is defined by quadrants crossing the city in a northeast–southwest direction, quasi-parallel to the shore and ensuring the oceanic influence is about the same all along the transect. The simulated surface temperature remained relatively constant with slight fluctuations around the city center's value of 27.5 °C, mainly due to changes in ISA density within the city (Figure 8). At approximately 10 km towards the northeast, the ISA expansion dominates and the temperature reaches a maximum of 27.8 °C. Further north around 20 km, the ISA tapers off at the city's boundaries and where rural areas begin, where the profile shows a slight decrease in surface temperature compared to the city center. Beyond these rural areas, approximately 30 km to the northeast, the temperature rises slightly, returning to around 27.5 °C—comparable to that observed in the city center. At this distance, Figure 8 shows that the data points overlap with the city of Mohammedia, which functions as a commuter hub for Casablanca and exhibits a similar surface temperature despite its smaller size. The southeastern side of the transect extends over rural areas where fallow croplands dominate the landscape and exhibit a temperature of 27.2 °C, indicating the existence of a UHS with an intensity of 0.4 °C.

Other studies have documented the UHI effect in Casablanca. For example, Ref. [39] analyzed the UHI in Casablanca using Landsat data from specific days of the year, rather than covering the entire season as is the case for our study. They indicated a strong correlation between UHI amplitude and vegetation cover, where areas with more greenery experienced LSTs. On the other hand, the study by [40] on UHI patterns in Casablanca utilized Landsat data by comparing LST observations and found that during summer, heat islands emerged in rural areas, while cool islands appeared in urban areas, similar to what was found in our parallel cross-section. Another study [36] analyzed UHI amplitude in Casablanca using MODIS LST data for the year 2013, comparing urban and rural temperatures during the vegetation growing season. The study found that Casablanca's daytime UHI amplitude was around 2 °C, while the nighttime amplitude was 2.1 °C. Combined, these results suggest that despite the moderating role of the ocean on LST changes, making them relatively small, localized land characteristics still play an important role in shaping the temperature structure.

3.1.4. Temperature Structure in Dakhla

Dakhla is situated on a narrow peninsula in the south of the country, stretching into the Atlantic Ocean (Figure 9). The city boasts a combination of desert landscapes and extensive oceanic surroundings, with an arid climate featuring hot, dry conditions, and a mild temperature range moderated by the oceanic breeze. A NE-SW cross-section analysis was performed to highlight the peninsula effect, along with a second NW-SE cross-section that captures the juxtaposition of the desert and the sea, illustrating the unique geographic and climatic characteristics present therein.

The NE-SW cross-section spans 20 km from Dakhla's city center towards the Atlantic Ocean in both directions, and shows a perfectly symmetrical and pronounced UHI with an amplitude of 8.7 °C (Figure 9). The city center temperature of 25.7 °C neatly contrasts with the 17.0 °C ocean surface temperature. In this context, Dakhla's unique peninsular geography and arid climate led to the formation of a quasi-perfect UHI with high amplitude indicative of the ISA's capacity to absorb and retain heat compared to large water bodies, while the region's arid environment, marked with high insolation and minimum vegetation, amplifies the UHI amplitude. However, and in line with other coastal cities, the urban impact on surface temperature at the city center is of 0.2 °C. This represents the additional increase in temperature due solely to the presence of the ISA in the peninsula, and shows that while the surrounding oceanic breezes provide a moderating influence, their cooling effect is overshadowed by the high thermal absorption capacity of Dakhla's urban surfaces.

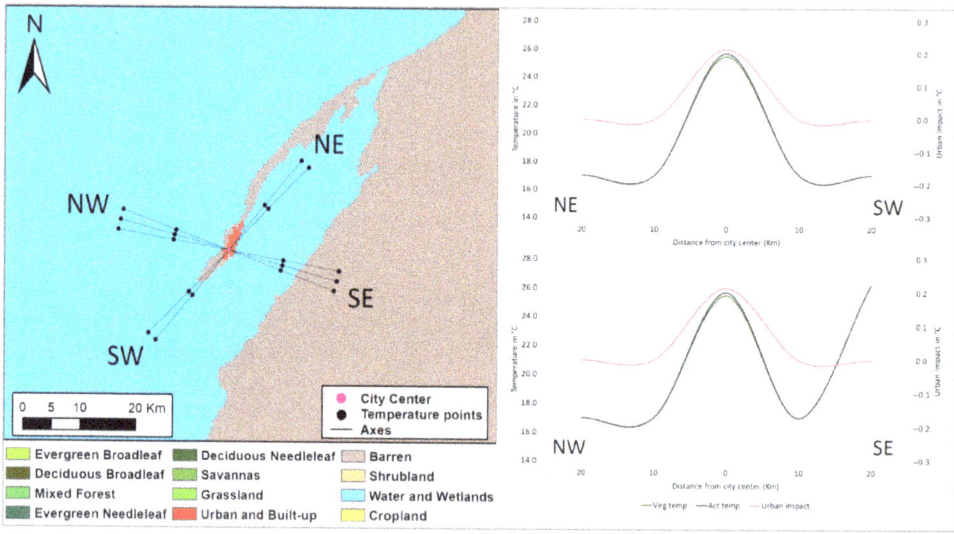

Figure 9. Same for Figure 3, except for simulated surface temperature cross-sections along two transects: a NE-SW transect highlighting the peninsula geography and a NW-SE transect crossing the city of Dakhla.

The NW-SE transect extends northwest towards the Atlantic seafront and southeast into the barren Sahara Desert (Figure 9). Similar to the first cross-section, the strong UHI of 8.7 °C is observed between the city and the surrounding oceanic water within the 10 km radius. However, further southeast over the desert, the simulated temperature is 26.2 °C, higher by 0.5 °C than that of the city center. This horizontal temperature gradient suggests that at these latitudes, the urban surface temperature is not much different from that of the desert despite the ocean's influence. The thermal structure of the city of Dakhla is complex, as when compared to the surrounding oceanic water, it forms a perfect UHI; however, when the comparison is extended to the mainland desert, an urban heat sink becomes apparent.

3.2. *Case of Inland Cities*

3.2.1. Temperature Structure in Marrakech

Located at the northern foothills of the High Atlas Mountains chain, Marrakech is characterized by a semi-arid climate, and experiences hot, dry conditions with pronounced diurnal temperature contrasts and irregular rainfall confined primarily to the winter months. Two temperature cross-sections were conducted: one oriented north–south and a second east–west.

The first cross-section extends northward into shrub and barren lands and southward into the grasslands of the Atlas Mountains (Figure 10). Marrakech's urban core recorded a simulated surface temperature of 30.4 °C, with an urban-induced increase in the city center of 0.2 °C. UHI effects are noted at 10 km north and south of the city, where the landscape is mainly comprised of shrublands, with respective amplitudes of 0.2 °C and 0.1 °C. A sharp decrease in temperature is observed further south of Marrakech, primarily driven by elevation change, which nearly doubles from around 450 m in the city to approximately 850 m at 30 km to the south (Figure 10). This decrease in temperature is also partly attributed to evapotranspiration cooling from irrigated agriculture in the Atlas Mountains, where snowmelt from higher elevations increases soil moisture and enhances agricultural productivity throughout the year. Towards the northern direction, a noticeable dip in surface temperature occurs at around 20 km distance, where temperatures drop to 29.6 °C over the northern grasslands. Further north, temperature rises again at the transect's edge but remains lower than that of the city center.

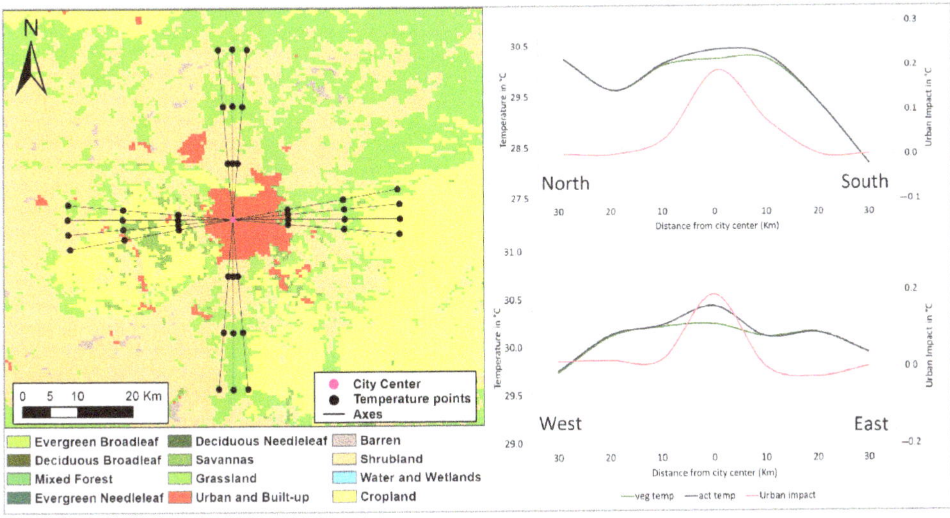

Figure 10. Same as Figure 3, except for simulated surface temperature cross-sections along a north–south transect and a west–east transect crossing the city of Marrakech.

The parallel cross-section expands westwards through a rural landscape comprised of a merge of croplands, shrublands, and mixed tree cover (10–20 kms) (Figure 10). This blend of vegetative cover, which is maintained through irrigation during the summer season [41], growing mostly subsistence crops, exhibits lower temperatures than that of the city, forming a slight UHI with a 0.2 °C amplitude. Eastwards, the transect spans mainly through rural croplands and grasslands past the 10 km mark, and at 10 km distance, the temperature profile reveals a second UHI of 0.6° C, further confirming that Marrakech city is warmer than all its rural surroundings in all four cardinal directions. In line with our findings, the study conducted by [36] to assess the UHI effect during the growing season found that Marrakech had a daytime UHI amplitude of 0.8 °C dropping at nighttime to 0.7 °C, with both amplitudes being some of the lowest found in the urban areas studied, which was attributed to the surrounding sparse vegetation and predominated agricultural use of the adjacent rural areas. On the other hand, Ref. [42] examined the UHI effect in the city, using mid-day air temperature measurements and humidity data, highlighting a maximum mean temperature difference that reached around 3.9 °C between various neighborhoods within the city. The latter study underscored that while vegetation helps reduce UHI amplitudes, large bare areas in arid regions like Marrakech play an opposite role in this dynamic, increasing daytime LSTs.

3.2.2. Temperature Structure in Errachidia

Errachidia is a gateway to the Sahara Desert and is located in eastern central Morocco. The city is bordered by the desert to the east and south and is characterized by a year-round arid climate, with extremely dry hot summers, and cool winters. Figure 11 shows the temperature profiles crossing the city.

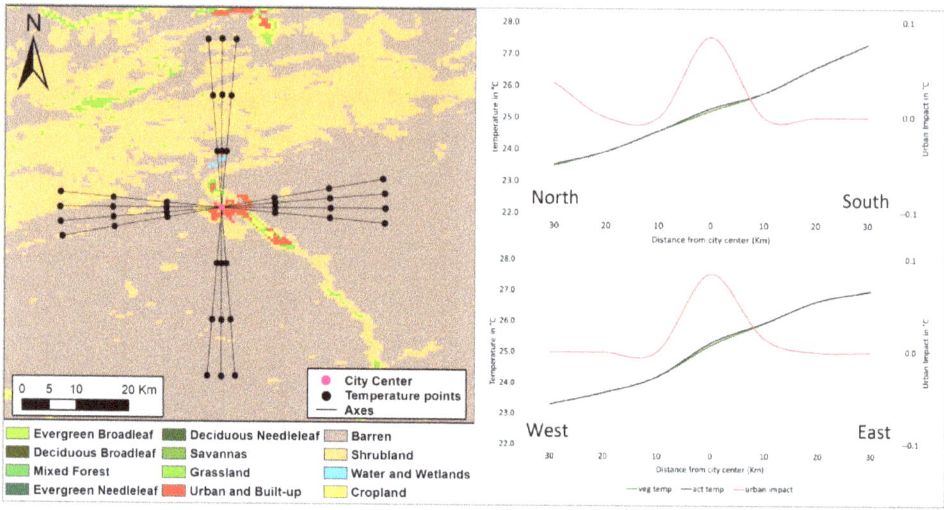

Figure 11. Same for Figure 3, except for simulated surface temperature cross-sections along a north–south transect and a west–east transect crossing the city of Errachidia.

This temperature profile reveals distinct thermal patterns across Errachidia's landscape. In the city center, the simulated surface temperature is 25.3 °C. A continuously increasing horizontal temperature gradient is simulated from north to south and from west to east. These contrasting effects can be clearly linked to the influence of changing land cover characteristics. Unlike Dakhla, which experiences a similar arid climate but benefits from the Ocean's moderating influence, Errachidia's landlocked location adjacent to the Sahara Desert results in more extreme temperature variations.

The urban impact on surface temperature illustrated by the pink curve reads around 0.1 °C in the city center (Figure 11), which is the lowest among the studied cities, largely due to the scarcity of vegetation within the city and the high temperature and water stresses that it undergoes during the summer daytime. In turn, the shrublands to the north and west do show cooler temperatures compared to the city center. Although sparse in their nature, these local shrublands are adapted to local climate and are mostly C4 plants composed of Alpha (*Panicum turgidum*) and Atriplex (*Atriplex halimus*). These species are native to the desert and are more water efficient and drought resistant as they assimilate about double the amount of carbon compared with a C3 plant for the same amount of water. The particularities of these plants located in the northern and western regions of Errachidia help maintain the UHI structure observed in these directions. Conversely, the barren desert to the south and east gets much warmer than the city core under the long periods of insolation during the summer, and in the absence of any heat mitigation mechanism, pronounced UHSs are observed. These findings align with previous research on arid regions' urban heat structures, which document the existence of UHSs in cities built in desert environments [28,43].

3.2.3. Temperature Structure in Oujda

The city of Oujda is located in northeastern Morocco and the Atlas Mountains extend across its southwestern boundary. Oujda's climate exhibits characteristics of both a Mediterranean and semi-arid climate, with hot, dry summers and significant diurnal temperature fluctuations, with sporadic rainfall occurring mainly in the winter months.

The city of Oujda has an almost circular shape and is surrounded by grasslands to the south and cropland to the north (Figure 12). However, our summer seasonal analysis precludes the existence of croplands as these are usually left fallow, except when irrigated. On the southern edge of the city lays an elevated terrain hosting the Sidi Mafaa forest. In this region, the predominant grassland vegetation exhibits significantly cooler surface temperatures compared to the fallow croplands north of the city. This temperature contrast is largely attributable to surface characteristics: the dry, dark fallow lands absorb more solar radiation, whereas the greener, vegetated areas in the south reflect more sunlight and benefit from evapotranspiration cooling, resulting in the simulated lower surface temperatures. This causes the temperature to decrease from north to south of the transect by about 2 °C within a radius of 30 km from the city center. A UHS is observed near the northern periphery, around 10 km from the city center, with an amplitude of −0.2 °C, and a UHI of 0.6 °C appears towards the southern outskirts, while in both directions, the amplitude of these effects increases progressively with distance along the transect. The west–east transect reveals that Oujda is consistently warmer than its immediate surrounding at a 10 km distance, with UHI effects extending in both directions and reaching amplitudes of approximately 0.3 °C. Among all urban areas analyzed, the city of Oujda exhibited the most pronounced impact of urbanization on surface temperature reaching 0.4 °C. Indeed, the impervious surfaces' temperature in Oujda's center was about 29.1 °C, which is 1.5 °C higher than that of the vegetation found in the city, a phenomenon driven by its unique climatic and urban morphological conditions. Unlike vertically dense cities where building shadows can mitigate surface heating, Oujda's predominantly horizontal urban expansion maximizes exposure to solar radiation, increasing heat storage in extensive impervious cover. This sprawl pattern—coupled with the city's semi-arid setting, reliance on ground-level irrigation without compensatory tree canopy, and widespread use of low-albedo materials—creates a thermal dipole between irrigated green zones and unshaded built surfaces.

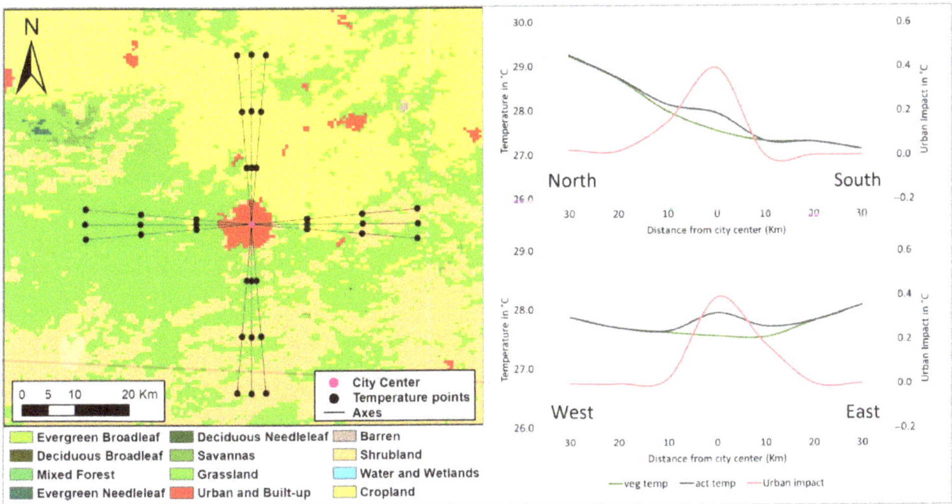

Figure 12. Same for Figure 3, except for simulated surface temperature cross-sections along a north–south and a west–east transect crossing the city of Oujda.

3.3. Inter-City Comparison

Studying the thermal structure of the seven selected cities sheds light on important insights regarding summer urban heat patterns in Morocco. When compared to maritime boundaries, coastal cities exhibited substantially strong UHI effects with an average amplitude of 6.18 ± 1.54 °C, underscoring how aquatic and terrestrial lands have different capacities to capture and retain heat, but more importantly, the role played by the Atlantic Ocean in shaping their thermal structure. Limited surface temperature differences were noted between urban areas and their surrounding rural landscape, suggesting that the effect of impervious surfaces is predominantly shadowed by the proximity to a large water body in the form of the ocean, and its related positive consequences on urban vegetation and heat dispersal. In contrast, during Morocco's hot and dry summer, urban vegetation in inland cities suffers from reduced soil moisture, transpiration, and high hydrological stress, which contributes to reduced UHI amplitudes, or the manifestation of UHS effects.

Out of the seven urban areas examined, Marrakech stood out as the only one where UHI effects were consistently observed across the surrounding rural areas located in the four cardinal directions. This is partly explained by the presence of large irrigated cropland perimeters in the immediate rural surroundings, where healthy vegetation moderates surface temperature through shading and evapotranspiration cooling, keeping a fresher surface temperature compared to that of the city's impervious surfaces which retain heat more effectively.

The impervious surfaces' effect on surface temperature was found to be the highest in the city of Oujda. This is mainly due to Oujda's compact core, where its concentrated ISA expands more horizontally, and limited urban vegetation produces sharp microclimatic gradients. Comparatively, other cities exhibited dampened thermal contrasts due to reasons such as coastal moderation or vertical building development (especially in the cases of Rabat, Casablanca, and Marrakech) offering more shading through urban canyons. This highlights how the urban growth trajectory can either exacerbate or mitigate surface temperature and heating caused by impervious surfaces.

Commuter urban agglomerations such as Salé and Temara near Rabat, Mohammedia near Casablanca, and Ait Melloul near Agadir provide more affordable housing compared to the main nearby urban centers. On the ground, this leads to more impervious surfaces at the expense of urban vegetation, in a clear contrast with the metropoles where economic sectors such as tourism warrant stricter urban planning policies that incorporate and enforce the inclusion of higher urban greenery. Consequently, and as highlighted in our results, commuter cities exhibited either similar or higher surface temperatures during the summer, a pattern consistent across the country. In these areas, the lack of urban parks and gardens in favor of a higher ISA and population densities exacerbates the vulnerability to thermal discomfort and associated health risks for residents. This makes addressing this issue a priority for urban planners, whose intervention becomes a must to put commuter cities on a sustainable development path that incorporates mitigating UHI effects through proactive urban vegetation expansion.

4. Concluding Remarks

Our analysis underscores the complex nexus between urbanization, geography, and environmental sustainability in Morocco, revealing critical spatial disparities in UHI dynamics and their cascading socio-ecological impacts. Morocco's most fertile agricultural plains lie directly beside rapidly expanding urban agglomerations concentrated in the northwest of the country, like Casablanca, Tangier, and Marrakech, which results in encroachment on prime cropland and natural habitats triggering arable-land loss, ecosystem disruption, increased runoff and pollution, and accelerated soil erosion. These findings highlight an urgent imperative to reframe urban planning as a multisector endeavor that balances growth with resource stewardship.

To mitigate these challenges, spatially tailored UHI interventions such as irrigated buffers, shade networks, and green infrastructure must be prioritized to enhance thermal equity across varied urban typologies. Second, robust regulatory frameworks are needed to curb sprawl, protect ecological zones, and incentivize compact, resource-efficient urban design. By embedding urban climate resilience into land-use policies and leveraging synergies between thermal mitigation and agricultural preservation. More attention must be directed towards satellite cities such as Salé and Ait Melloul, found to experience higher ISAs and LSTs. Promoting green space ratio frameworks, validated globally for their environmental and socio-economic benefits, can help mitigate thermal disparities. In the Moroccan context, such frameworks should incorporate native planting considerations and explore community stewardship programs to support sustainability and social engagement. Green roofing initiatives also present a valuable opportunity, especially where private gardens are limited, as is often the case in lower-income urban areas. Such integrative approaches will not only alleviate immediate heat risks but also fortify the nation's efforts against climatic pressures.

Additionally, future work could extend this analysis by deploying high-resolution remote-sensing and urban-scale climate modeling to evaluate the long-term effectiveness of potential greening strategies, assess socio-economic trade-offs, and explore nature-based solutions tailored to the country's varied climatic zones. Such integrative research will be essential for guiding evidence-based policy and ensuring equitable, livable, and productive urban futures.

Author Contributions: Conceptualization, M.A.L., L.B., N.E.-d. and M.Y.K.; methodology, M.A.L. and L.B.; software, M.A.L. and L.B.; validation, M.A.L. and L.B.; formal analysis, M.A.L. and L.B.; investigation, M.A.L.; resources M.A.L. and L.B.; data curation, M.A.L. and L.B.; writing—original draft preparation, M.A.L. and L.B.; writing—review and editing, M.A.L., L.B., N.E.-d. and M.Y.K.; visualization, M.A.L.; supervision, M.A.L. and L.B.; project administration, M.A.L., L.B., N.E.-d. and M.Y.K.; funding acquisition, M.A.L. and L.B. All authors have read and agreed to the published version of the manuscript.

Funding: This work was carried out with the support of the Moroccan National Center for Scientific and Technical Research (CNRST) under the PhD-Associate Scholarship—PASS Program.

Data Availability Statement: Data from this study are available on request from the authors.

Conflicts of Interest: The authors declare no conflict of interest.

References

1. Chestnut, L.G.; Breffle, W.S.; Smith, J.B.; Kalkstein, L.S. Analysis of Differences in Hot-Weather-Related Mortality across 44 U.S. Metropolitan Areas. *Environ. Sci. Policy* **1998**, *1*, 59–70. [CrossRef]
2. Carnahan, W.H.; Larson, R.C. An Analysis of an Urban Heat Sink. *Remote Sens. Environ.* **1990**, *33*, 65–71. [CrossRef]
3. Imhoff, M.L.; Zhang, P.; Wolfe, R.E.; Bounoua, L. Remote Sensing of the Urban Heat Island Effect across Biomes in the Continental USA. *Remote Sens. Environ.* **2010**, *114*, 504–513. [CrossRef]
4. Mohajerani, A.; Bakaric, J.; Jeffrey-Bailey, T. The Urban Heat Island Effect, Its Causes, and Mitigation, with Reference to the Thermal Properties of Asphalt Concrete. *J. Environ. Manag.* **2017**, *197*, 522–538. [CrossRef] [PubMed]
5. Smithers, R.J.; Doick, K.J.; Burton, A.; Sibille, R.; Steinbach, D.; Harris, R.; Groves, L.; Blicharska, M. Comparing the Relative Abilities of Tree Species to Cool the Urban Environment. *Urban Ecosyst.* **2018**, *21*, 851–862. [CrossRef]
6. Saranaathan, V.; Firoz, A.; Sridhar, K.; Govindaraju, M. Comprehensive Analysis of Urban Heat Island and Climate Change Impact on the Environment—An Overview. *J. Basic Sci.* **2024**, *24*, 326–334.
7. Steeneveld, G.J.; Koopmans, S.; Heusinkveld, B.G.; van Hove, L.W.A.; Holtslag, A.A.M. Quantifying Urban Heat Island Effects and Human Comfort for Cities of Variable Size and Urban Morphology in the Netherlands. *J. Geophys. Res. Atmos.* **2011**, *116*, D20129. [CrossRef]
8. Mohan, M.; Kikegawa, Y.; Gurjar, B.R.; Bhati, S.; Kolli, N.R. Assessment of Urban Heat Island Effect for Different Land Use–Land Cover from Micrometeorological Measurements and Remote Sensing Data for Megacity Delhi. *Theor. Appl. Clim.* **2013**, *112*, 647–658. [CrossRef]
9. Ghous, M.; Khalida, K.; Basit, A.; Ougahi, J. Temporal analysis of urbanization and resulting local weather change: A case study of Lahore, Punjab, Pakistan. *Sci. Int.* **2015**, *27*, 1281–1287.

10. Gherraz, H.; Guechi, I.; Alkama, D. Quantifying the Effects of Spatial Patterns of Green Spaces on Urban Climate and Urban Heat Island in a Semi-Arid Climate. *Bull. Soc. R. Sci. Liege* **2020**, *89*, 164–185. [CrossRef]
11. Founda, D.; Santamouris, M. Synergies between Urban Heat Island and Heat Waves in Athens (Greece), during an Extremely Hot Summer (2012). *Sci. Rep.* **2017**, *7*, 10973. [CrossRef]
12. Zhao, L.; Oppenheimer, M.; Zhu, Q.; Baldwin, J.W.; Ebi, K.L.; Bou-Zeid, E.; Guan, K.; Liu, X. Interactions between Urban Heat Islands and Heat Waves. *Environ. Res. Lett.* **2018**, *13*, 034003. [CrossRef]
13. Bounoua, L.; Thome, K.; Nigro, J. Cities Exacerbate Climate Warming. *Urban Sci.* **2021**, *5*, 27. [CrossRef]
14. Koomen, E.; Diogo, V. Assessing Potential Future Urban Heat Island Patterns Following Climate Scenarios, Socio-Economic Developments and Spatial Planning Strategies. *Mitig. Adapt. Strateg. Glob. Change* **2017**, *22*, 287–306. [CrossRef]
15. Ahmed, S. Assessment of Urban Heat Islands and Impact of Climate Change on Socioeconomic over Suez Governorate Using Remote Sensing and GIS Techniques. *Egypt. J. Remote Sens. Space Sci.* **2018**, *21*, 15–25. [CrossRef]
16. Hong, J.-W.; Hong, J.; Kwon, E.E.; Yoon, D.K. Temporal Dynamics of Urban Heat Island Correlated with the Socio-Economic Development over the Past Half-Century in Seoul, Korea. *Environ. Pollut.* **2019**, *254*, 112934. [CrossRef]
17. Oke, T.R. The Energetic Basis of the Urban Heat Island. *Quart. J. R. Meteorol. Soc.* **1982**, *108*, 1–24. [CrossRef]
18. Roth, M. Review of Urban Climate Research in (Sub)Tropical Regions. *Int. J. Climatol.* **2007**, *27*, 1859–1873. [CrossRef]
19. Rajagopalan, P.; Lim, K.C.; Jamei, E. Urban Heat Island and Wind Flow Characteristics of a Tropical City. *Sol. Energy* **2014**, *107*, 159–170. [CrossRef]
20. Tiafack, O.; Prisca, N.; Dian, T. Understanding Urban Growth through Heat Islands Using Remotely Sensed Data: Yaounde Case Study, Cameroon. *Curr. Urban Stud.* **2022**, *10*, 163–187. [CrossRef]
21. Simwanda, M.; Ranagalage, M.; Estoque, R.C.; Murayama, Y. Spatial Analysis of Surface Urban Heat Islands in Four Rapidly Growing African Cities. *Remote Sens.* **2019**, *11*, 1645. [CrossRef]
22. Isioye, O.A.; Ikwueze, H.U.; Akomolafe, E.A. Urban Heat Island Effects and Thermal Comfort in Abuja Municipal Area Council of Nigeria. *FUTY J. Environ.* **2020**, *14*, 19–34.
23. Haut Commissariat au Plan (HCP). *Indicateurs Recensement général RGPH, Taux d'urbanisation*; Site institutionnel du Haut-Commissariat au Plan du Royaume du Maroc: Rabat, Morocco, 2024. Available online: https://www.hcp.ma/Taux-d-urbanisation_a3243.html (accessed on 23 December 2024).
24. Bounoua, L.; Nigro, J.; Zhang, P.; Thome, K.; Lachir, A. Mapping Urbanization in the United States from 2001 to 2011. *Appl. Geogr.* **2018**, *90*, 123–133. [CrossRef]
25. Tilton, J.C.; De Colstoun, E.B.; Wolfe, R.E.; Tan, B.; Huang, C. Generating Ground Reference Data for a Global Impervious Surface Survey. In Proceedings of the 2012 IEEE International Geoscience and Remote Sensing Symposium, Munich, Germany, 22–27 July 2012; IEEE: Munich, Germany, 2012; pp. 5993–5996. [CrossRef]
26. Friedl, M.; Sulla-Menashe, D. MODIS/Terra+Aqua Land Cover Type Yearly L3 Global 500 m SIN Grid. SIN Grid V061. 2022. Available online: https://lpdaac.usgs.gov/products/mcd12q1v061/ (accessed on 28 August 2024). [CrossRef]
27. Sellers, P.J.; Randall, D.A.; Collatz, G.J.; Berry, J.A.; Field, C.B.; Dazlich, D.A.; Zhang, C.; Collelo, G.D.; Bounoua, L. A Revised Land Surface Parameterization (SiB2) for Atmospheric GCMS. Part I: Model Formulation. *J. Clim.* **1996**, *9*, 676–705. [CrossRef]
28. Bounoua, L.; Safia, A.; Masek, J.; Peters-Lidard, C.; Imhoff, M.L. Impact of Urban Growth on Surface Climate: A Case Study in Oran, Algeria. *J. Clim.* **2009**, *48*, 217–231. [CrossRef]
29. Collatz, G.J.; Ribas-Carbo, M.; Berry, J.A. Coupled Photosynthesis-Stomatal Conductance Model for Leaves of C4 Plants. *Funct. Plant Biol.* **1992**, *19*, 519–538. [CrossRef]
30. Zhang, C.; Dazlich, D.A.; Randall, D.A.; Sellers, P.J.; Denning, A.S. Calculation of the Global Land Surface Energy, Water and CO_2 Fluxes with an off-line Version of SiB2. *J. Geophys. Res. Atmos.* **1996**, *101*, 19061–19075. [CrossRef]
31. Denning, A.S.; Collatz, G.J.; Zhang, C.; Randall, D.A.; Berry, J.A.; Sellers, P.J.; Colell0, G.D.; Dazlich, D.A. Simulations of Terrestrial Carbon Metabolism and Atmospheric CO_2 in a General Circulation Model, Part 1: Surface Carbon Fluxes. *Tellus B Chem. Phys. Meteorol.* **1996**, *48*, 521–542. [CrossRef]
32. Denning, A.S.; Randall, D.A.; Collatz, G.J.; Sellers, P.J. Simulations of Terrestrial Carbon Metabolism and Atmospheric CO_2 in a General Circulation Model, Part 2: Simulated CO_2 Concentrations. *Tellus B Chem. Phys. Meteorol.* **1996**, *48*, 543–567. [CrossRef]
33. Al-Hamdan, M.Z.; Quattrochi, D.A.; Bounoua, L.; Lachir, A.; Zhang, P. Using Landsat, MODIS, and a Biophysical Model to Evaluate LST in Urban Centers. *Remote Sens.* **2016**, *8*, 952. [CrossRef]
34. Gelaro, R.; McCarty, W.; Suárez, M.J.; Todling, R.; Molod, A.; Takacs, L.; Randles, C.A.; Darmenov, A.; Bosilovich, M.G.; Reichle, R.; et al. The Modern-Era Retrospective Analysis for Research and Applications, Version 2 (MERRA-2). *J. Clim.* **2017**, *30*, 5419–5454. [CrossRef] [PubMed]
35. Parastatidis, D.; Mitraka, Z.; Chrysoulakis, N.; Abrams, M. Online Global Land Surface Temperature Estimation from Landsat. *Remote Sens.* **2017**, *9*, 1208. [CrossRef]
36. Fathi, N.; Bounoua, L.; Messouli, M. A Satellite Assessment of the Urban Heat Island in Morocco. *Can. J. Remote Sens.* **2019**, *45*, 26–41. [CrossRef]

37. Haut Commissariat au Plan (HCP). *Population Legale du Royaume du Maroc Répartie par Régions, Provinces et Préfectures et Communes Selon les Résultats du Recensement Général de la Population et de L'habitat 2024*; Site institutionnel du Haut-Commissariat au Plan du Royaume du Maroc: Rabat, Marocco. Available online: https://www.hcp.ma/Population-legale-du-Royaume-du-Maroc-repartie-par-regions-provinces-et-prefectures-et-communes-selon-les-resultats-du_a3974.html (accessed on 1 December 2024).
38. El Ghazouani, L.; Bounoua, L.; Nigro, J.; Mansour, M.; Radoine, H.; Souidi, H. Combining Satellite Data and Spatial Analysis to Assess the UHI Amplitude and Structure within Urban Areas: The Case of Moroccan Cities. *Urban Sci.* **2021**, *5*, 67. [CrossRef]
39. Azmi, R.; Saadane, A.; Kacimi, I. Estimation of Spatial Distribution and Temporal Variability of Land Surface Temperature over Casablanca and the Surroundings of the City Using Different Landat Satellite Sensor Type (TM, ETM+ and OLI). *Int. J. Innov. Appl. Stud.* **2015**, *11*, 49–57.
40. Bahi, H.; Rhinane, H.; Bensalmia, A.; Fehrenbach, U.; Scherer, D. Effects of Urbanization and Seasonal Cycle on the Surface Urban Heat Island Patterns in the Coastal Growing Cities: A Case Study of Casablanca, Morocco. *Remote Sens.* **2016**, *8*, 829. [CrossRef]
41. Centre Royal de Télédétection Spatiale (CRTS). *Cartographie des Cultures Irriguees et Suivi de leur Evolution a Partir des Images Satellite*; Centre Royal de Télédétection Spatiale: Rabat, Morocco, 2014. Available online: https://crts.gov.ma/files/Rapport_final_cartograhie_irrigation_nationale.pdf (accessed on 25 December 2024).
42. Gourfi, A.; Taïbi, A.N.; Salhi, S.; Hannani, M.E.; Boujrouf, S. The Surface Urban Heat Island and Key Mitigation Factors in Arid Climate Cities, Case of Marrakesh, Morocco. *Remote Sens.* **2022**, *14*, 3935. [CrossRef]
43. Shepherd, J.M. Evidence of Urban-Induced Precipitation Variability in Arid Climate Regimes. *J. Arid. Environ.* **2006**, *67*, 607–628. [CrossRef]

Disclaimer/Publisher's Note: The statements, opinions and data contained in all publications are solely those of the individual author(s) and contributor(s) and not of MDPI and/or the editor(s). MDPI and/or the editor(s) disclaim responsibility for any injury to people or property resulting from any ideas, methods, instructions or products referred to in the content.

land

MDPI

Article

Redefining Urban Boundaries for Health Planning Through an Equity Lens: A Socio-Demographic Spatial Analysis Model in the City of Rome

Elena Mazzalai [1], Susanna Caminada [1], Lorenzo Paglione [2,*] and Livia Maria Salvatori [1]

[1] Health District III, ASL Roma 1, 00193 Rome, Italy; elena.mazzalai@aslroma1.it (E.M.); livia.salvatori@aslroma1.it (L.M.S.)
[2] Department of Prevention, ASL Roma 1, 00193 Rome, Italy
* Correspondence: lorenzo.paglione@aslroma1.it

Abstract: Urban health planning requires a multi-scalar understanding of the territory, capable of capturing socio-economic inequalities and health needs at the local level. In the case of Rome, current administrative subdivisions—Urban Zones (*Zone Urbanistiche*)—are too large and internally heterogeneous to serve as effective units for equitable health planning. This study presents a methodology for the territorial redefinition of Rome's Municipality III, aimed at supporting healthcare planning through an integrated analysis of census sections. These were grouped using a combination of census-based socio-demographic indicators (educational attainment, employment status, single-person households) and real estate values (OMI data), alongside administrative and road network data. The resulting territorial units—21 newly defined Mesoareas—are smaller than Urban Zones but larger than individual census sections and correspond to socio-territorially homogeneous neighborhoods; this structure enables a more nuanced spatial understanding of health-related inequalities. The proposed model is replicable, adaptable to other urban contexts, and offers a solid analytical basis for more equitable and targeted health planning, as well as for broader urban policy interventions aimed at promoting spatial justice.

Keywords: urban health planning; spatial equity; health inequalities; spatial justice; urban governance; health service planning

1. Introduction

Social inequalities in health constitute a quantifiable issue with a profound impact on population health. Traditionally, such inequalities have been examined—within the framework of social epidemiology—as disparities in exposure, and thus in disease incidence, as well as in outcomes [1]. A substantial body of the literature addresses these phenomena. Concerning inequalities in outcomes, attention has focused particularly on factors related to healthcare facility accessibility [2], as well as on the pathways of diagnostic and clinical appropriateness—elements investigated under the auspices of evaluative epidemiology [3]. With respect to inequalities in exposure, since the initial definition and systematization of the social determinants of health [4,5], research has concentrated on individual-level variables such as educational attainment [6] and employment status [7], as well as on collective exposure metrics, including pollution [8] and other forms of environmental stressors [9].

There is growing interest in the literature in studying collective variables as well; whereas individual factors generally concern the characteristics of the person (the so-called "who you are" [10]), additional dimensions are now emerging forcefully in the scientific debate, within conceptual frameworks such as Urban Health and One Health [11,12]. These area-level variables pertain to the "where you live" and include material anthropic factors—such as the quality of the built environment, both indoor [13] and outdoor [14]; the interaction between the built environment and its surroundings, for example with respect to climate change–related phenomena [15]—as well as immaterial aspects linked, for instance, to walkability [16], characteristics of public space [17], and the presence of facilities in terms of services and functions [2].

In this sense, at the intersection of these two dimensions—particularly in urban contexts—the housing market plays a central role as a filter, mediating individuals' economic ability to choose *where to live* [18,19]. Despite its relevance, this factor remains largely underexplored in the biomedical literature as an indicator of social inequality.

Social inequalities in health therefore affect disease occurrence and pose a significant challenge for health service planning. In Italy, following the reform enacted by Law 833/1978 [20], a process of managerialization was introduced with the establishment of Local Health Units (LHUs, from *Azienda Sanitaria Locale* in Italian) and a new organizational structure divided into Hospital, District, and Prevention levels [21]. Although the debate on issues related to managerialization falls outside the scope of this study, it is important to highlight that, to date, the existence of the so-called "territory"—namely the Health Districts and the Department of Prevention—has become central to assessing the health needs of the reference population. In this regard, these two structures now require—especially in light of the recent additional reform introduced by Ministerial Decree 77/2022 (DM 77/2022) [22], which endows LHUs with competencies in "epidemiological intelligence"—to understand, study, and act based on needs-driven planning rather than demand-driven approaches, thereby operating proactively to promote health equity by countering social inequalities in health [23].

The Italian Ministerial Decree 77/2022 further elaborates on the concept of so-called "catchment areas", which are intended as planning tools for the structuring of healthcare services financed through the Next Generation EU framework, rather than instruments for assessing existing accessibility conditions. In this context, catchment areas are defined primarily based on population thresholds, thus delineating residential territories according to the minimum set of services to be provided to a given population.

Within the broader reorganization of territorial healthcare outlined by DM 77/2022, the catchment areas established for general practitioners (*Medici di Medicina Generale*, approximately 1500 residents per physician) and family counseling centers (*Consultori Familiari*, up to 20,000 residents) represent the elementary building blocks of the system. These thresholds form the foundational units upon which broader and more complex organizational levels—such as the Case della Comunità (Community Health Centers), serving catchment areas of approximately 40,000 inhabitants—are structured. Ensuring coherence across these different levels requires a territorial reading capable of aggregating urban areas that are homogeneous in socio-demographic and functional terms. This is essential to support a more targeted and equitable healthcare planning approach. Complex urban contexts thus pose a significant challenge for LHUs, particularly given the difficulties in studying and delineating territories due to existing administrative overlaps in their very definition [24]. The context of the Municipality of Rome (hereafter, Rome) represents an additional layer of complexity, both because of its extensive area (over 1200 km^2) and the social and urban stratification present across its neighborhoods [25]. Although targeted health-promotion interventions—leveraging urban studies that integrate GIS tools, surveys, administrative data, and qualitative territorial knowledge—have yielded promising methodological advancements [26], substantial work remains to harmonize the frameworks for identifying, studying, and intervening within the territory, especially in the context of Rome.

Rome indeed exhibits numerous territorial, administrative, historical, and statistical subdivisions: the city is partitioned into 15 Municipalities (*Municipi* in Italian), with populations ranging from 128,048 in *Municipio VIII* (47.1 km^2) to 313,164 in *Municipio VII* (47.6 km^2) [27]. An additional administrative layer comprises the 155 Urban Zones (UZs) [28], which are essential for a more nuanced understanding of the city's urban fabric [29]. This schema, devised by the Statistical Office of the Municipality of Rome Capital, was introduced in 1977 for purposes of statistical analysis, urban planning, and territorial management. The delineation of these zones is grounded in criteria of urban homogeneity and follows the principal discontinuities within the built environment. Further historical subdivisions include the so-called "*Rioni*" in the Historic Centre [30] and the "*Quartieri*" in the Historic Periphery [31].

The UZs of the Municipality of Rome Capital are, in turn, uniquely divisible into census sections, which correspond approximately to individual blocks in the consolidated urban fabric and to settlement clusters in the dispersed Urban–Rural Fringe/Rurban Landscape (or *città-campagna* in Italian) fabric [32]. These census sections are likewise delineated based on the principal discontinuities of the urban structure. They serve as the geographic unit through which the Italian National Institute of Statistics (ISTAT) disseminates data from the Permanent Population and Housing Census, and since the 2011 Census, Rome has comprised 13,506 such sections [24].

Recently, thanks to a renewed drive by the Municipality of Rome to implement the so-called 15-Minute City [33], the subdivision of UZ has been called into question, as it now functions more as a historical partition that, in certain contexts marked by social and urban dynamism, poorly mirrors the true differentiation between neighborhoods.

In this regard, it is worth noting that a more detailed territorial subdivision than the one currently in use enables a greater capacity to analyze the socio-economic and urban fabric in terms of inequalities. Assessments in this direction have highlighted virtuous examples, such as the so-called "Microareas of Trieste" [34], although such an approach remains difficult to implement in a complex and metropolitan urban context like that of Rome.

The debate remains ongoing, but the Administration's objective is to equip itself with an effective tool for territorial governance grounded in a deep understanding of the area.

The III Municipality of Roma Capitale

The present study thus aligns with this research strand, enabled by the "epidemiological intelligence" initiative developed by LHU Roma 1 under the mandate of Ministerial Decree 77/2022. Accordingly, the research was conducted in a pilot district of LHU Roma 1—Municipio III Montesacro (hereafter, III Municipality)—which nonetheless effectively encapsulates the complexity of Rome's broader territory. Covering nearly 100 km^2, Rome's III Municipality is divided into 13 Urban Zones and has a resident population of 203,396 (2023 data) (Figure 1) [27]. It occupies the north-western sector of Rome, bounded by the banks of the River Aniene and the municipalities of Guidonia, Mentana, and Monterotondo. Its characteristic wedge-shaped form—common to many of Rome's Municipalities—reflects a spectrum of urban and natural environments, as follows:

- Central consolidated fabric: Densely built and populated neighbourhoods typical of the historic inner-ring periphery (Conca D'Oro, Montesacro, Tufello) [32].
- Mid-density "*palazzine*" sectors: Residential blocks, often grouped into complexes (Talenti, Nuovo Salario) [32].
- Post-war public housing estates: Large residential developments from the 1950s to 1960s (Vigne Nuove, Serpentara, Colle Salario) [32].

- Recent expansion zones: Neighbourhoods developed since the 1990s (Casale Nei, Bufalotta).
- Peripheral *"borgate"* and new gated communities: Formerly informal settlements outside the *Grande Raccordo Anulare*, the motorway ring that surrounds the inner city of Rome [35], now subject to major urban interventions (Castel Giubileo, Fidene, Settebagni, Cinquina) [32].

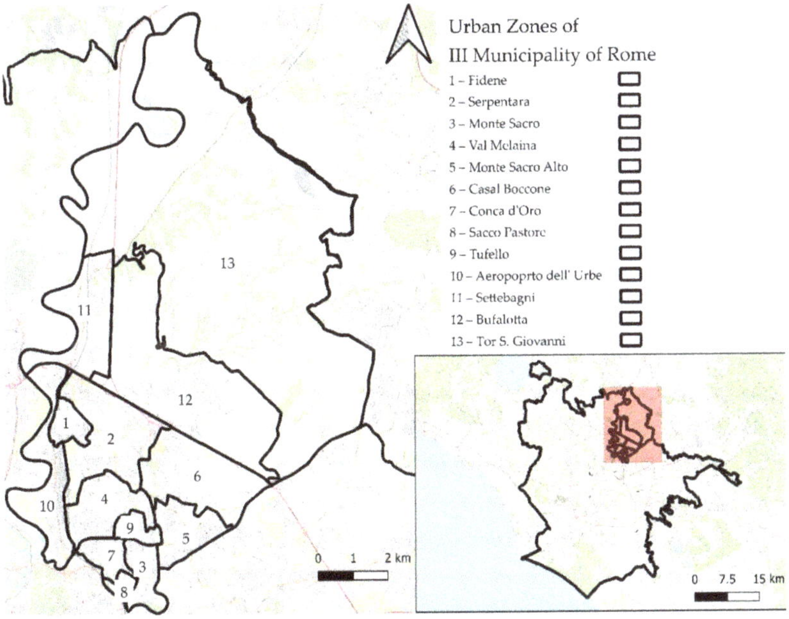

Figure 1. Urban zones of III Municipality of Rome. The red box in the overview map corresponds to the geographical extent displayed in the main map.

Except along the Via Salaria corridor—where the mixed industrial and residential settlement of Settebagni lies—the outermost areas of III Municipality encompass a vast green reserve, featuring scattered hamlets and largely unspoiled tracts of the classic *"Campagna Romana"* landscape within the Marcigliana Reserve [36].

The III Municipality experienced several distinct phases of urban development during the 20th century, beginning with the original settlement of Montesacro/Garden City, situated adjacent to the site of the plebeian secession in Republican Rome (494 BC) and later the location of Simón Bolívar's oath (1805) [37,38] before he embarked on his South American political career. Subsequent growth saw the establishment of the "historic borgata" at Tufello, followed by the construction of the densely built Conca D'Oro neighborhood during the post-war building boom [32]. Early public housing projects and later urbanization initiatives [32], as well as further developments along the Motorway Ring (*Grande Raccordo Anulare—GRA*, in Italian), largely mirrored broader patterns across Rome—alternating between planned, state-led interventions and speculative construction—underscoring the considerable social and spatial complexity of the territory itself [39,40].

The study and classification of areas—aimed at responding to the LHU's strategic planning requirements—therefore constitute a primary necessity to ensure that service levels are appropriately aligned with needs, within an equity framework [41].

The aim of this study was to develop a subdivision of the territory of Rome Capital's Municipality III, starting from the existing Urban Zones, by identifying areas (Mesoareas) with a population ranging between 1500 and 20,000 inhabitants, that are as homogeneous as possible in terms of socioeconomic status.

2. Materials and Methods

The software tools used for this study were QGIS Desktop 3.16.3 (QGIS Project, open-source, International), Microsoft Excel 2016 (Microsoft Corporation, Redmond, Washington, USA) (step 1–5) and STATA MP 13, (StataCorp LLC, 4905 Lakeway Drive, College Station, 322 Texas, USA (step 6). The following layers were imported into the QGIS project (Table 1):

- A vector layer containing the 2021 census section boundaries of the Municipalities of Rome (geographical data), provided by ISTAT in the WGS 84/UTM Zone 32N projection, available both as point and polygon geometries. This layer was joined with the database containing data from the 2021 ISTAT Census, the most recent census available.
- A vector layer of the administrative subdivisions (Municipalities), created and made available by IPTSAT S.r.l. through the DatiOpen.it platform [42].
- A vector layer representing the Urban Zones of the Municipalities of Rome, in the WGS 84 geographic coordinate system [43].
- A vector layer delineating the boundaries of the *Osservatorio del Mercato Immobiliare* (OMI—Real Estate Market Observatory) zones for the year 2021, provided by the Italian *Agenzia delle Entrate*. The OMI zones are defined as territorially homogeneous areas for which annual real estate valuations are available. These valuations provide, for each OMI zone, a minimum and maximum price range per square meter in euros, for both market value and rental value, differentiated by property type [44]. For this study, the average sale price of residential properties intended for civilian use was calculated and assigned to each OMI zone in the corresponding vector layer. Figure 5 shows the OMI zones covering the area of Municipality III, along with the corresponding average sale prices of residential properties.
- Linear vector road network provided by Anas S.p.A., the national agency responsible for managing Italy's road infrastructure [45].

Table 1. Overview of the main data sources used for the delineation of Mesoareas and the analysis of socioeconomic indicators within the Urban Zones of Municipality III. For each dataset, the type of data and the corresponding source are reported.

Dataset Description	Year	Source
Vector layer of 2021 census sections of the Municipality of Rome, including both point and polygon geometries.	2021	ISTAT (Istituto Nazionale di Statistica)
Database containing demographic and socioeconomic data from the 2021 ISTAT Census, joined to the spatial layer of census sections.	2021	ISTAT (Istituto Nazionale di Statistica)
Vector layer of administrative subdivisions (Municipalities).	Accessed May 2025	IPTSAT s.r.l., via DatiOpen.it platform
Vector layer of Urban Zones (Zone Urbanistiche) of the Municipality of Rome in WGS 84 coordinate system.	Accessed May 2025	Roma Capitale/Geoportale
Vector layer of OMI (Osservatorio del Mercato Immobiliare) Zones, with minimum and maximum property sale and rental prices per square meter. Used to calculate average residential sale prices.	2021	Agenzia delle Entrate—OMI
Linear vector layer of the national road network.	Accessed May 2025	ANAS S.p.A. (Azienda Nazionale Autonoma delle Strade)

Figure 2 illustrates the methodological process flowchart.

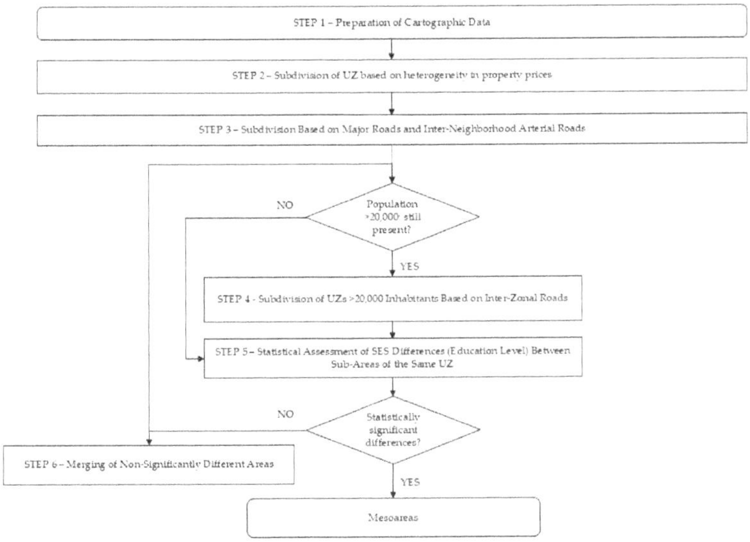

Figure 2. Methodological process for the subdivision of Urban Zones (UZs) and identification of Mesoareas. The flowchart illustrates the sequential steps based on property market heterogeneity, major road networks, population thresholds, and statistical comparisons to define meaningful intra-urban sub-areas.

2.1. Step 1—Preparation of Cartographic Data: Generation of a Vector Layer of Census Sections with Linked Attributes from the 2011 Census and Urban Zones

As a first step, to associate each census section (CS) with its corresponding Municipality, the intersection function was used, with the point vector layer of the census geographic data as the input layer, and the vector layer of the Municipalities of Rome as the output layer. Therefore, only the CSs belonging to III Municipality were selected and exported. Through a table join tool, it was also possible to isolate the CSs of III Municipality within the polygonal census geographical data layer.

Subsequently, to associate each CS from the previously generated layer with its corresponding UZ, the intersection function was again applied, this time with an arbitrary buffer of −5 m. The output layer in this case was the vector layer of Rome's Urban Zones.

From the resulting polygon vector layer of the CSs for III Municipality of Roma Capitale, now enriched with both 2021 census information and UZs identifiers, a series of operations was carried out with the goal of identifying UZs that had a very large population, or a high degree of heterogeneity in terms of average real estate sale prices. The subdivision of UZs was carried out using the polygon vector layer of census sections as the base spatial unit within QGIS. CSs were aggregated into new groupings to define sub-areas based on spatial criteria aimed at maximizing internal socio-demographic homogeneity and adhering to a population range between 1500 and 20,000 residents, in line with the territorial guidelines for the primary healthcare catchment areas outlined in Ministerial Decree 77 [22]. The choice of CSs as the base unit was supported by two factors: (1) each CS is entirely contained within a single UZ, avoiding overlap, and (2) in most urban contexts, CS boundaries align with road infrastructure. This spatial correspondence enabled the use of major road axes and arterial streets as effective criteria for delineating Mesoarea boundaries within each UZ, as described in Steps 3–4.

2.2. *Step 2—Subdivision of UZ Based on Heterogeneity in Property Prices (OMI Areas)*

The identification of Urban Zones intersected by multiple OMI areas was carried out through the application of the spatial intersection function between the point vector layer representing the centroids of the census sections and the polygon vector layer of the OMI areas. This procedure allowed us to determine which Urban Zones are affected by more than one OMI area, based on the spatial location of CS centroids falling within different OMI boundaries.

Census sections belonging to the same Urban Zone and intersecting different OMI areas were aggregated to define sub-areas, conditional on a combined population size greater than 1500 residents. This subdivision aimed to capture intra-ZU socioeconomic heterogeneity based on real estate values.

2.3. *Step 3—Subdivision Based on Major Roads and Inter-Neighborhood Arterial Roads*

Subsequently, the Urban Zones (or sub-areas resulting from Step 1) were further subdivided based on the main road axes. The 'General Urban Traffic Plan' (*Piano Generale del Traffico Urbano*—PGTU) of Roma Capitale, approved by City Council Resolution No. 70/2014, and specifically its annex 'Road Regulation and Functional Classification', classifies the city's roads into five categories: A (motorways), S (high-speed roads), IQ (inter-neighborhood roads), IZ (inter-zonal roads), and Q (neighborhood roads). Roads within Municipality III falling under categories A, S, and IQ were identified: the first two correspond to high-speed roads, while the third refers to roads connecting different neighborhoods and characterized by specific urban planning functions.

Using the ANAS (the Italian national authority on road administration, see above) road network graph, the aforementioned road axes were identified. Among these, those already delineating the boundaries of pre-existing urban zones or of the sub-areas derived from Step 1 were excluded. The remaining axes, which largely coincide with the boundaries of census tracts, were used to separate census tracts belonging to the same urban zone into distinct sub-areas.

2.4. *Step 4—Subdivision of UZs > 20,000 Inhabitants Based on Inter-Zonal Roads*

Step 4 aimed to further subdivide only those urban zones or sub-areas (identified in the previous steps) that still had a resident population exceeding 20,000 inhabitants. To this end, road axes classified as 'interzonal' (IZ) in the aforementioned General Urban Traffic Plan (PGTU) of Roma Capitale were considered. According to the document, interzonal roads represent the classification level immediately below 'inter-neighborhood' roads and are defined as axes connecting areas of the city that, although belonging to the same neighborhood, represent distinct urban areas.

Accordingly, interzonal road axes were identified within the areas resulting from the earlier subdivisions, in which the resident population remained above 20,000 inhabitants even after the adjustments made in Steps 2 and 3.

2.5. *Step 5—Statistical Assessment of SES Differences (Education Level) Between Sub-Areas of the Same UZ*

To assess whether socioeconomic differences exist among the zones subdivided based on real estate sale prices or population size, chi-square (χ^2) tests were conducted using data from the 2021 census. Specifically, the percentage of individuals with a university degree among the population aged over 25 years was compared across Mesoareas within the same UZs.

For UZs that were divided into more than two Mesoareas, Bonferroni correction was applied.

2.6. Step 6—Merging of Non-Significantly Different Areas

Sub-areas resulting from the previous steps that, at Step 5, did not exhibit statistically significant differences in educational level were merged, provided they were adjacent. Conversely, where significant differences were observed, the corresponding areas were classified as Mesoareas.

2.7. Step 4bs—Subdivision of UZs > 20,000 Inhabitants Based on Inter-Zonal Roads

Where the merging process described in the previous step resulted in areas with a population exceeding 20,000 inhabitants, the methodology detailed in Step 4—Subdivision of UZs > 20,000 Inhabitants Based on Inter-Zonal Roads—was reapplied.

2.8. Step 5b—Statistical Assessment of SES Differences (Education Level) Between Sub-Areas of the Same UZ

Subsequently, a Statistical Assessment of SES Differences (Education Level) Between Sub-Areas of the Same UZ, as described in Step 5, was conducted for the subdivisions resulting from Steps 6 and 4b. Where statistically significant differences were identified, the corresponding areas were classified as Mesoareas.

2.9. Step 6—Final Statistical Assessment

For each meso-area and Urban Zone, descriptive statistics for the education level indicator were computed, including the mean, median, interquartile range (IQR), and variance. The distributions of the proportion of university graduates within Mesoareas and Urban Zones were visualized using box-and-whisker plots, highlighting the median and IQR to summarize central tendency and variability.

Subsequently, the distribution of the aforementioned variables was assessed using the Shapiro–Wilk test. Differences in the percentage of university graduates among Mesoareas within each Urban Zone were examined using the Kruskal–Wallis test. In total, five separate tests were conducted (one for each Urban Zone subdivided into Mesoareas). Upon obtaining a significant Kruskal–Wallis result, pairwise comparisons between subgroups were performed using Dunn's test with Bonferroni correction to control for Type I error associated with multiple comparisons.

For each classification level (Urban Zones and Mesoareas), the average within-group variance was also calculated to explore the degree of internal heterogeneity within each grouping level.

Socioeconomic differences among the Mesoareas resulting from the previous steps were further tested using additional proxy indicators of socioeconomic status. Due to the unavailability of a validated composite deprivation index based on the 2021 Italian census data, the dimensions considered in the deprivation index by Rosano et al. [46] were analyzed separately. Specifically, using the chi-square test on 2021 census data, the following variables were examined within Mesoareas belonging to the same UZ:

- Percentage of individuals aged over 25 with a university degree;
- Percentage of employed individuals aged 15–65;
- Percentage of single-person households out of the total number of families.

Population density, included in the Rosano et al. [46] index, was not tested because ISTAT had not yet released the necessary data to calculate this dimension at the census tract level.

3. Results

3.1. Step 1—Preparation of Cartographic Data

From the first step, 791 CSs were associated with Municipality III. The total population in 2021 was 201,900 individuals. Seventy-three CSs from the ISTAT shapefile did not match the census information database. A visual inspection of their locations suggests these are uninhabited areas, such as squares, parks, or natural zones. It was assumed that these CSs have a population of 0 (Figure A1). Each census section was assigned to one of the 13 Urban Zones within Municipality III. Based on this association, the distribution of the 201,900 residents, as shown in Table 2 and Figure 3, was obtained.

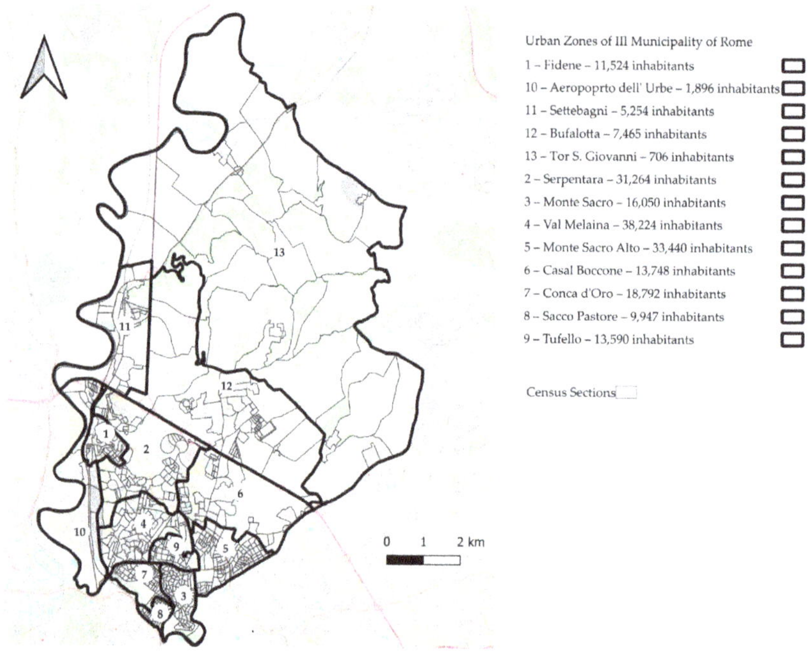

Figure 3. Urban Zones (UZs) of Municipality III subdivided by census sections (CSs). The association between CSs and UZs was established using the methods described in Step 1 of the methodological process.

Table 2. This table reports the description of the Urban Zones within III Municipality, including their area and 2021 ISTAT population, as derived from the procedures described above.

Urban Zone	**Area (km^2)**	**Inhabitants (ISTAT 2021)**
Aeroporto dell' Urbe	4.4	1896
Bufalotta	13.9	7465
Casal Boccone	5.7	13,748
Conca d'Oro	1.2	18,792
Fidene	1.1	11,524
Monte Sacro	1.7	16,050
Monte Sacro Alto	2.6	33,440
Sacco Pastore	0.5	9947

Table 2. *Cont.*

Urban Zone	Area (km²)	Inhabitants (ISTAT 2021)
Serpentara	5.9	31,264
Settebagni	4.8	5254
Tor S. Giovanni	51.9	706
Tufello	0.9	13,590
Val Melaina	3.2	38,224
III Municipality of Rome	97.9	201,900

The outcomes of the sub-area splitting and merging process, carried out from the initial Urban Zones to the final Mesoareas, as detailed in Steps 2 to 6, are summarized in Figure 4.

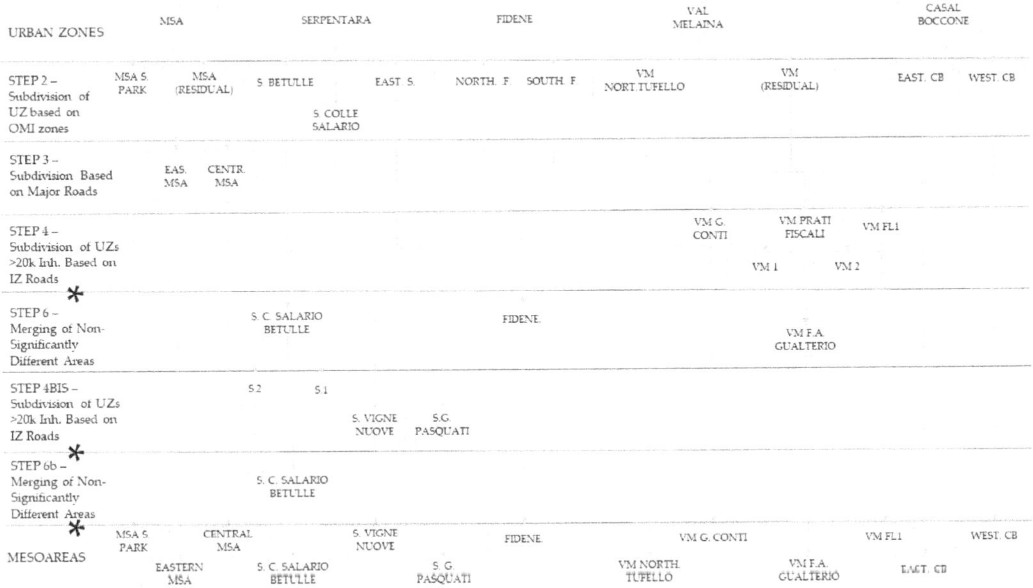

Figure 4. Results of the subdivision of Urban Zones (UZs) based on real estate market heterogeneity and road axes. Asterisks (*) indicate the steps at which statistical analyses were conducted to assess differences in the percentage of university graduates across Mesoareas within the same UZ. Abbreviations: S = Serpentara; VM = Val Melaina; CB = Casal Boccone; MSA = Monte Sacro Alto; F = Fidene. Tufello, Aeroporto dell'Urbe, Tor San Giovanni, Bufalotta, Monte Sacro, Settebagni and Conca d'Oro UZs were not subject to further subdivision.

3.2. Step 2—Subdivision of UZ Based on Heterogeneity in Property Prices

As illustrated in Table 3 and Figure 5, 12 Urban Zones (UZs) contain census sections that fall within different OMI zones. However, in only five UZs, these divisions can be used to identify sub-areas with populations exceeding 1500 inhabitants, in line with the objectives of this study: Casal Boccone, Fidene, Monte Sacro Alto, Val Melaina, and Serpentara, as detailed below.

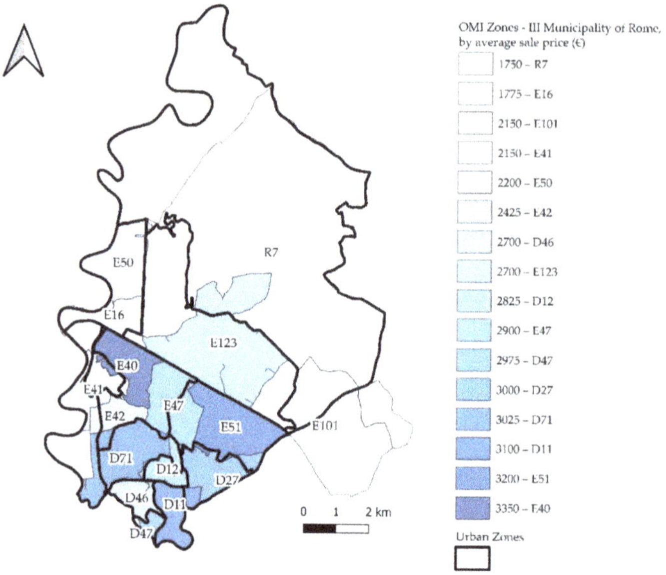

Figure 5. The map shows, in shades of blue, the OMI zones covering the area of Municipality III. The bold black boundaries represent the subdivision into Urban Zones.

Table 3. The table presents the OMI zones falling within each Urban Zone (UZ), identified through the intersection between the centroid layer of the 2021 census sections within each UZ and the polygon vector layer of the OMI zones, as described in Step 2. The table also reports the corresponding population figures based on the 2021 census data.

Urban Zones	OMI Zones	Inhabitants	Actions
Aeroporto dell'Urbe	D71	178	None
	E16	1589	None
	E41	129	None
	E42		None
Bufalotta	E101	5	None
	E123	7425	None
	R7	35	None
Casal Boccone	D27	536	Considered part of Casal Boccone
	E47	6192	Led to the definition of Western Casal Boccone
	E51	7020	Led to the definition of Eastern Casal Boccone
Conca d'Oro	D46	18,784	None
	D47	8	None
	D11	0	None
Fidene	E40	2288	Led to the definition of Northern Fidene
	E41	9236	Led to the definition of Southern Fidene
Monte Sacro	D11	15,462	None
	D46	588	None
	d47	0	None

Table 3. *Cont.*

Urban Zones	OMI Zones	Inhabitants	Actions
Monte Sacro Alto	D11	2673	Led to the definition of Monte Sacro Alto Sannazzaro Park
	D27	29,807	Led to the definition of Monte Sacro Alto (Residual Area)
	E51	960	Considered part of Monte Sacro Alto (Residual Area)
Sacco Pastore	D46	2	None
	D47	9945	None
	D11	0	None
Settebagni	E16	590	None
	E50	4664	None
Serpentara	E40	9681	Led to the definition of Serpentara Colle Salario
	E41	1351	Considered part of Serpentara Betulle
	E42	10,346	Led to the definition of Serpentara Betulle
	E47	9886	Led to the definition of Eastern Serpentara
	D71	0	Led to the definition of Serpentara Colle Salario
Tor S. Giovanni	E101	146	None
	E123	180	None
	R7	380	None
Val Melaina	D12 (eastern area)	2164	Led to the definition of Northern Tufello
	D12 (central-southern area)	1239	Considered part of Val Melaina (Residual Area)
	D71	34,821	Led to the definition of Residual Val Melaina
Tufello	D12	13,590	None
			None

The Fidene UZ, located adjacent to the GRA, in the western part of III Municipality, includes a north-eastern section where real estate prices are higher, suggesting different socio-economic characteristics compared to the rest of the UZ. For this reason, the Fidene Urban Zone was subdivided into two sub-areas: Fidene North and Fidene South (Figure 6).

The Serpentara UZ, which surrounds the Fidene UZ on its eastern border, displays internal heterogeneity in terms of average property sale prices. Broadly speaking, its territory can be divided into three areas characterized by distinct average sale prices, as shown in Figure 7. As indicated in Step 2 of the Methods Section, CSs were used to delineate three separate Mesoareas: Serpentara Colle Salario (the area with the highest average price), Eastern Serpentara, and Serpentara Betulle (the area with the lowest average price). As shown in Figure 7 and Table 3, near the borders of the UZ, there are areas that fall within different OMI zones. However, since these zones include populations of fewer than 1500 inhabitants, they were not classified as separate sub-areas and were instead assimilated into the surrounding areas. Specifically, the southernmost area falling within OMI zone D71 has no residents, while the central-western area located in OMI zone E41, with a population of 1351 inhabitants, was incorporated into the Serpentara Betulle sub-area.

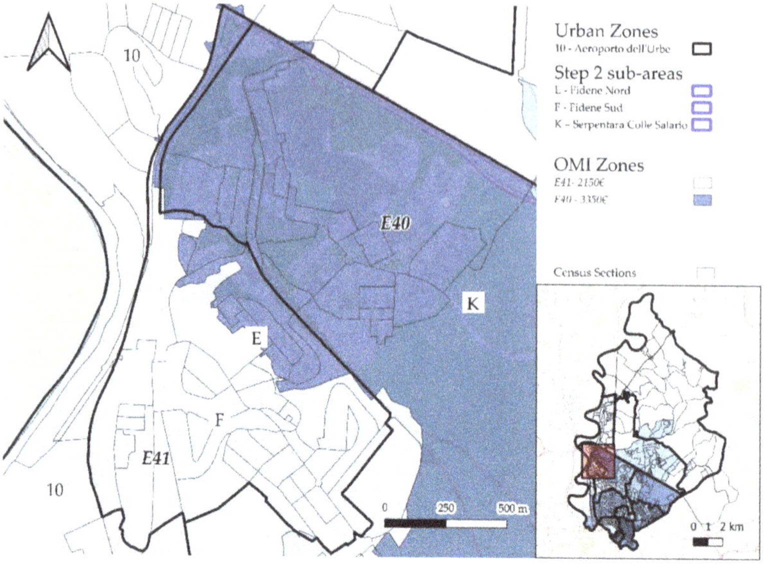

Figure 6. The image illustrates the Step 2 subdivision process of the Fidene (UZ). In the Urban Zone of Fidene, two OMI zones are present: E40, with higher average property sale prices, and E41. The census sections (CSs) whose centroids fall within OMI zone E40 were grouped to form the sub-area Fidene North (E), while those falling within OMI zone E41 constitute the sub-area Fidene South (F). The red box in the overview map corresponds to the geographical extent displayed in the main map.

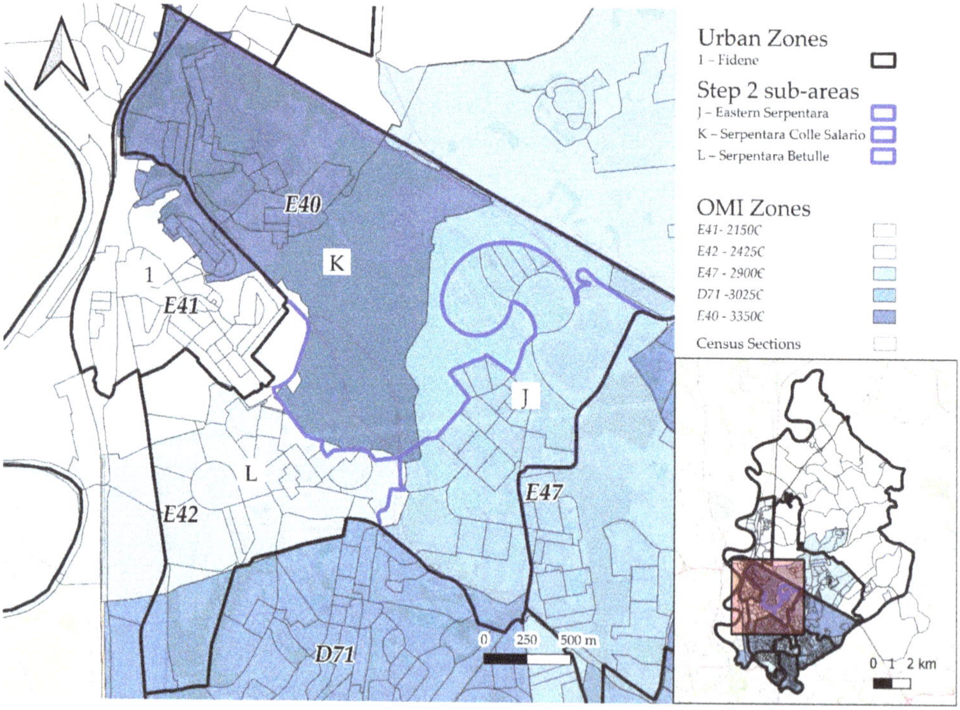

Figure 7. The image illustrates the Step 2 subdivision process of the Serpentara Urban Zone (UZ). Three OMI zones fall within this UZ: E40, with the highest average property sale price; E47, with an

intermediate value; and E42, with the lowest value. Census sections (CSs) whose centroids fall within OMI zone E40 were grouped into the sub-area Serpentara Colle Salario (K); those falling in E47 formed Serpentara Est (J); and those in E42 formed Serpentara Betulle (L). In the northern part of Serpentara, bordering the Fidene UZ (1), some CSs fall into OMI zone E41. These CSs, comprising 1351 inhabitants, were included in Serpentara Betulle. CSs located in the southernmost portion of the UZ, falling within OMI zone D71 and containing zero inhabitants, were also assigned to Serpentara Betulle. The red box in the overview map corresponds to the geographical extent displayed in the main map.

The Val Melaina UZ is largely homogeneous in terms of the average sale price, except for an eastern area that falls within an OMI zone characterized by a lower price. Therefore, CSs were used to separate this area from the rest of the UZ. This Mesoarea has thus been renamed "Northern Tufello" due to its urban continuity with the Tufello UZ, with which it shares a similar average real estate sale price (Figure 8). Similarly, in the area bordering the Tufello Urban Zone (UZ), some census sections fall within OMI zone D12, with a combined population of 1239 inhabitants; however, these sections were not considered for further subdivision (Figure 8 and Table 3).

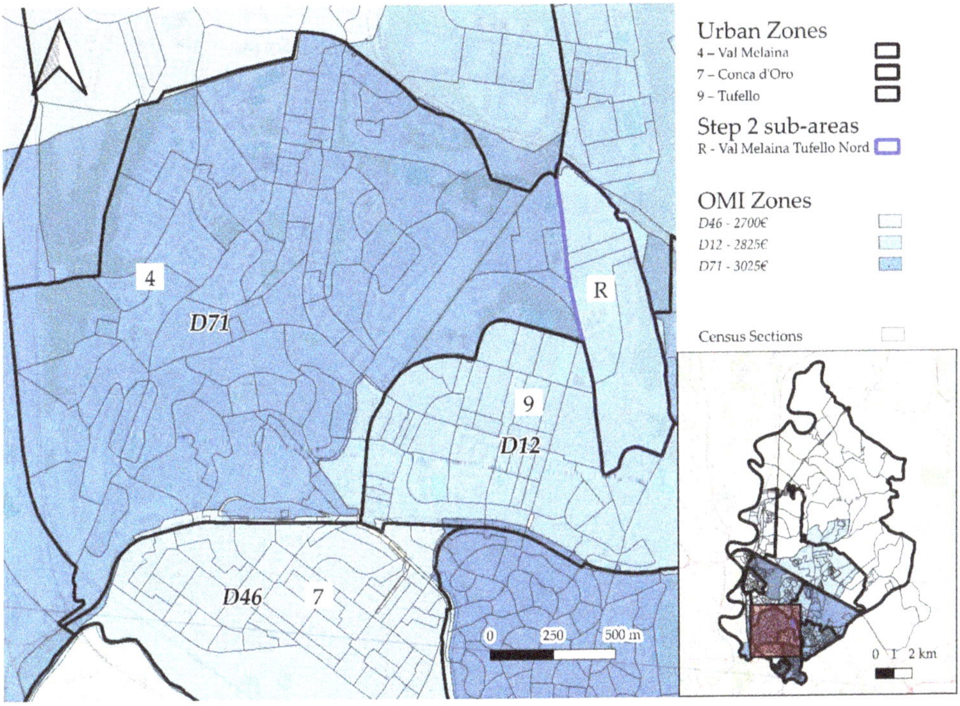

Figure 8. The image shows the subdivision process of the Val Melaina Urban Zone (UZ) based on Step 2. Two OMI zones intersect this UZ: D71, with a higher average residential property sale price, and D12. The census sections (CSs) whose centroids fall within OMI zone D12 were grouped into the sub-area Northern Tufello (E). Some CSs along the border with the adjacent Tufello UZ also fall within D12; as these CSs account for fewer than 1500 residents, they were included in the portion of Val Melaina that was not subdivided, further designated as the Val Melaina Residual Area. The red box in the overview map corresponds to the geographical extent displayed in the main map.

Most of the CSs within the Casal Boccone UZ fall into two different OMI areas: a smaller western portion with lower average prices, consistent with the adjacent Serpentara–Vigne Nuove area (E47), and a larger eastern portion with higher prices (E51). Consequently, the

Casal Boccone UZ was subdivided into two Mesoareas: Eastern Casal Boccone and Western Casal Boccone (Figure 9). Additionally, the Casal Boccone UZ includes a small southwestern area falling within a different OMI zone (D27); this area, with a population of 536 inhabitants, was incorporated into the Casal Boccone East sub-area (Figure 9 and Table 3).

The Monte Sacro Alto UZ includes a relatively small area in the southwest characterized by a slightly higher average sale price compared to the rest of the UZ. Census sections were used to isolate this area, which has been renamed 'Monte Sacro Alto Sannazzaro Park', due to the presence of a green space that occupies nearly half of its surface (Figure 10). Additionally, some small areas in the northern part, near the border with the Casal Boccone UZ, fall within an OMI zone with a different average sale price (E51); however, since the population in this portion is fewer than 1500 inhabitants (960), no separate Mesoarea was defined based on this criterion (Figure 10 and Table 3).

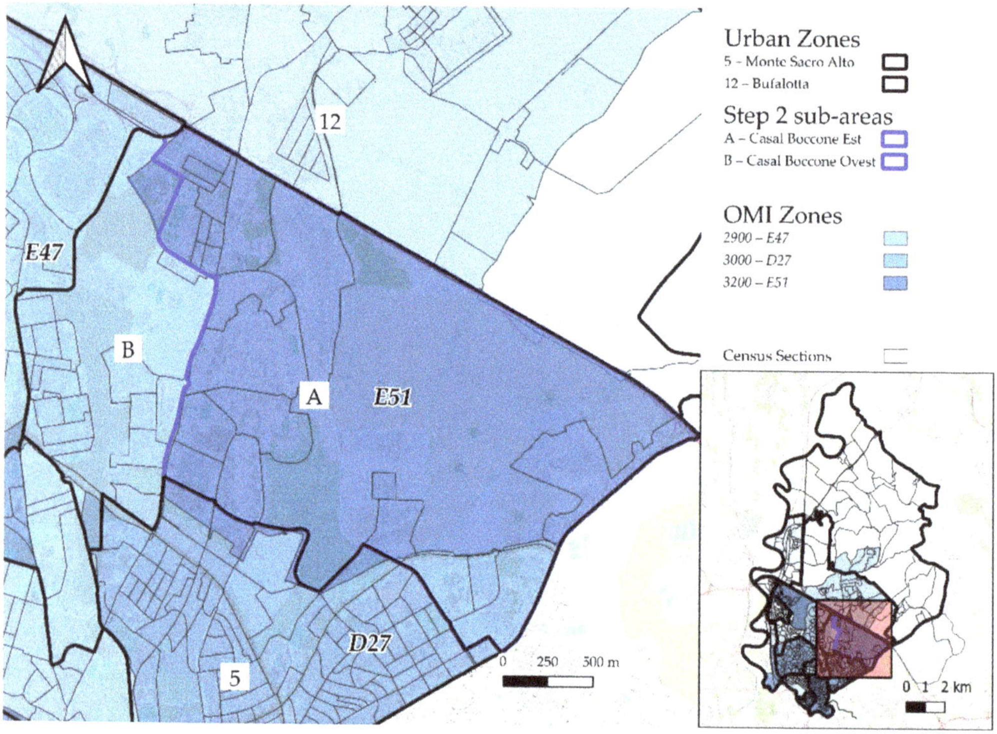

Figure 9. The figure shows the subdivision process of the Casal Boccone Urban Zone (UZ) based on Step 2. Two main OMI zones intersect this UZ: E51, with the highest average residential property sale price, and E47, with a lower price. The census sections (CSs) whose centroids fall within OMI zone E51 were grouped into the sub-area Casal Boccone East (A), while those falling within OMI zone E47 formed the sub-area Casal Boccone West (B). In the northern portion of Casal Boccone, at the border with Monte Sacro Alto UZ (5), some CSs fall within OMI zone D27; these CSs, with a total population of 536 inhabitants, were included in the Serpentara Betulle area. The southernmost CSs, falling within OMI zone D71 and containing no inhabitants, were also considered part of Casal Boccone East. The red box in the overview map corresponds to the geographical extent displayed in the main map.

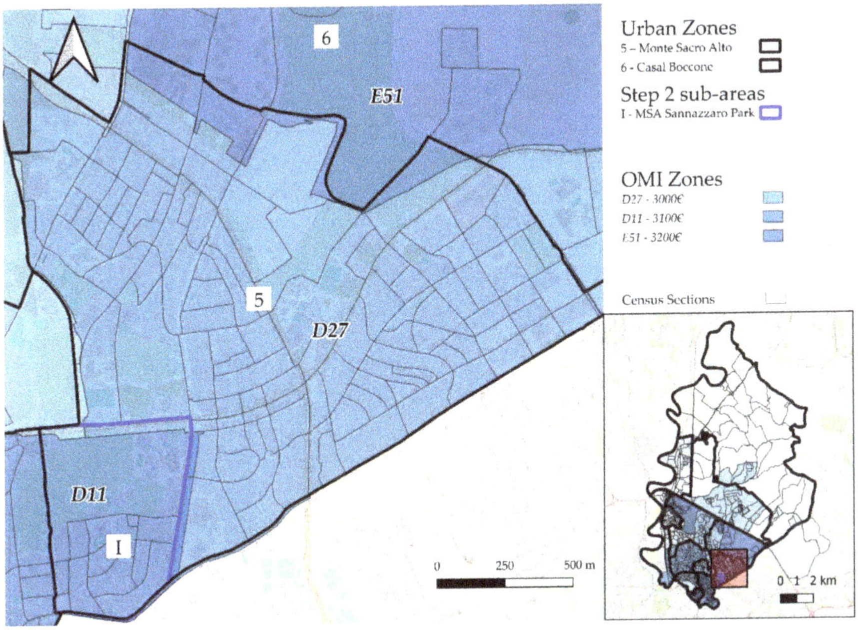

Figure 10. The image illustrates the subdivision process of the Monte Sacro Alto (MSA) Urban Zone (UZ) based on Step 2. Two main OMI zones intersect this UZ: D11, with the highest average residential property sale price, and D27, with the lowest. The census sections (CSs) whose centroids fall within OMI zone D11 were grouped into the sub-area MSA Sannazzaro Park (I). In the southern part of MSA, near the border with Casal Boccone UZ (6), some CSs fall within OMI zone E51. These CSs, with a combined population of 960 inhabitants, were included in the portion of MSA that was not further subdivided, designated as the MSA Residual Area. The red box in the overview map corresponds to the geographical extent displayed in the main map.

3.3. *Step 3—Subdivision Based on Major Roads and Inter-Neighborhood Arterial Roads*

Roads within Municipality III corresponding to the A (motorways), S (high-speed roads), and IQ (inter-neighborhood roads) classifications were identified on the ANAS road network graph. These include the following:

1. Via Fucini Renato;
2. Via Graf Antonio;
3. Viale Jonio;
4. Via Nomentana Nuova (Nomentana—Sempione);
5. Via dei Prati Fiscali;
6. Viadotto Gronchi;
7. Viadotto Sandro Pertini;
8. Viadotto Saragat;
9. Viadotto Segni.

Among these, three road axes already serve as boundaries between previously identified Urban Zones (UZs) or sub-areas, as follows:

1. The road axis formed by Viale dei Prati Fiscali and Viale Jonio separates the Urban Zones (UZs) of Conca d'Oro and Monte Sacro to the south from those of Val Melaina and Conca d'Oro. In its easternmost section, it delineates the boundary between the Monte Sacro Alto UZ and its sub-area 'Monte Sacro Alto Sannazzaro', which was defined during Step 2 (Figure A2).

2. The high-speed road axis composed of roads no. 6–9 constitutes the so-called 'Viadotto dei Presidenti' (Presidents' Viaduct) and separates the Fidene UZ from the Serpentara UZ to the northeast. As it continues, it crosses census sections (CSs) with low population density within the Serpentara UZ, which form the boundary between sub-areas identified in Step 1—specifically, between Serpentara Colle Salario and Serpentara Betulle along the Viadotto Giuseppe Saragat. Further along, it crosses the boundary CSs between the sub-area Serpentara Vigne Nuove and the Val Melaina UZ, in the section corresponding to Viadotto Antonio Segni. In its final section, named Viadotto Giovanni Gronchi, it traverses the boundary CSs between the Casal Boccone Est sub-area (outlined in Step 1) and the Monte Sacro Alto UZ (Figure A3).
3. Via Nomentana, located within the 'Nomentana–Sempione' section— the only portion of Via Nomentana within Municipality III classified as an inter-neighborhood road by the General Urban Traffic Plan of Roma Capitale— constitutes a short segment of the larger consular road, Via Nomentana. It runs through the Sacco Pastore Urban Zone (UZ), effectively dividing it into eastern and western sectors. The eastern sector, situated along the banks of the Aniene River, counts 859 inhabitants and was therefore not identified as a distinct sub-area (Figure A4).

The remaining road axis, Via Renato Fucini–Via Arturo Graf, was instead used to further subdivide the Monte Sacro Alto UZ. Consequently, two additional sub-areas were identified: Central Monte Sacro Alto Centro and Eastern Monte Sacro Alto. Overall, the Monte Sacro Alto UZ was divided into three sub-areas: Central Monte Sacro Alto (16,334 inhabitants), Eastern Monte Sacro Alto (14,433 inhabitants), and Monte Sacro Alto Sannazzaro Park (2673 inhabitants) (Figure 11).

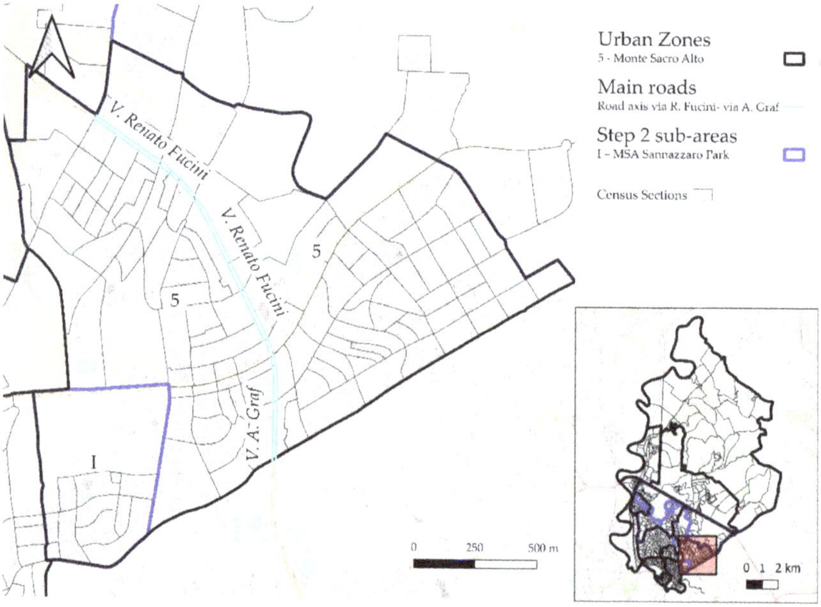

Figure 11. The image shows the road axis formed by Via Renato Fucini (to the north) and Via Arturo Graf (to the south). This road axis crosses the residual area of Monte Sacro Alto (MSA) that remained after the subdivision carried out in Step 2. It was used as a reference to further divide the census sections of MSA into two sub-areas: one located east of the axis, named MSA Est (14,433 inhabitants), and one to the west, between the road axis and MSA Sannazzaro Park, named MSA Centrale (16,334 inhabitants). The red box in the overview map corresponds to the geographical extent displayed in the main map.

The subdivision using the indicated methods resulted in 20 sub-areas. (Table 4).

Table 4. Sub-areas resulting from the subdivisions due to heterogeneity in average real estate sale prices or main road axis.

Mesoareas	Inhabitants (ISTAT 2021)
Aeroporto dell'Urbe	1896
Bufalotta	7465
Eastern Casal Boccone	7556
Western Casal Boccone	6192
Conca d'Oro	18,792
Northern Fidene	2288
Southern Fidene	9236
Monte Sacro	16,050
Central Monte Sacro Alto	16,334
Monte Sacro Alto Sannazzaro Park	2673
Eastern Montesacro Alto	14,433
Saccro Pastore	9947
Serpentara Vigne Nuove	9886
Serpentara Colle Salario	9681
Serpentara Betulle	11,697
Settebagni	5254
Northern Tor S. Giovanni	299
Southern S. Giovanni	407
Tufello	13,590
Northern Tufello	2164
Val Melaina (residual)	36,060
III Municipality	201,900

3.4. Step 4—Subdivision of UZs > 20,000 Inhabitants Based on Inter-Zonal Roads

Among the sub-areas delineated in the previous steps, the only sub-area with a population exceeding 20,000 inhabitants is Val Melaina, with 36,060 inhabitants (in Step 2, the Northern Tufello sub-area, with 2164 inhabitants, was separated from the Val Melaina UZ).

To identify the road axes along which further subdivisions of this area could be performed, roads classified as 'inter-zonal' by Roma Capitale were identified on the road network graph, as follows:

1. Via Cavriglia;
2. Via Vaglia;
3. Via Filippo Antonio Gualterio;
4. Piazza Minucciano;
5. Via Molazzana;
6. Via Seggiano;
7. Via Comano;
8. Piazza della Filattiera;
9. Via Pian di Sco;
10. Via Monte Cervialto.

These roads enable the subdivision of the Val Melaina area along four axes (Figure 12), as follows:

1. The first, located furthest north, follows Piazza della Filattiera, Via Pian di Sco, and Via Comano.
2. The second runs along Piazza Minucciano, Via Cavriglia, Via Molazzano, Via Seggiano, and Via Vaglia.
3. The third follows Via Filippo Antonio Gualterio.
4. The fourth follows Via Monte Cervialto.

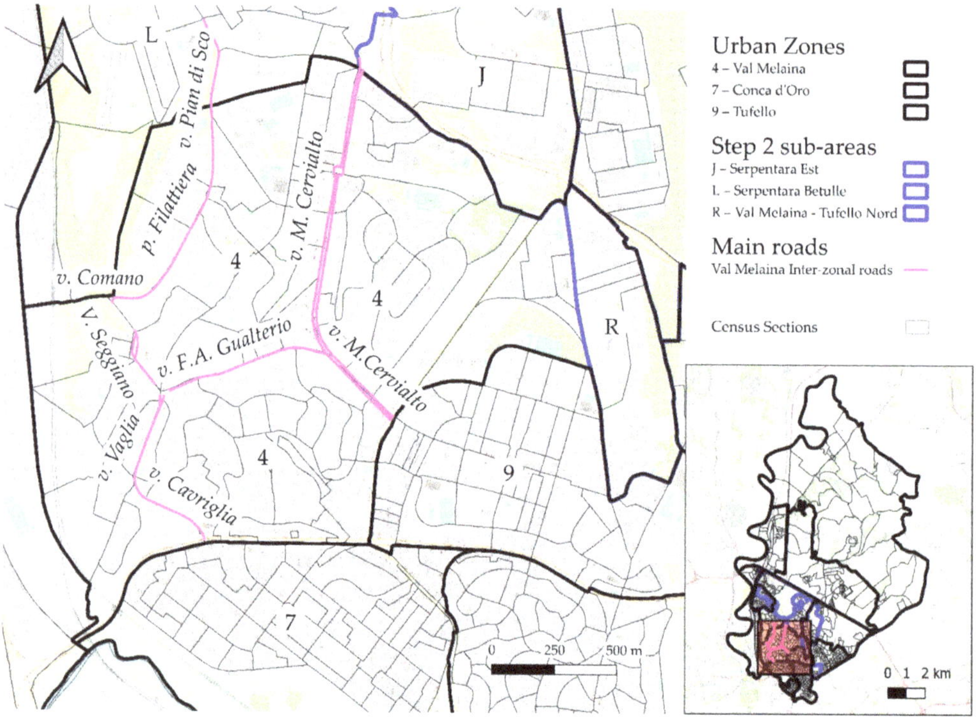

Figure 12. The image illustrates the inter-zonal roads within the Urban Zone (UZ) of Val Melaina. These road axes were used to delineate five additional sub-areas within Val Melaina: one to the west of the Seggiano–Vaglia–Cavriglia axis, named 'Val Melaina FL1'; one to the northwest of the Comano–Filattiera–Pian di Sco axis, named 'Val Melaina 1'; one southeast of this axis and bounded by Via Seggiano, Via Gualterio, Via Monte Cervialto, and the border with the sub-areas of Serpentara (L, J), named 'Val Melaina 2'; one east of Via Monte Cervialto, named 'Val Melaina Giovanni Conti'; and one bordered to the northeast by Via Cavriglia, Via Vaglia, Via Gualterio, and Via Monte Cervialto, and to the southwest by the boundaries with Conca d'Oro (7) and Tufello (9), named 'Val Melaina Prati Fiscali'. The red box in the overview map corresponds to the geographical extent displayed in the main map.

Through this step, the Val Melaina Urban Zone (UZ) was further subdivided into five sub-areas: Val Melaina 1 (1883 inhabitants), Val Melaina Giovanni Conti (9927 inhabitants), Val Melaina 2 (8602 inhabitants), Val Melaina FL1 (3905 inhabitants), and Val Melaina Prati Fiscali (11,743 inhabitants). Therefore, by including the previously identified Mesoarea of Northern Tufello (2164 inhabitants), Val Melaina was ultimately divided into six sub-areas (Figure 4).

The subdivision derived from Step 2–Sept 4 thus resulted in 24 sub-areas, with a resident population below 20,000 inhabitants each (Table 5).

Table 5. Sub-areas resulting from the Step 2–4 subdivisions.

Mesoareas	Inhabitants (ISTAT 2021)
Aeroporto dell'Urbe	1896
Bufalotta	7465
Eastern Casal Boccone	7556
Western Casal Boccone	6192
Conca d'Oro	18,792
Northern Fidene	2288
Southern Fidene	9236
Monte Sacro	16,050
Central Monte Sacro Alto	16,334
Monte Sacro Alto Sannazzaro Park	2673
Eastern Montesacro Alto	14,433
Saccro Pastore	9947
Serpentara Est	9886
Serpentara Colle Salario	9681
Serpentara Betulle	11,697
Settebagni	5254
Tor S. Giovanni	706
Tufello	13,590
Val Melaina Northern Tufello	2164
Val Melaina Giovanni Conti	9927
Val Melaina Prati Fiscali	11,743
Val Melaina1	1883
Val Melaina2	8602
Val Melaina FL1	3905
III Municipality	201,900

3.5. Step 5—Statistical Assessment of SES Differences (Education Level) Between Sub-Areas of the Same UZ

Almost all Mesoareas within the same UZ—obtained either by OMI index or road axis subdivision—show statistically significant differences in the percentage of university graduates, used as a proxy for socio-economic status (SES) (Table 6).

No statistically significant differences were found between Northern/Southern Fidene, between Val Melaina Prati Fiscali/Val Melaina NorthernTufello, Val Melaina Prati Fiscali/Val Melaina2, Val Melaina1/Val Melaina2, Val Melaina Prati Fiscali/Val Melaina1, Val Melaina Northern Tufello/Val Melaina2, and between sub-areas derived from Serpentara UZ.

Table 6. Results from the chi2 analysis of differences in educational level per sub-area defined during Steps 2–4. Statistically significant results ($p < 0.05$) are shown in bold.

UZ		University Degree or More N (%)	Lower than University Degree N(%)	p
Casal Boccone	eastern western	2119 (35.8) 1069 (20.4)	3796 (64) 4165 (89.6)	**<0.001**
Fidene	northern southern	310 (16.4) 1316 (17.4)	1584 (83.6) 6228 (82.6)	0.267
Monte Sacro Alto	Central Sannazzaro Park	5320 (39.5) 1057 (48.0)	8139 (60.5) 1144 (52.0)	**<0.001**
Monte Sacro Alto	Central Eastern	5320 (39.5) 4507 (37.9)	8139 (60.5) 7373 (62.1)	**0.010**
Monte Sacro Alto	Sannazzaro Park Eastern	1057 (48.0) 4107 (32.5)	1144 (52.0) 8514 (67.5)	**<0.001**
Serpentara	Eastern Colle Salario	2149 (26.8) 2076 (26.0)	5859 (73.2) 5897 (74.0)	0.253
Serpentara	Eastern Betulle	2149 (26.8) 2553 (25.4)	5859 (73.2) 7515 (74.6)	0.024
Serpentara	Colle Salario Betulle	2076 (26.0) 2553 (25.4)	5897 (74.0) 7515 (74.6)	0.300
Val Melaina	Northern Tufello Giovanni Conti	583 (32.5) 2117 (25.7)	1212 (67.5) 6111 (74.3)	**<0.001**
Val Melaina	Northern Tufello Prati Fiscali	583 (32.5) 3388 (35.2)	1212 (67.5) 6248 (64.8)	0.029
Val Melaina	Northern Tufello 1	583 (32.5) 575 (38.3)	1212 (67.5) 926 (61.7)	**<0.001**
Val Melaina	Northern Tufello 2	583 (32.5) 2559 (35.6)	1212 (67.5) 4639 (64.4)	0.015
Val Melaina	Giovanni Conti Prati Fiscali	2117 (25.7) 3388 (35.2)	6111 (74.3) 6248 (64.8)	**<0.001**
Val Melaina	Giovanni Conti 1	2117 (25.7) 575 (38.3)	6111 (74.3) 926 (61.7)	**<0.001**
Val Melaina	Giovanni Conti 2	2117 (25.7) 2559 (35.6)	6111 (74.3) 4639 (64.4)	**<0.001**
Val Melaina	Prati Fiscali 1	3388 (35.2) 575 (38.3)	6248 (64.8) 926 (61.7)	0.018
Val Melaina	Prati Fiscali 2	3388 (35.2) 2559 (35.6)	6248 (64.8) 4639 (64.4)	0.599
Val Melaina	1 2	575 (38.3) 2559 (35.6)	926 (61.7) 4639 (64.4)	0.043
Val Melaina	FL1 Northern Tufello	1317 (40.5) 583 (18.1)	1938 (59.5) 2630 (81.9)	**<0.001**
Val Melaina	FL1 Giovanni Conti	1317 (40.5) 2117 (25.7)	1938 (59.5) 6111 (74.3)	**<0.001**
Val Melaina	FL1 Prati Fiscali	1317 (40.5) 3388 (35.2)	1938 (59.5) 6248 (64.8)	**<0.001**
Val Melaina	FL1 1	1317 (40.5) 575 (22.4)	1938 (59.5) 1995 (77.6)	**<0.001**
Val Melaina	FL1 2	1317 (40.5) 2559 (35.6)	1938 (59.5) 4639 (64.4)	**<0.001**

3.6. Step 6—Merging of Non-Significantly Different Areas

Due to the lack of statistical significance in the analysis, the following areas were merged:

- Fidene Nord/Sud = Fidene;
- Val Melaina 1/Val Melaina 2/Val Melaina Prati Fiscali = Val Melaina F.A. Gualterio;
- Serpentara Betulle/Serpentara Colle Salario = Serpentara Colle Salario Betulle.

After this step, a total of 20 sub-areas were identified, as reported in Table 7.

Table 7. Mesoareas resulting from the subdivisions due to extensive dimensions, heterogeneity in average real estate sale prices or main road axis, or large number of inhabitants.

Mesoareas	Inhabitants (ISTAT 2021)
Aeroporto dell'Urbe	1896
Bufalotta	7465
Eastern Casal Boccone	7556
Western Casal Boccone	6192
Conca d'Oro	18,792
Fidene	11,524
Monte Sacro	16,050
Central Monte Sacro Alto	16,334
Monte Sacro Alto Sannazzaro Park	2673
Eastern Montesacro Alto	14,433
Saccro Pastore	9947
Serpentara Est	9886
Serpentara Colle Salario Betulle	21,378
Settebagni	5254
Tor S. Giovanni	706
Tufello	13,590
Val Melaina Northern Tufello	2164
Val Melaina Giovanni Conti	9927
Val Melaina F.A. Gualterio	22,228
Val Melaina FL1	3905
III Municipality	201,900

3.7. Step 4b—Subdivision of UZs > 20,000 Inhabitants Based on Inter-Zonal Roads

Among the sub-areas delineated in the previous steps, two sub-areas still had a population exceeding 20,000 inhabitants: Val Melaina F.A. Gualterio and Serpentara Colle Salario Betulle.

Since the Urban Zone (UZ) of Val Melaina had already been subdivided based on inter-zonal road axes, no further subdivisions were carried out in this area. Instead, the inter-zonal roads classified by Roma Capitale within the UZ of Serpentara were identified on the street network graph, including the following:

1. Via Amalia Bettini;
2. Viale Gino Cervi;
3. Via Pian di Sco;
4. Via Titina de Filippo.

These roads enable a further subdivision of the Serpentara area along two axes: the first, in the southwest, follows via Pian di Sco and via Titina de Filippo; the second, in the southeast, follows via Amalia Bettini and Viale Gino Cervi (Figure 13).

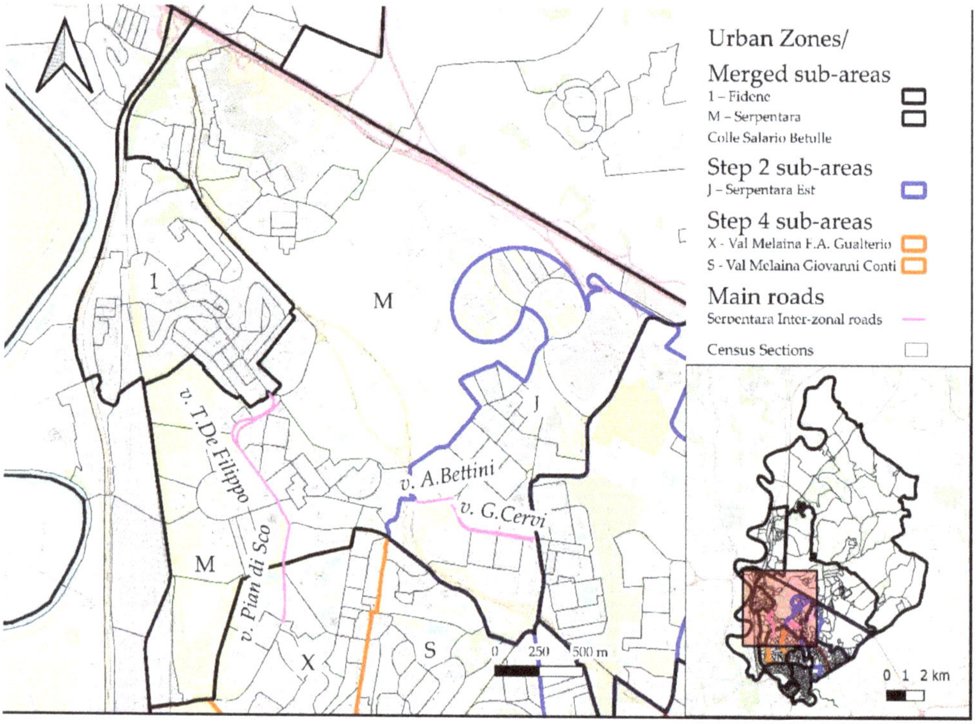

Figure 13. The image displays the inter-zonal roads within the Urban Zone of Serpentara. The road axis formed by Via Titina De Filippo and Via Pian di Sco allows for the subdivision of the broader Serpentara Colle Salario Betulle area—created through the merging operations in Step 6—into two sub-areas: one to the east of the axis, named Serpentara 2, and one to the west, named Serpentara 3. Additionally, the road axis composed of Via Amalia Bettini and Viale Gino Cervi enables the subdivision of the Serpentara Est area into two further sub-areas: a smaller southern portion called Serpentara Giulio Pasquati, and a larger northern portion referred to as Serpentara Vigne Nuove. The red box in the overview map corresponds to the geographical extent displayed in the main map.

At the conclusion of this step, the Serpentara Urban Zone is divided into four sub-areas: Serpentara Giulio Pasquati, Serpentara Vigne Nuove, Serpentara 2, and Serpentara 3 (Figure 4).

3.8. Step 5b—Statistical Assessment of SES Differences (Education Level) Between Sub-Areas of the Same UZ

Table 8 reports the results of the chi-squared test carried out to assess differences in educational attainment across the Serpentara sub-areas identified in the previous steps.

Table 8. The table shows the results of the Chi-squared tests conducted between the sub-areas derived from the Serpentara Urban Zone through the subdivisions carried out in Step 4b. Statistically significant results ($p < 0.05$) are shown in bold.

UZ		University Degree or More n (%)	Lower than University Degree n (%)	*p*
Serpentara	3 2	3462 (25.5) 1167 (26,1)	10,116 (74.5) 4463 (73,9)	0.387
Serpentara	3 Giulio Pasquati	3462 (25.5) 339 (14)	10,116 (74.5) 2082 (86)	**<0.001**
Serpentara	2 Giulio Pasquati	1167 (26.1) 339 (14)	1167 (26,1) 2082 (86%)	**<0.001**
Serpentara	Vigne Nuove 2	1810 (32,4) 1167 (26,1)	3777 (67.6) 4463 (73.9)	**<0.001**
Serpentara	Vigne Nuove Giulio Pasquati	1810 (32,4) 339 (14)	3777 (67.6) 2082 (86%)	**<0.001**
Serpentara	3 Vigne Nuove	3462 (25.5) 1810 (32.4)	10,116 (74.5) 3777 (67.6)	**<0.001**

3.9. Step 6b—Merging of Non-Significantly Different Areas

Given the lack of statistically significant differences between the two adjacent sub-areas Serpentara 2 and Serpentara 3, they were merged again into a single sub-area named Serpentara Colle Salario Betulle (referred to as Serpentara CS Betulle in the tables). In contrast, the sub-areas Serpentara Giulio Pasquati and Serpentara Vigne Nuove were retained (Figure 4).

3.10. Step 5c—Statistical Assessment of SES Differences (Education Level) Between Sub-Areas of the Same UZ

Table 9 presents the results of the chi-squared test performed to evaluate statistically significant differences in educational attainment across the Serpentara sub-areas delineated in the preceding methodological steps.

Table 9. The table shows the results of the Chi-squared tests conducted between the sub-areas derived from the Serpentara Urban Zone through the merging of sub-areas carried out in Step 6b. Statistically significant results ($p < 0.05$) are shown in bold.

UZ		University Degree or More n (%)	Lower than University Degree n (%)	*p*
Serpentara	CS Betulle Giulio Pasquati	4629 (25.7) 339 (14)	13,412 (74.3) 2082 (86)	**<0.001**
Serpentara	Vigne Nuove CS Betulle	1810 (32,4) 4629 (25.7)	3777 (67.6) 13,412 (74.3)	**<0.001**
Serpentara	Vigne Nuove Giulio Pasquati	1810 (32,4) 339 (14)	3777 (67.6) 2082 (86%)	**<0.001**

3.11. Statistical Assessment of Differences Between Mesoareas

As a result of the previous steps, a total of 21 Mesoareas were identified (see Table 10, Figures 14 and 15). The adopted methodology allowed the delineation of Mesoareas with population sizes ranging from 706 inhabitants (Tor San Giovanni) to 22,228 inhabitants (Val Melaina F.A. Gualterio). Two Mesoareas—Val Melaina F.A. Gualterio (22,228 inhabitants) and Serpentara Colle Salario Betulle (21,278 inhabitants)—exceed the 20,000-resident threshold. Considering the lack of statistically significant differences within their internal census sections and the impossibility of applying further subdivisions based on the road network structure, no additional splitting was performed.

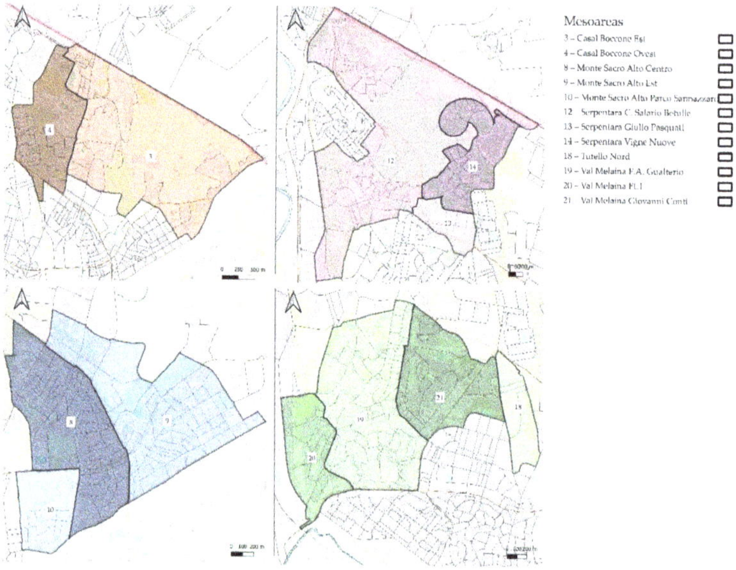

Figure 14. This figure illustrates the Mesoareas derived from the subdivision of Urban Zones (UZs), produced and retained following the sequence of methodological steps outlined earlier.

Table 10. III Municipality of Rome, Mesoareas.

Mesoareas	Inhabitants (ISTAT 2021)
Aeroporto dell'Urbe	1896
Bufalotta	7465
Eastern Casal Boccone	7556
Western Casal Boccone	6192
Conca d'Oro	18,792
Fidene	11,524
Monte Sacro	16,050
Central Monte Sacro Alto	16,334
Monte Sacro Alto Sannazzaro Park	2673
Eastern Montesacro Alto	14,433
Saccro Pastore	9947
Serpentara Vigne Nuove	7092
Serpentara Colle Salario Betulle	21,378
Serpentara Giulio Pasquati	2794
Settebagni	5254

Table 10. *Cont.*

Mesoareas	Inhabitants (ISTAT 2021)
Tor S. Giovanni	706
Tufello	13,590
Northern Tufello	2164
Val Melaina Giovanni Conti	9927
Val Melaina FL1	3905
Val Melaina F.A. Gualterio	22,228
III Municipality	201,900

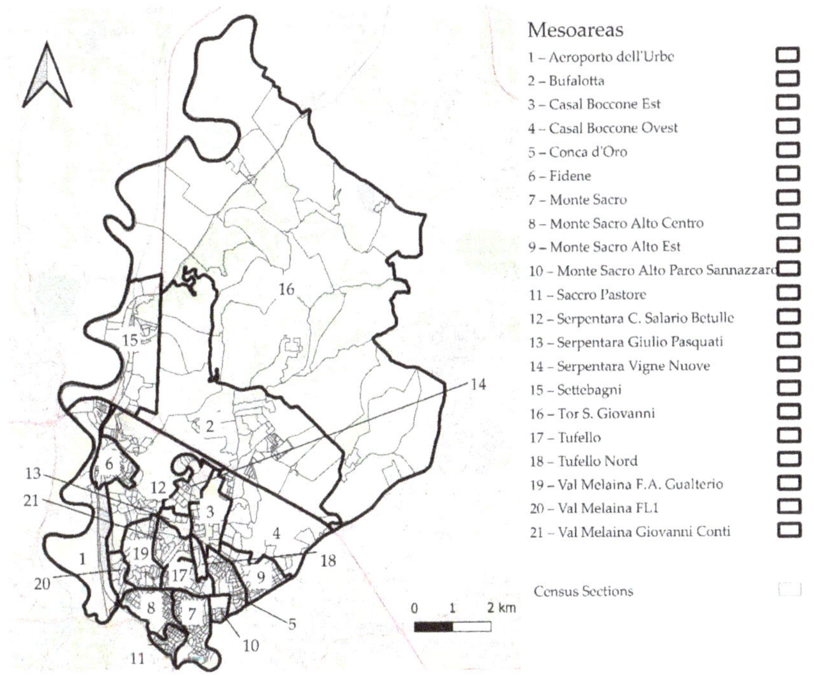

Figure 15. Mesoareas of the III Municipality of Rome.

3.12. Summary Measures of Educational Attainment by Urban Classification

Table 11 summarizes the mean, median, interquartile range, and variance of the percentage of university graduates within the Urban Zones. Figures A5 and A6 show the box-and-whisker plots of the distribution of the percentage of university graduates across Urban Zones and Mesoareas, respectively.

Table 12 presents the same set of statistics for the the Mesoareas defined in the previous sections.

Regarding Urban Zones, the mean percentage of graduates varies notably across UZs, ranging from 16.5% in Fidene to 45.4% in Monte Sacro. Similarly, the median values confirm this trend, with Monte Sacro (46.5%), Monte Sacro Alto (40.1%), and Sacco Pastore (39.1%) showing the highest concentrations of university-educated residents. In contrast, the lowest median percentages are found in Fidene (16.7%) and Tufello (16.7%). Variance remains relatively low across zones (0.002–0.015), indicating limited internal dispersion. However, IQR values—ranging from 0.059 in Fidene to 0.216 in Casal Boccone—suggest that in some zones, particularly Casal Boccone and Tor S. Giovanni (IQR = 0.173), there is considerable heterogeneity in educational attainment levels across Mesoareas. Overall, these results underscore the spatial variability in the distribution of university graduates within the studied urban zones.

Table 11. Descriptive statistics (mean, variance, median, and interquartile range) for educational attainment across all Urban Zones within Municipality III.

Urban Zones	Mean	Variance	Median	IQR
Fidene	0.165	0.002	0.167	0.059
Sacco Pastore	0.389	0.005	0.391	0.075
Conca d'Oro	0.382	0.006	0.381	0.087
Monte Sacro Alto	0.400	0.007	0.401	0.089
Settebagni	0.237	0.008	0.222	0.076
Monte Sacro	0.454	0.009	0.465	0.11
Bufalotta	0.212	0.009	0.183	0.151
Tufello	0.187	0.009	0.167	0.13
Tor S. Giovanni	0.183	0.010	0.182	0.173
Val Melaina	0.335	0.011	0.345	0.157
Serpentara	0.261	0.011	0.275	0.137
Aeroporto dell' Urbe	0.188	0.013	0.164	0.098

Table 12. Descriptive statistics (mean, variance, median, and interquartile range) for educational attainment across all Mesoareas within Municipality III.

Urban Zones	Mean	Variance	Median	IQR
Fidene	0.165	0.002	0.167	0.059
Monte Sacro Alto Sannazzaro Park	0.493	0.003	0.482	0.074
Val Melaina FL1	0.408	0.004	0.425	0.074
Serpentara Giulio Pasquati	0.140	0.004	0.123	0.08
Monte Sacro Alto East	0.386	0.005	0.371	0.108
Saccro Pastore	0.389	0.005	0.391	0.075
Conca d'Oro	0.382	0.006	0.381	0.087
Northern Tufello	0.286	0.007	0.295	0.138
Serpentara Vigne Nuove	0.326	0.007	0.314	0.106
Monte Sacro Alto Central	0.395	0.007	0.405	0.07
Settebagni	0.237	0.008	0.222	0.076
Val Melaina F.A. Gualterio	0.354	0.008	0.35	0.134
Monte Sacro	0.454	0.009	0.465	0.11
Bufalotta	0.212	0.009	0.183	0.151
Tufello	0.187	0.009	0.167	0.13
Serpentara Colle Salario Betulle	0.238	0.010	0.257	0.145
Tor S. Giovanni	0.183	0.010	0.182	0.173
Aeroporto dell'Urbe	0.188	0.013	0.164	0.098
Casal Boccone East	0.296	0.013	0.278	0.232
Val Melaina Giovanni Conti	0.272	0.015	0.256	0.184
Casal Boccone West	0.233	0.016	0.221	0.248

The results confirm and expand the variability already observed at the Urban Zone level. The mean percentage of university graduates ranges from a minimum of 14.0% in Serpentara Giulio Pasquati to a maximum of 49.3% in Monte Sacro Alto Sannazzaro Park. The median values follow a similar pattern, with the lowest value found again in Serpentara Giulio Pasquati (12.3%) and the highest in Monte Sacro Alto Sannazzaro Park (48.2%).

The IQR, which captures the spread of the middle 50% of observations, highlights considerable internal heterogeneity in some Mesoareas, particularly in Casal Boccone West (0.248) and Casal Boccone East (0.232). By contrast, Mesoareas such as Monte Sacro Alto Sannazzaro Park (0.074) and Val Melaina FL1 (0.074) show more homogeneous distributions. The variance values remain generally low, yet slightly higher in areas such as Casal Boccone West (0.016) and Val Melaina Giovanni Conti (0.015), indicating greater dispersion in those territories.

Overall, the Mesoarea-level data underline pronounced socio-spatial disparities in educational attainment, with certain Mesoareas—especially those in the Monte Sacro Alto and Val Melaina clusters—exhibiting notably higher and more homogeneous percentages of university graduates, while others reflect both lower levels and greater internal variability.

The average variance of the percentage of university graduates is 0.009 across Urban Zones and 0.008 across Mesoareas. This slight difference may reflect the smaller spatial scale of Mesoareas, although the variation between the two levels remains limited.

3.13. Non-Parametric Analysis of Educational Differences Among Mesoareas Within Urban Zones

Differences in the percentage of university graduates among Mesoareas within each Urban Zone (UZ) were assessed using Kruskal–Wallis tests, due to the non-normal distribution of the variable (Shapiro–Wilk $p < 0.05$). The analysis was conducted on the five UZs that were subdivided into Mesoareas; among these, four showed statistically significant differences between Mesoareas ($p < 0.05$) (Table 13).

Table 13. Results of Kruskal–Wallis tests performed on the percentage of university graduates across subdivided Mesoareas within Urban Zones. Where three or more groups were compared, post-hoc pairwise comparisons using Dunn's test with Bonferroni correction are presented. Statistically significant differences ($p < 0.05$) between specific Mesoareas are highlighted, illustrating localized variations in educational attainment. Statistically significant results ($p < 0.05$) are shown in bold.

Urban Zone	No. of Mesoareas	Kruskal–Wallis Test (χ^2, df)	Global p-Value	Significant Pairwise Comparisons (Dunn–Bonferroni)	Adjusted p-Value
Casal Boccone	2	2.364 with 1 d.f.	0.1242		
MSA	3	17.493 with 2 d.f.	**0.0002**	MSA Sannazzaro Park vs. Central MSA	**0.0012**
				MSA Sannazzaro Park vs. Eastern MSA	**<0.001**
Serpentara	3	15.674 with 2 d.f.	**0.0004**	Serpentara Vigne Nuove vs. Serpentara CS Betulle	**0.0018**
				Serpentara Vigne Nuove vs. Serpentara Giulio Pasquati	**0.0023**
Val Melaina	4	13.848 with 3 d.f.	**0.0031**	VM Giovanni Conti vs. VM F.A. Gualterio	**0.02**
				VM Giovanni Conti vs. VM FL1	**0.0032**

Specifically, in the UZ Monte Sacro Alto, differences among Mesoareas were statistically significant ($\chi^2 = 117.493$, df = 2, $p = 0.0002$). Post hoc Dunn tests with Bonferroni correction revealed significant differences between Monte Sacro Alto Sannazzaro Park and

Monte Sacro Alto Central ($p = 0.0012$), and between Monte Sacro Alto Sannazzaro Park and Monte Sacro Alto East ($p < 0.001$); no significant difference emerged between Monte Sacro Alto Central and Monte Sacro Alto East.

Similarly, in Serpentara ($\chi^2 = 15.674$, df = 2, $p = 0.0004$), the Mesoarea Serpentara Vigne Nuove showed significant differences when compared to both Serpentara Colle Salario Betulle ($p = 0.0018$) and Serpentara Giulio Pasquati ($p = 0.0023$).

In Val Melaina, the Kruskal–Wallis test was also significant ($\chi^2 = 13.848$, df = 3, $p = 0.0031$). The Val Melaina Giovanni Conti Mesoarea differed significantly from both Val Melaina F.A. Gualterio ($p = 0.02$) and Val Melaina FL1 ($p = 0.0032$).

In Casal Boccone, no statistically significant difference was found between the two Mesoareas ($\chi^2 = 2.364$, df = 1, $p = 0.1242$).

3.14. *Results of Chi-Square Analyses on Educational Attainment, Employment Status, and Single-Person Households*

Table 14 reports the results of the chi-squared tests comparing employment status variables across Mesoareas belonging to the same Urban Zones.

Table 14. Results from the chi2 analysis of differences in occupational status. Statistically significant results ($p < 0.05$) are shown in bold.

UZ		Employed (%)	Unemployed (%)	p
Casal Boccone	eastern western	3495 (67.8) 2435 (61.5)	1658 (32.2) 1525 (38.5)	**<0.001**
Monte Sacro Alto	Central Sannazzaro Park	6433 (65.4) 1081 (65.2)	3409 (34.6) 578 (34.8)	0.872
Monte Sacro Alto	Central Eastern	6433 (65.4) 6037 (66.7)	3409 (34.6) 3008 (33.3)	0045
Monte Sacro Alto	Sannazzaro Park Eastern	1081 (65.2) 6037 (66.7)	578 (34.8) 3008 (33.3)	0.209
Serpentara	Vigne Nuove CS Betulle	3515 (71.4) 8612 (64.9)	1406 (28.6) 4649 (35.1)	**<0.001**
Serpentara	Vigne Nuove Giulio Pasquati	3515 (71.4) 925 (56.8)	1406 (28.6) 704 (43.2)	**<0.001**
Serpentara	CS Betulle Giulio Pasquati	8612 (64.9) 2553 (25.4)	4649 (35.1) 7515 (74.6)	**<0.001**
Val Melaina	Giovanni Conti Northern Tufello	3550 (62) 961 (67.8)	2176 (38) 457 (32.2)	**<0.001**
Val Melaina	F.A. Gualterio Northern Tufello	8263 (60.8) 961 (67.8)	5331 (39.2) 457 (32.2)	**<0.001**
Val Melaina	Northern Tufello FL1	961 (67.8) 1465 (66.5)	457 (32.2) 738 (33.5)	0.427
Val Melaina	Giovanni Conti F.A. Gualterio	3550 (62) 8263 (60.8)	2176 (38) 5331 (39.2)	0.114
Val Melaina	Giovanni Conti FL1	3550 (62) 1465 (66.5)	2176 (38) 738 (33.5)	**<0.001**
Val Melaina	F.A. Gualterio FL1	8263 (60.8) 1465 (66.5)	5331 (39.2) 738 (33.5)	**<0.001**

In the majority of pairwise comparisons, Mesoareas with a higher percentage of graduates also exhibited a higher employment rate. An exception is observed in the comparison between Casal Boccone East and Casal Boccone West, where, despite a significantly higher proportion of graduates in Casal Boccone East, Casal Boccone West shows a significantly higher employment rate (39% vs. 32%). Similarly, within all comparisons among the Mesoareas of Monte Sacro Alto, the employment rate distribution is inverted, with none of the differences in employment rates reaching statistical significance. A comparable pattern is also evident within the Val Melaina Urban Zone, specifically the following:

1. The Northern Tufello Mesoarea shows a significantly higher employment rate compared to Val Melaina F.A. Gualterio (68% vs. 61%), despite Val Melaina F.A. Gualterio having a higher percentage of graduates (20% vs. 18% in Northern Tufello).
2. The Northern Tufello Mesoarea exhibits a higher employment rate compared to Val Melaina FL1 (68% vs. 67%), while Val Melaina FL1 shows a higher percentage of graduates (24% vs. 18% in Northern Tufello). However, the difference in employment status is not statistically significant.
3. The Val Melaina Giovanni Conti Mesoarea shows a higher employment rate compared to Val Melaina F.A. Gualterio (62% vs. 61%), while Val Melaina F.A. Gualterio has a higher percentage of graduates (20% vs. 15% in Giovanni Conti). However, the difference in employment status is not statistically significant.

Most of the tested differences in employment status are statistically significant, except for the previously mentioned comparisons within the Monte Sacro Alto Mesoarea, as well as those between Northern Tufello and Val Melaina FL1, and between Val Melaina Giovanni Conti and Val Melaina FL1.

Table 15 reports the results of the chi-square test on the variable "percentage of single-person households" across Mesoareas within the same Urban Zones.

Table 15. Results from the chi2 analysis of differences in single-person households. Statistically significant results ($p < 0.05$) are shown in bold.

UZ		Single-Person Households (%)	Not-Single-Person Households (%)	p
Casal Boccone	eastern western	1528 (43,2) 1069 (37.1)	2007 (56.8) 1810 (62.9)	**<0.001**
Monte Sacro Alto	Central Sannazzaro Park	3615 (45%) 549 (43,4)	4425 (55) 716 (56.6)	0.299
Monte Sacro Alto	Central Eastern	3615 (45%) 3130 (44.5)	4425 (55) 3906 (55.5)	0.556
Monte Sacro Alto	Sannazzaro Park Eastern	549 (43,4) 3130 (44.5)	716 (56.6) 3906 (55.5)	0.474
Serpentara	Vigne Nuove CS Betulle	1672 (47.1%) 3995 (39.3)	1879 (52.9) 6159 (60.7)	**<0.001**
Serpentara	Vigne Nuove Giulio Pasquati	1672 (47.1%) 417 (33.1)	1879 (52.9) 841 (66.9)	**<0.001**
Serpentara	CS Betulle Giulio Pasquati	3995 (39.3) 417 (33.1)	6159 (60.7) 841 (66.9)	**<0.001**
Val Melaina	Giovanni Conti Northern Tufello	1760 (38.9) 544 (49.3)	2765 (61.1) 559 (50.7)	**<0.001**
Val Melaina	F.A. Gualterio Northern Tufello	6044 (52.2) 544 (49.3)	5544 (47.8) 559 (50.7)	0.071
Val Melaina	Northern Tufello FL1	544 (49.3) 807 (42.3)	559 (50.7) 1099 (57.7)	**<0.001**

Table 15. *Cont.*

UZ		Single-Person Households (%)	Not-Single-Person Households (%)	*p*
Val Melaina	Giovanni Conti F.A. Gualterio	1760 (38.9) 6044 (52.2)	2765 (61.1) 5544 (47.8)	**<0.001**
Val Melaina	Giovanni Conti FL1	1760 (38.9) 807 (42.3)	2765 (61.1) 1099 (57.7)	0.01
Val Melaina	F.A. Gualterio FL1	6044 (52.2) 807 (42.3)	5544 (47.8) 1099 (57.7)	**<0.001**

For this variable, a distribution differing from that of educational attainment emerges in only two cases: the comparison between the Monte Sacro Alto Central and MSA Sannazzaro Park Mesoareas, where the former shows a higher percentage of single-person households (45% Monte Sacro Alto Central vs. 43% MSA Sannazzaro Park), although not statistically significant, unlike the pattern observed for educational attainment (40% Monte Sacro Alto Central vs. 48% MSA Sannazzaro Park). The second case involves the Northern Tufello and Val Melaina FL1 Mesoareas (single-person households: 49% Northern Tufello vs. 42% Val Melaina FL1; percentage of graduates: 18% Northern Tufello vs. 24% Val Melaina FL1), where the difference in the distribution of single-person households is statistically significant.

Most differences in the distribution of single-person households tested are significant, except for those within Monte Sacro Alto Mesoareas and those between Northern Tufello/Val Melaina F.A. Gualterio and Val Melaina Giovanni Conti/Val Melaina F.A. Gualterio Mesoareas.

4. Discussion

4.1. Subdivision of Urban Zones into Mesoareas Based on OMI Zones and Road Axes, and Chi-Square Tests on Educational Attainment

The methodology developed in this study enabled a redefinition of the boundaries within III Municipality, dividing the existing UZs into 21 Mesoareas, with population sizes ranging from 706 (Tor San Giovanni) to 22,228 (Val Melaina F.A. Gualterio). Two Mesoareas—Val Melaina F.A. Gualterio and Serpentara Colle Salario Betulle—exceed the target population threshold of 20,000 residents, which had been established based on the catchment areas for healthcare services outlined in the Introduction. The adopted method, which involved splitting densely populated areas based on major interzonal roadways or previously identified heterogeneity in real estate markets, was not sufficient to isolate sub-areas within these larger Mesoareas that showed statistically significant differences in the distribution of university-level educational attainment. Further work will therefore be necessary to develop a more refined methodology capable of identifying potential internal subdivisions characterized by socioeconomic differentiation.

As part of the methodological process, five UZs were initially subdivided into subareas based on heterogeneity in the real estate market. In four of these UZs, such subdivisions were retained due to the statistically significant differences observed in the distribution of educational attainment between the resulting Mesoareas:

- In Val Melaina, the Mesoarea Northern Tufello was delineated from the rest of the territory. The percentage of university graduates in this Mesoarea is significantly higher than in the adjacent one, Val Melaina Giovanni Conti, yet lower than in the other two Mesoareas derived from the same UZ, Val Melaina F.A. Gualterio and Val Melaina FL1.

- In Monte Sacro Alto, the Mesoarea Monte Sacro Alto Sannazzaro Park—bordering the UZ with the highest percentage of graduates, Monte Sacro—was separated based on a significantly higher proportion of university graduates compared to the rest of Monte Sacro Alto. Notably, this newly defined Mesoarea displays the highest share of graduates among all Mesoareas within III Municipality.
- Casal Boccone UZ was split into two Mesoareas that show a statistically significant difference in the distribution of university-level education: Casal Boccone East (36%) and Casal Boccone West (20%). Within these Mesoareas, the population distribution is highly uneven, with residents in Casal Boccone West concentrated in the westernmost section, contiguous with the Serpentara Mesoareas, with which it seems to share similar socioeconomic characteristics. Conversely, the population of Casal Boccone East is concentrated in the southern portion of the Mesoarea, adjacent to Monte Sacro Alto East, and appears to reflect similar socioeconomic patterns.

By contrast, two proposed subdivisions of UZ based on differences in the real estate market were discarded, as the subsequent statistical tests did not reveal significant differences in the socioeconomic level between the resulting sub-areas.

In the case of Fidene, the absence of statistically significant differences between the two Mesoareas—defined based on the presence of two distinct OMI zones (Colle Salario E40 and Fidene E41)—may be attributable to the asymmetric distribution of property transaction prices within the higher-value zone, Colle Salario E40 (minimum: €2800/m^2; maximum: €3900/m^2) (Table A1). This asymmetry may have led to an upward distortion of the average price, potentially influenced by a limited number of high-end properties [47], without necessarily indicating a true socioeconomic distinction between the populations of the two areas. In this context, the use of average property values may overestimate SES differences due to the particular dynamics of the local real estate market. It is also important to note that the Italian Revenue Agency (Agenzia delle Entrate) does not publish median property values, making it impossible to assess the internal distribution of prices within each OMI zone. Furthermore, in this case, the spatial intersection between census section centroids and OMI zones did not allow for a clean delineation of sub-area boundaries that precisely matched OMI borders. It is worth noting, however, that the area bordering the two sub-areas (Northern and Southern Fidene) is characterized by relatively low residential density (Figure A1), primarily due to the presence of green spaces. The misalignment between census section boundaries and OMI zones resulted in Northern Fidene also encompassing parts of the Fidene OMI zone (E41). This, combined with the relatively small population size of Northern Fidene (2288 residents) compared to the remaining area of Fidene (9236 residents), may have limited the ability to detect significant SES differences between the two groups.

Similarly, no statistically significant differences in the average level of education were found among the three Mesoareas derived from the Serpentara UZ, which had been delineated based on average property prices. The Mesoarea Serpentara Colle Salario was defined according to the boundaries of the OMI zone Colle Salario E40—the same real estate zone that includes Northern Fidene. As such, the considerations discussed above regarding the specific real estate dynamics of this area may also be relevant here. The other two Mesoareas—Serpentara Vigne Nuove and Serpentara Betulle—were delineated according to the internal boundaries of the Serpentara UZ, corresponding to the OMI zones Vigne Nuove E47 and Serpentara E42, respectively. These OMI zones also exhibit a wide range in property transaction prices, with differences of €1000 for E47 and €850 for E42 (Table A1), suggesting potential price variability that may not be fully captured by average values alone. Moreover, it is important to note that the two indicators used in this analysis—property prices and educational attainment—may reflect distinct dimensions of socioeconomic status and may not always align spatially or conceptually.

The literature shows that education level is independently associated with mortality and is minimally affected by other socioeconomic position (SEP) indicators, while property prices are also associated with mortality independently of individual educational level [48,49]. Residential area and real estate market values can be considered as synthetic indicators of socioeconomic position, urban quality, and area-level deprivation [50,51], encompassing both material and immaterial factors (e.g., functions, safety, services), particularly in the absence of individual-level data [52]. This rationale may also explain the opposite scenario—i.e., significant differences in educational levels among Mesoareas that are homogeneous in terms of OMI indicators—as observed between Central Monte Sacro Alto and Eastern Monte Sacro Alto, or between some of the Mesoareas within Val Melaina and Serpentara.

In the territories of Monte Sacro Alto, Val Melaina, and Serpentara, the subdivision based on main road axes made it possible to reveal differences in the distribution of educational attainment that would not have been identified through real estate market data alone. Within the Monte Sacro Alto UZ, for instance, the Mesoareas of MSA Central and MSA East were delineated from a territorially homogeneous area in terms of average property values. Nonetheless, these Mesoareas exhibit a statistically significant, albeit slight, difference in the proportion of university degree holders (40% in MSA Central vs. 38% in MSA East), suggesting the presence of a possible center–periphery gradient in educational attainment.

In the Val Melaina UZ, the subdivision based on major road axes enabled the identification of three Mesoareas: Val Melaina Giovanni Conti, Val Melaina FL1, and Val Melaina F.A. Gualterio. The Mesoarea of Giovanni Conti shows a significantly lower percentage of university degree holders (26%) compared to the other two Mesoareas (40% in FL1 and 36% in F.A. Gualterio). The area of Val Melaina Giovanni Conti, which appears to concentrate a population with more disadvantaged SES indicators, could not have been distinguished based solely on real estate market data. This may be explained not only by the different SES dimensions captured by property values and education level, but also by the potential exclusion of specific housing types—such as public housing—from the real estate market. This observation points to the possible need for integrating additional sources of information, such as expert knowledge, qualitative assessments of the built environment, or data from urban development plans, to achieve a finer-grained territorial segmentation.

In the western part of Serpentara UZ, even the subdivision based on interzone road axes did not reveal significant differences in the distribution of educational attainment across areas. In contrast, in the eastern sector of Serpentara UZ, this same method enabled the identification of the Serpentara Giulio Pasquati Mesoarea, which shows a significantly lower percentage of university graduates compared to the rest of the Urban Zone—and indeed the lowest percentage among all Mesoareas considered. As observed in the case of Val Melaina Giovanni Conti, the real estate market alone would not have captured this area's more disadvantaged profile in terms of educational attainment. A potential explanation for this may again lie in the exclusion of certain housing types—such as public housing—from the real estate market, which could obscure underlying socioeconomic differences when relying exclusively on market data.

Finally, it is worth noting that, where statistically significant differences were observed, Mesoareas derived from OMI zones with higher average property sale prices within the same UZ tended to exhibit a higher proportion of residents with a university degree—an association consistent with the existing literature on the relationship between real estate value and SES. This finding supports the idea that property sale prices can serve as a proxy for the SES of the resident population [53]. Property value has been used in the

United States as a proxy for SES to investigate the distribution of chronic conditions such as obesity, where low-SES areas report higher obesity rates [54], even after adjusting for age, race/ethnicity, household size, employment status, and home ownership [55].

4.2. Intra-Mesoareas Variability in Educational Attainment: Descriptive Patterns and Statistical Testing

Descriptive statistics computed for each Mesoarea and UZ further illustrate the internal variability of educational attainment within and among areas. While the overall distribution of university degree attainment tends to increase from some peripheral areas (e.g., Serpentara Giulio Pasquati, Fidene) toward more central or affluent zones (e.g., MSA, Conca d'Oro), the spread of values within areas remains heterogeneous. The interquartile range (IQR) and variance, in particular, offer insight into internal disparities.

Notably, nearly all UZ that were subdivided into Mesoareas—such as Casal Boccone, Serpentara, Val Melaina, and Tufello—show some of the highest variances, suggesting that the decision to further disaggregate these areas was empirically well grounded. At the Mesoarea level, high-variance areas like Casal Boccone West and Val Melaina Giovanni Conti may benefit from more refined analytical methodologies capable of identifying finer-grained spatial heterogeneity. The Mesoarea-based breakdown also allowed for the identification of the areas with the lowest (Serpentara Giulio Pasquati) and highest (MSA Sannazzaro Park) proportions of university graduates across the entire municipality. Interestingly, the two areas with the lowest population density in the district—Aeroporto dell'Urbe and Tor San Giovanni—exhibit among the highest variance values, further suggesting that heterogeneity may persist even in sparsely populated zones and highlighting the importance of localized analysis.

At the UZ level, measures of dispersion such as variance appear slightly higher overall than in Mesoareas, consistent with the notion that Mesoareas may capture more internally homogeneous subunits. However, differences in variance values between levels remain modest, warranting cautious interpretation.

To further investigate the differences in the percentage of university graduates across Mesoareas, Kruskal–Wallis tests were conducted within selected UZ. These analyses, consistent with the chi-square test results, confirmed the presence of statistically significant variation in several cases. In Monte Sacro Alto (χ^2 = 17.493, df = 2, p = 0.0002), pairwise post-hoc comparisons using Dunn's test with Bonferroni correction identified significant differences between Sannazzaro Park and both the Central (p = 0.0012) and Eastern (p < 0.001) Mesoareas, indicating clear internal educational disparities within the zone. Similarly, in Serpentara (χ^2 = 15.674, df = 2, p = 0.0004), Vigne Nuove differed significantly from both Colle Salario Betulle (p = 0.0018) and Giulio Pasquati (p = 0.0023), suggesting a gradient in educational attainment across the Mesoareas. In Val Melaina (χ^2 = 13.848, df = 3, p = 0.0031), significant differences were found between Giovanni Conti and both F.A. Gualterio (p = 0.02) and FL1 (p = 0.0032).

In contrast, no statistically significant differences were observed in Casal Boccone (χ^2 = 2.364, df = 1, p = 0.1242), despite the chi-square test having indicated significant differences in the distribution of education levels between its two Mesoareas. This discrepancy may reflect the different nature of the tests; while the chi-square test captures overall distributional differences, the Kruskal–Wallis test focuses on differences in central tendency and ranking. These results reinforce the presence of intra-zone heterogeneity in educational attainment and underscore the usefulness of combining multiple statistical approaches to capture different aspects of variability.

4.3. Educational, Occupational, and Household Structure Differences Across Mesoareas

The chi-square tests performed to assess differences between Mesoareas within the same UZ revealed several statistically significant differences in the distribution of university degree attainment, employment status, and single-person households. These dimensions were selected because they are included in the deprivation index developed by Rosano et al. [46], which was originally calculated using 2011 census data. However, the complete deprivation index was not applied in this analysis, as it has not been validated for use with the 2021 census data. Moreover, housing density—another component of the Rosano index—was excluded from the present analysis due to the unavailability of corresponding 2021 data from ISTAT. As such, the focus was limited to educational attainment, employment status, and household composition, which were deemed the most robust and interpretable indicators available for assessing socio-economic differences at the Mesoarea level.

All chi-square tests for differences in educational attainment between Mesoareas yielded statistically significant results, indicating consistent intra-UZ variation in the proportion of university graduates. This different distribution of educational attainment is often accompanied by significant differences in employment levels and, in some cases, household composition.

Particularly noticeable are the results in Serpentara, where all three dimensions—education, employment, and household structure—show highly significant and concordant differences among the three Mesoareas. For example, Serpentara Giulio Pasquati, which exhibits the lowest percentage of university graduates (14%), also shows the lowest employment rate (57%) and the highest proportion of single-person households (67%), in contrast with neighboring Mesoareas such as Vigne Nuove and Colle Salario Betulle. This alignment of disadvantage across multiple indicators suggests consistent socio-spatial disparities within the same UZ.

A similar pattern is observed in the Val Melaina group. Val Melaina Giovanni Conti, characterized by a relatively low percentage of graduates (15%) and moderate employment levels (62%), differs significantly from Val Melaina FL1, which records higher educational attainment (24%) and a more favorable employment profile (67%). In this case, both the chi-square tests on education and employment are statistically significant ($p < 0.001$), reinforcing the interpretation of unequal conditions across Mesoareas.

In contrast, some UZs, such as Monte Sacro Alto, show significant differences in educational attainment (e.g., between Sannazzaro Park and other Mesoareas) without corresponding differences in employment status or household composition. This may indicate that not all dimensions of socio-demographic differentiation necessarily co-occur, or that the variables considered may not fully capture socio-economic disparities between the Mesoareas in question.

In certain comparisons, particularly within the UZ of Val Melaina, the distribution of single-person households does not align with the patterns observed in educational attainment or, when significant, in employment status. For instance, Val Melaina Giovanni Conti, despite having the lowest percentage of university graduates (15%) and a relatively low employment rate (62%), shows a smaller proportion of single-person households (39%) compared to Val Melaina F.A. Gualterio (52%), which has higher levels of both education (20%) and employment (61%). Similarly, Val Melaina FL1, with the highest percentage of graduates in the group (24%) and a higher employment rate (67%), has a lower share of single-person households (42%) than Gualterio. These inconsistencies suggest that the presence of single-person households may not directly reflect socio-economic disadvantage in the same way as educational or occupational variables. Rather, household composition may be shaped by additional demographic or cultural factors not captured by the indicators used in this analysis. It is important to note that the 2021 census data do not include specific information on single-parent households. As a result, the analysis was limited to considering the percentage of single-person households, regardless of the presence of dependent children. This constraint may affect the interpretation of household composition indicators, as single-person households represent only one aspect of potentially vulnerable family structures.

Overall, these results highlight substantial heterogeneity within UZ, with some Mesoareas consistently exhibiting more disadvantaged profiles across multiple indicators. The convergence of significant differences in both education and employment in several cases reinforces the relevance of Mesoarea-level analysis for identifying localized patterns of socio-economic inequality.

4.4. Interpretation of Findings in Light of Policy Frameworks and the Scientific Literature

This study demonstrates the feasibility of reconstructing new forms of territorial analysis starting from socioeconomic statistical indicators, which can be classified as social determinants of health, for the purpose of health planning. This territorial subdivision was also informed by the planning requirements set forth in the reform of territorial healthcare introduced by Ministerial Decree 77/2022 [22], particularly regarding the definition of minimum catchment areas for structuring family counseling services and general medical practice. These two services represent the most elementary units within the organizational framework outlined by the decree, with catchment thresholds of approximately 20,000 and 1500 residents, respectively. Understanding the catchment areas of general practitioners can help identify which segments of the population remain isolated or underserved and enable more informed planning of practice openings or the reprogramming of healthcare services provision in particularly disadvantaged areas. On the other hand, identifying a more homogeneous territorial subdivision in terms of population size and socio-economic conditions provides a more informative picture of the territory. Indeed, since that the epidemiology of several illnesses is strictly influenced by SES [4], knowing in advance the characteristics of an area can help to better anticipate potential health demands, as well as prevent or respond more promptly to the needs of its inhabitants, knowing.

More specifically, the subdivision of the territory of III Municipality proposed in this study—providing a finer level of spatial granularity than the official Urban Zones—could support relevant institutions in more effectively identifying the catchment areas of health and social services that, according to the territorial reorganization outlined in Ministerial Decree 77/2022, operate on smaller population thresholds. These include services such as the community nurse (*infermiere di comunità*, 1 per 3000 inhabitants) or Family Counseling Centers (*Consultori Familiari*, 1 per 20,000 inhabitants). As a result, this Mesoarea-based territorialization may facilitate the identification of underserved areas according to current service planning and established standards. Furthermore, it may help to pinpoint areas that, while technically covered by existing service catchments, exhibit socioeconomically disadvantaged profiles that could translate into more complex and specific health needs—calling for better-prepared and more responsive services. For instance, the planned Community Health Center (*Casa della Comunità*) on Via Dina Galli, located in the southeastern portion of the Val Melaina Giovanni Conti mesoarea, could be envisioned as serving a cluster of surrounding mesoareas including Val Melaina Giovanni Conti, Northern Tufello, Tufello, and Serpentara Giulio Pasquati—areas where indicators of socioeconomic vulnerability are most concentrated. By contrast, the other two planned Community Health Centers are located in the more central and eastern areas of III Municipality, where the population displays generally more favorable characteristics: the center on Via Lampedusa is situated in Monte Sacro, adjacent to the Mesoareas of Conca d'Oro, Sacco Pastore, and MSA Sannazzaro Park, while the center on Via Paolo Monelli is located in Eastern Casal Boccone and may serve surrounding areas including Casal Boccone and Central and Eastern MSA. This distribution of facilities could pose challenges—such as an unequal distribution of catchment complexity and service burden—but it also presents opportunities. In particular, it enables institutions to strategically concentrate proactive public health interventions and community initiatives in the specific sections of the territory that demonstrate the greatest need. Using the methodology presented, a new territorial subdivision can be developed—one that more closely reflects the socioeconomic characteristics and population density, not only for the III Municipality, but for the whole area of Rome Capital. The innovative contribution of this study lies in the integration of multiple layers of analysis: tools traditionally associated with economic valuation—such as the OMI zoning—elements of territorial and administrative planning—such as the General Urban Traffic Plan (PGTU) and Urban Zones—and more structured instruments for interpreting social health inequalities, such as individual-level socioeconomic determinants.

Regarding real estate values, which have already been tested as a proxy in the Roman context [50], these can be interpreted not only as *immobilized capital*, but also, in a broader sense, as *potential capital*—particularly in a context like Rome, where a high rate of residential homeownership prevails [56]. In this sense, dwellings function as dormant patrimonial assets: they may not generate immediate income, yet they embody a mobilizable market value and can therefore be considered part of a household's wealth and socioeconomic position [56]. Due to the high number of residential property owners—driven more by clientelist and family-oriented political strategies that promote a family-centered welfare model than by market forces or globalization—spatial segregation has been shown to be reduced [57], compared to the US, where Black and White residentially are clearly separated [58]. On the other hand, since public housing is poor and mainly for natives, immigrants often turn to the private rented sector, ending up in overcrowded apartments in middle and upper-middle class areas [57].

While the use of urban, architectural, and real estate characteristics to assess socioeconomic vulnerability is not new within the Italian literature [59], the present study represents the first attempt to systematically integrate these dimensions into a coherent framework for sociosanitary needs planning, with the aim of supporting health policy strategies grounded in equity.

The analysis shows that in areas with higher OMI index values, the resident population exhibits higher levels of educational attainment, suggesting a potential association between property prices and socioeconomic status. This finding highlights the internal heterogeneity within the current UZs and supports the conclusion that the existing subdivision—established in 1977 [43]—is outdated and no longer adequate for understanding local dynamics. It underscores the need for a more detailed territorial breakdown to better capture the reality of the district.

Nonetheless, the UZ classification remains the most used territorial framework in the Municipality of Rome for providing and describing population characteristics [60,61]. Therefore, a finer-grained subdivision that can still be linked to the existing UZ would allow for improved territorial analysis while maintaining comparability with previously available or officially published data. In this regard, both national and international experiences have employed small-area approaches for territorial analysis—comparable to the census sections used as foundational units in the present study. However, such approaches cannot be directly applied to the complexity of the Roman context, which requires scalable tools capable of capturing territorial dynamics at an intermediate level—between the small-area scale and the *Municipio*, which currently represents the core unit of territorial governance and sociosanitary planning in Rome. While these tools are widely used globally to assess service accessibility, they generally serve a retrospective function, addressing outcomes *ex post* [62], or are used as foundational units for the construction of deprivation indices or exposure models [63], in a manner that is conceptually analogous to the Italian context [46].

Identifying homogeneous areas is crucial for effective healthcare planning based on the actual needs of the population. The implementation of an efficient territorial analysis tool is especially timely given the ongoing reform of territorial healthcare in Italy, initiated with the drafting of the National Recovery and Resilience Plan (Piano Nazionale Ripresa e Resilienza, PNRR) and its operational articulation (Ministerial Decree 77/2022) [22], which re-establishes the centrality of territorial healthcare, primary health care, proactive medicine, and health promotion and prevention. Effective planning cannot disregard a precise understanding of the territory [64,65].

In particular, with the implementation of the *Casa della Comunità* (Community Health Center, CdC) model, the integrated analysis of patients' health and social healthcare needs becomes central to the stratification of the population for service planning purposes [66]. This stratification must necessarily consider indicators of socioeconomic status. Furthermore, in defining the catchment areas of CdCs, disparities between different territories and contexts must be considered [67].

A revision of the territorial subdivisions of the Health District could integrate additional layers of complexity through variables that describe other dimensions of territorial deprivation—such as the availability of health and social-health services and facilities [68–72], elements evaluating the quality of the urban environment [73], environmental vulnerability [9] (e.g., the presence of urban green spaces in their various forms [74–76], average temperature levels [77], or even factors influencing health-related choices, such as the presence of so-called "food deserts" [78]. All of this contributes to a multidimensional assessment of socioeconomic vulnerability through the lens of intersectionality [79–81].

As of now, the idea of redrawing the Urban Zones of Rome Capital has not yet been considered by political decision-makers and those involved in the governance of territorial healthcare. Other Italian initiatives have adopted a proximity-based approach, such as the experience of the 'Habitat Microareas' project, implemented in the city of Trieste. This initiative aimed to strengthen experimental interventions promoting well-being and social cohesion in areas with a significant presence of public housing. Initially targeting the tenants of a few buildings in the city, the project was later extended to include families living in neighborhoods surrounding the public housing complexes, through collaboration with various local stakeholders [82]. The project yielded positive results in terms of population inclusion and enabled the testing of empowerment models for the individuals involved [83]. Other examples from the cities of Turin [84], Bologna [85], and Milan [86] showed how a more detailed and context-specific subdivision of the territory can reveal health inequalities that would otherwise remain hidden, thereby enabling the development of more effective health planning to address such conditions.

The analytical process proposed in this study is inherently scalable and could be extended to the entire territory of Rome, given the availability of the same data sources used herein. At the same time, the framework would benefit from the integration of more refined tools, such as the deprivation indices already developed for the Italian context, although these have not yet been validated using the data employed in the present study—particularly those from the 2021 census.

In this sense, our work aims to develop and introduce a new methodology to update UZ boundaries and constitutes an initial step toward the effective multidimensional integration of parameters, with the aim of developing a more reliable tool for the planning of healthcare needs. Through the proposed methodology, we seek to contribute to the ongoing scientific debate on how to integrate diverse types of information—each in some way related to the domain of health needs—into coherent planning frameworks. This aligns with recent efforts in the city of Rome to define forms of *"double vulnerability"*, encompassing both environmental and social dimensions [9], and more broadly with the advancement of an intersectional approach to health. Such an approach would enhance the capacity of healthcare systems to perform multidimensional assessments of sociosanitary vulnerability at the population level.

5. Limitations and Conclusions

This work represents one of the first attempts at the national level in Italy, and the first in the city of Rome, to define or revise the administrative territorial divisions from the perspective of healthcare planning. It proposes a method based on open-access data available to all Health Districts, making it potentially replicable in other contexts. Moreover, it allows to overcome the obstacle of population stratification without using individual-level data, which is problematic due to privacy concerns. However, to date, no pilot actions or projects have been proposed or discussed with policy makers, even though the current methodology to reshape UZs has been presented and integrated into the Local Healthcare Unit Roma 1 Corporate Organizational Act as a possible adoptable future strategy to rethink the territorial subdivisions. This work it to be intended as an exploratory methodological framework that could inform future planning processes.

Despite offering a pragmatic basis for the initial subdivision of Urban Zones, the use of real estate market heterogeneity—specifically OMI data—presents several limitations when aiming to delineate sub-areas that meaningfully reflect differences in SES. First, real estate markets are influenced by a range of factors beyond SES, such as urban development trends, planning regulations, and neighborhood amenities. While the study area does not include territories subject to significant housing market distortions, such as university campuses or tourist districts, other localized influences may still affect property values. Second, the structure of OMI data itself poses methodological challenges: values are provided only as minimum and maximum transaction prices within each zone, without details on the volume or distribution of sales at different price points. This means that averages can be skewed by outliers—e.g., exceptionally high- or low-priced transactions—making them a potentially unreliable proxy for the actual distribution of property values. Furthermore, the absence of median values limits the ability to assess internal price symmetry. Lastly, the spatial boundaries of OMI zones do not always align with census tract boundaries, the administrative units used in this study to reconstruct the Mesoarea subdivisions. This misalignment can lead to inconsistencies in how demographic and real estate data overlap, reducing the precision of any SES inferences drawn from real estate heterogeneity alone.

Finally, using the education level alone as a proxy for SES—although it is the most representative factor and most strongly associated with health outcomes [87,88]—may in some cases be limiting or insufficiently informative, and could underlie the lack of statistical significance. The decision to rely on educational level was primarily driven by the unavailability of publicly accessible data on other individual-level socio-economic variables, largely due to privacy concerns or adequate spatial resolution. For instance, it was not possible to collect the data required to compute the deprivation index proposed by Rosano et al. [46] for the Italian context, as such information was not available, and the development and validation of a new deprivation index was beyond the scope of this study. While we consider educational level to be an adequate proxy—based on the arguments previously outlined—for conducting a preliminary analysis aimed at redefining Urban Zones (UZs), we also recognize that a more comprehensive socio-economic status (SES) indicator could be highly valuable for the identification of Mesoareas.

The forthcoming update of the socioeconomic deprivation index [46], calibrated on the 2021 census data, will—once available—provide a further advancement in the development of a multidimensional and integrated model for health planning. This update, previously validated using 2011 data within the Italian context, is expected to enhance the accuracy and applicability of territorial vulnerability assessments, as anticipated. A subsequent phase of analysis will certainly address the evaluation of accessibility, once the specific socio-healthcare facilities to be included in the territorial planning framework have been identified. The high variance observed in some Mesoareas—including those that were delineated through the methodological subdivision of the Urban Zones proposed in this study—suggests the need to complement quantitative data with qualitative assessments. Where intra-area heterogeneity remains high despite efforts to achieve greater internal homogeneity, additional methodologies may be required to refine territorial segmentation. These could include expert-based approaches, such as structured elicitation of local knowledge from professionals working in urban planning, public health, or social services. Moreover, visual surveys of the built environment and housing stock, or the analysis of planning documents such as the General Regulatory Plan (*Piano Regolatore Generale*) of the Municipality of Rome, may provide further insights into socio-spatial dynamics that are not fully captured by census or real estate data. Integrating these sources can support a more nuanced and context-sensitive identification of sub-areas, particularly where standard indicators fall short in representing the lived reality of urban populations.

The aim of the study was to present a potential model for redefining administrative boundaries within the territory of Rome, while simultaneously improving understanding of the area, its urban and social dynamics, and consequently the health needs of the population from an equity perspective. In this sense, these Mesoareas represent a possible methodological advancement, especially in light of the current discussions taking place within the Municipality of Rome.

Author Contributions: Conceptualization, methodology, validation, E.M., S.C., L.P. and L.M.S.; formal analysis, E.M., S.C. and L.M.S.; investigation, E.M., S.C., L.P. and L.M.S.; resources, E.M. and L.M.S.; data curation, E.M., S.C., L.P. and L.M.S.; writing—original draft preparation, E.M., S.C., L.P. and L.M.S.; writing—review and editing, E.M., S.C., L.P. and L.M.S.; visualization, E.M., S.C., L.P. and L.M.S.; supervision, L.M.S.; project administration, L.M.S. All authors have read and agreed to the published version of the manuscript.

Funding: This research received no external funding.

Data Availability Statement: All data presented in this study are available on request from the corresponding author.

Conflicts of Interest: The authors declare no conflicts of interest.

Appendix A

Table A1. OMI Zones in III Municipality and surroundings. Sale price of residential properties intended for civilian, year 2021. Agenzia delle Entrate.

OMI Zone	Zone Code	Min Sale Price (€)	Max Sale Price (€)	Average Sale Price (€)
'Riserva Della Marcigliana (Via Di Santa Colomba)'	R7	1400	2100	1750
'Castel Giubileo-Bel Poggio (Via Castorano)'	E16	1500	2050	1775
'Fidene-Villa Spada (Via Radicofani)'	E41	1750	2550	2150
'Casal Monastero (Via Belmonte In Sabina)'	E101	1800	2500	2150
'Fidene-Villa Spada (Via Radicofani)'	E41	1750	2550	2150
'Serpentara (Viale Lina Cavalieri)'	E42	2000	2850	2425
'Conca D'Oro (Via Val Di Lanzo)'	D46	2200	3200	2700
'La Cinquina-Bufalotta (Via Feo Belcari)'	E123	2200	3200	2700

Table A1. *Cont.*

OMI Zone	Zone Code	Min Sale Price (€)	Max Sale Price (€)	Average Sale Price (€)
'Valmelaina-Tufello (Via Delle Isole Curzolane)'	D12	2350	3300	2825
'Vigne Nuove-Porta Di Roma (Via Delle Vigne Nuove)'	E47	2400	3400	2900
'Sacco Pastore (Via Val Trompia)'	D47	2450	3500	2975
'Talenti (Via Ugo Ojetti)'	D27	2500	3500	3000
'Nuovo Salario-Prati Fiscali (Via Monte Cervialto)'	D71	2450	3600	3025
'Montesacro (Viale Adriatico)'	D11	2600	3600	3100
'Casal Boccone Bufalotta (Via Paolo Monelli)'	E51	2700	3700	3200
'Colle Salario (Via Monte Giberto)'	E40	2800	3900	3350

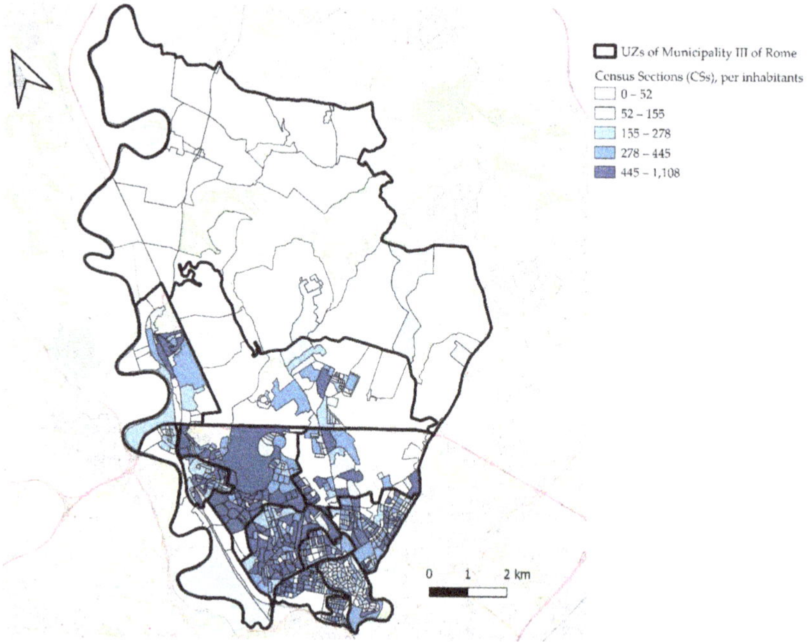

Figure A1. The map shows the subdivision of Rome's III Municipality into census sections, shaded according to population density. The bold black boundaries represent the Urban Zones (UZs). Census sections in the western portion of the area within the GRA (Grande Raccordo Anulare) show higher population density. In contrast, sparsely populated areas are located in the northern periphery—mainly within the Riserva Naturale della Marcigliana—and in the southwestern sector beyond Via Salaria, along the left bank of the Tiber River.

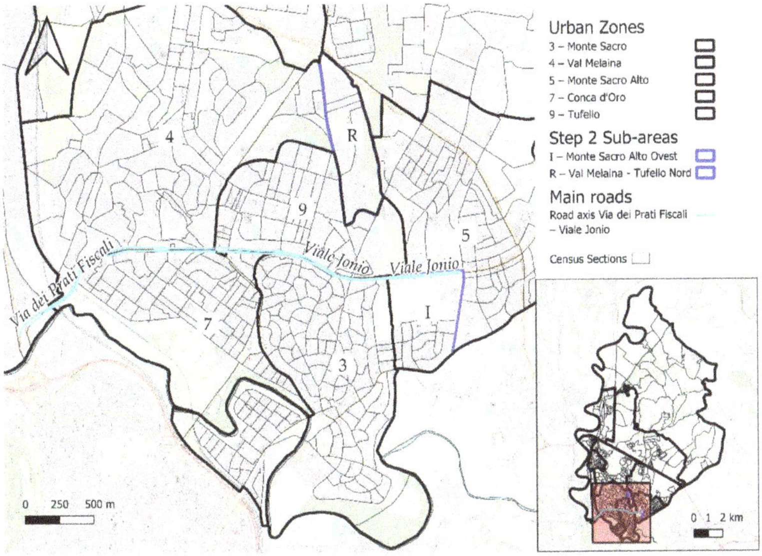

Figure A2. The image shows the road axis formed by Via dei Prati Fiscali (to the west) and Viale Jonio (to the east). This road axis already serves as the boundary between several Urban Zones (UZs) within Municipality III, and in its easternmost segment, it separates the UZs of Tufello (9) and Monte Sacro Alto (5) from the sub-area Monte Sacro Alto—Parco Sannazzaro, as defined in Step 2. The red box in the overview map corresponds to the geographical extent displayed in the main map.

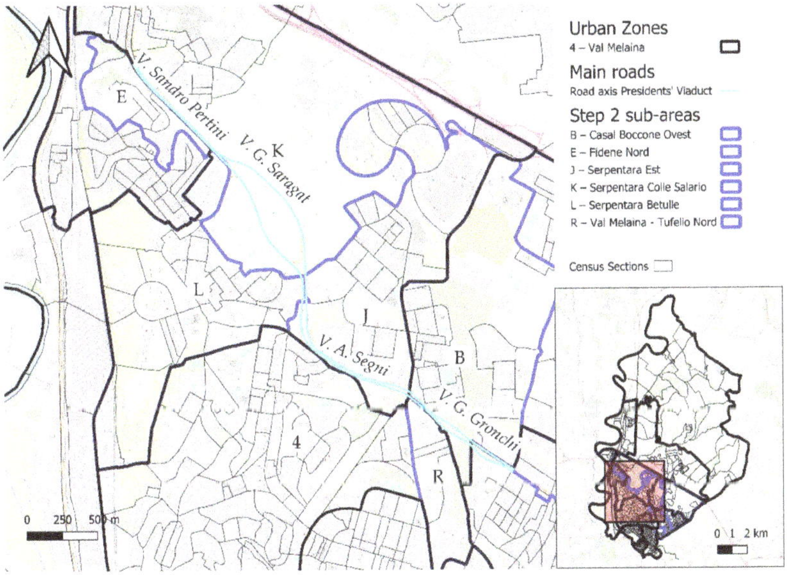

Figure A3. The image shows the road axis known as the Viadotto dei Presidenti. This road crosses the boundary census sections between the sub-areas created in Step 2 from the following Urban Zones: Fidene (E), Serpentara (J, K, L), Val Melaina (4, R), and Casal Boccone (B). The red box in the overview map corresponds to the geographical extent displayed in the main map.

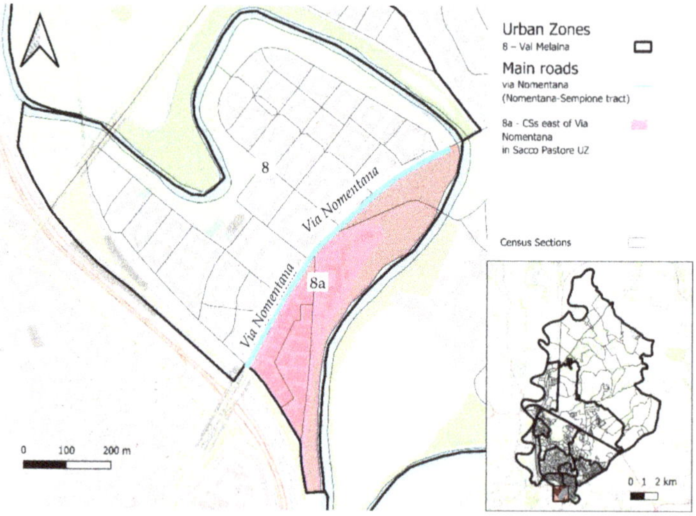

Figure A4. The image shows the section of Via Nomentana between Via Nomentana Nuova and Piazza Sempione (Ponte Tazio), the only segment designated as an 'interquartiere' road by the General Urban Traffic Plan of Roma Capitale. This road crosses the Urban Zone of Sacco Pastore, dividing it into a larger northwestern area and a smaller southeastern portion along the banks of the Aniene River. Due to its low population (859 residents), the latter area was not defined as a separate sub-area. The red box in the overview map corresponds to the geographical extent displayed in the main map.

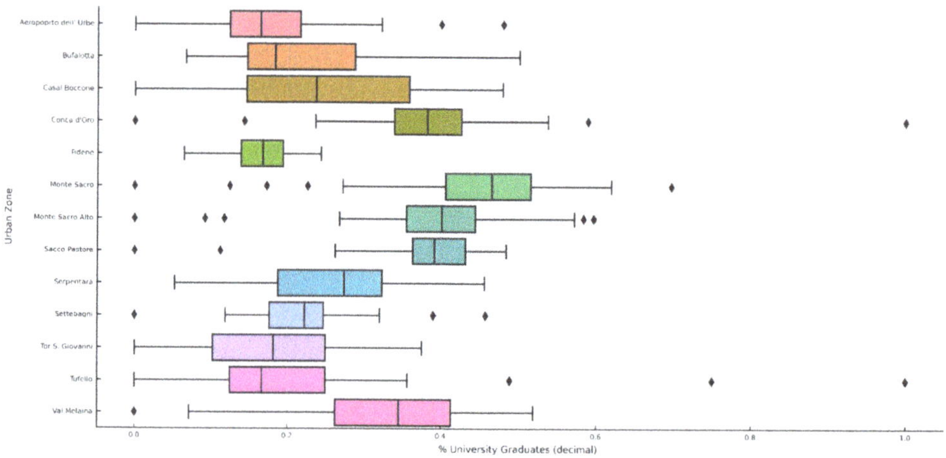

Figure A5. Box plots showing the distribution of selected variables across the Urban Zones within Municipality III. The plots illustrate the median, interquartile range (IQR), and potential outliers.

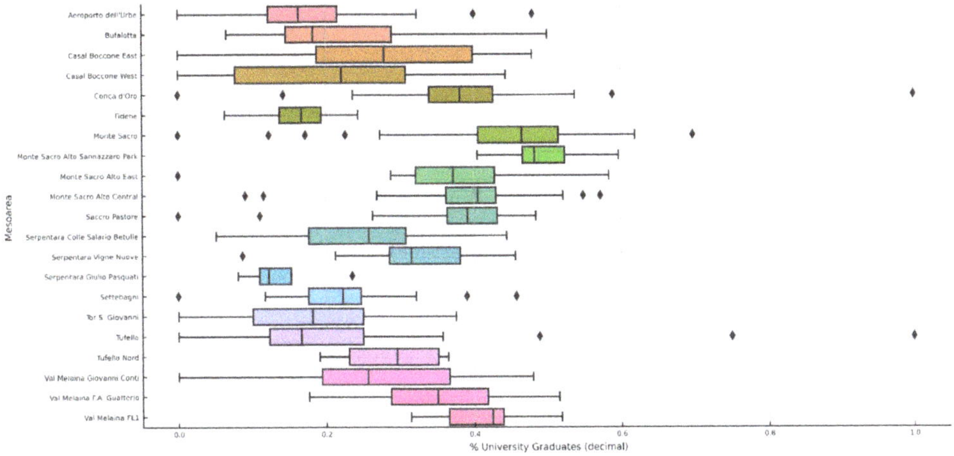

Figure A6. Box plots showing the distribution of selected variables across the Mesoareas identified within Municipality III. The plots illustrate the median, interquartile range (IQR), and potential outliers, highlighting the variability and skewness of socio-demographic and territorial indicators within and between Mesoareas.

References

1. Diderichsen, F.; Hallqvist, J.; Whitehead, M. Differential vulnerability and susceptibility. In *Challenging Inequities in Health: From Ethics to Action*; Evans, T., Whitehead, M., Diderichsen, F., Bhuiya, A., Eds.; Oxford University Press: New York, NY, USA, 2001; pp. 58–71. [CrossRef]
2. Calandrini, E.; Paglione, L.; Bargagli, A.M.; Agabiti, N.; Battisti, A.; Salvatori, L.M.; Marceca, M.; Brandimarte, M.A.; Di Rosa, E.; Iorio, S.; et al. Does Urbanization Correlate with Health Service Assistance? An Observational Study in Rome, Italy. In *Equity in Health*; Green Energy and Technology; Springer: New York, NY, USA, 2023; pp. 95–111, ISBN 978-3-031-16181-0. [CrossRef]
3. Ventura, M.; Fusco, D.; Bontempi, K.; Colais, P.; Davoli, M. Regional Outcome Evaluation Program (P.Re.Val.E.): Reduction of inequality in access to effective health care in the Lazio region of Italy (2012–2015). *PLoS ONE* **2018**, *13*, e0194972. [CrossRef] [PubMed]
4. Marmot, M. Social determinants of health inequalities. *Lancet* **2005**, *365*, 1099–1104. [CrossRef] [PubMed]
5. Dahlgren, G.; Whitehead, M. *Policies and Strategies to Promote Social Equity in Health*; Institute for Futures Studies: Stockholm, Sweden, 1991.
6. Cacciani, L.; Bargagli, A.M.; Cesaroni, G.; Forastiere, F.; Agabiti, N.; Davoli, M. Education and Mortality in the Rome Longitudinal Study. *PLoS ONE* **2015**, *10*, e0137576. [CrossRef] [PubMed]
7. Nardi, A.; Dei Bardi, L.; Davoli, M.; Agabiti, N.; Cesaroni, G. Differences in mortality between temporary and permanent workers: Results from the Rome Longitudinal Study. *BMJ Open* **2022**, *12*, e058594. [CrossRef]
8. Bauleo, L.; Giannini, S.; Ranzi, A.; Nobile, F.; Stafoggia, M.; Ancona, C.; Iavarone, I.; the EpiCovAir Study Group. A Methodological Approach to Use Contextual Factors for Epidemiological Studies on Chronic Exposure to Air Pollution and COVID-19 in Italy. *Int. J. Environ. Res. Public Health* **2022**, *19*, 2859. [CrossRef]
9. Badaloni, C.; De Sario, M.; Caranci, N.; De' Donato, F.; Bolignano, A.; Davoli, M.; Leccese, L.; Michelozzi, P.; Leone, M. A spatial indicator of environmental and climatic vulnerability in Rome. *Environ. Int.* **2023**, *176*, 107970. [CrossRef]
10. Costa, G.; Caiazzo, A.; Marinacci, C.; Spadea, T. Individual and Contextual Determinants of Inequalities in Health: The Italian Case. In *The Political and Social Contexts of Health*; Routledge: Abingdon, UK, 2020; pp. 115–146. [CrossRef]
11. D'Alessandro, D. Urban public health: A multidisciplinary approach. In *Urban Public Health: A Multidisciplinary Approach*; Springer Nature: Cham, Switzerland, 2020; pp. 1–8. [CrossRef]
12. Brown, H.L.; Pursley, I.G.; Horton, D.L.; La Ragione, R.M. One health: A structured review and commentary on trends and themes. *One Health Outlook* **2024**, *6*, 17. [CrossRef]
13. Settimo, G.; Gola, M.; Capolongo, S. The Relevance of Indoor Air Quality in Hospital Settings: From an Exclusively Biological Issue to a Global Approach in the Italian Context. *Atmosphere* **2020**, *11*, 361. [CrossRef]
14. D'Alessandro, D.; Appolloni, L. Housing and Health: Here We Go Again. *Int. J. Environ. Res. Public Health* **2021**, *18*, 12060. [CrossRef]

15. Cofone, L.; Sabato, M.; Di Paolo, C.; Di Giovanni, S.; Donato, M.A.; Paglione, L. Urban, Architectural, and Socioeconomic Factors Contributing to the Concentration of Potential Arbovirus Vectors and Arbovirosis in Urban Environments from a One Health Perspective: A Systematic Review. *Sustainability* **2025**, *17*, 4077. [CrossRef]
16. Paglione, L.; Donato, M.A.; Cofone, L.; Sabato, M.; Appolloni, L.; D'Alessandro, D. The Healthy City Reimagined: Walkability, Active Mobility, and the Challenges of Measurement and Evaluation. *Urban Sci.* **2024**, *8*, 157. [CrossRef]
17. Cornu, T.; Marchal, B.; Renmans, D. How do urban green spaces influence heat-related mortality in elderly? A realist synthesis. *BMC Public Health* **2024**, *24*, 457. [CrossRef] [PubMed]
18. Smith, N. Gentrification and the rent-gap. *Ann. Assoc. Am. Geogr.* **1987**, *77*, 462–465. [CrossRef]
19. Bier, T.; Moving Up, Filtering Down: Metropolitan Housing Dynamics and Public Policy. Discussion Paper Prepared for the Brookings Institution Center on Urban and Metropolitan Policy. 2001. Available online: https://www.brookings.edu/wp-content/uploads/2016/06/bier.pdf (accessed on 15 June 2025).
20. Italian Republic. Law 23 December 1978, n. 833, Istituzione Del Servizio Sanitario Nazionale; Gazzetta Ufficiale Della Repubblica Italiana, Serie Generale, n. 360, 28 December 1978; Suppl. Ordinario. Available online: https://www.gazzettaufficiale.it/eli/id/1978/12/28/078U0833/sg (accessed on 10 June 2025).
21. Italian Republic. Law 8 November 1992, n. 502, Riordino Della Disciplina in Materia Sanitaria E Istituzione Del Servizio Sanitario Nazionale; Gazzetta Ufficiale della Repubblica Italiana, Serie Generale, n. 303, 30 December 1992; Suppl. Ordinario. Available online: https://www.gazzettaufficiale.it/eli/id/1992/12/30/092G0502/sg (accessed on 10 June 2025).
22. Italian Republic. Ministero della Salute. Decreto 23 Maggio 2022, n. 77. Regolamento Recante La Definizione Di Modelli E Standard per Lo Sviluppo Dell'assistenza Territoriale Nel Servizio Sanitario Nazionale. Gazzetta Ufficiale Della Repubblica Italiana, Serie Generale, n. 144, 22 June 2022; Entrata in Vigore 07 July 2022. Available online: https://www.gazzettaufficiale.it/eli/id/2022/06/22/22G00085/SG (accessed on 10 June 2025).
23. He, Z.; Pfaff, E.; Guo, S.J.; Guo, Y.; Wu, Y.; Tao, C.; Stiglic, G.; Bian, J. Enriching Real-world Data with Social Determinants of Health for Health Outcomes and Health Equity: Successes, Challenges, and Opportunities. *Yearb. Med. Inform.* **2023**, *32*, 253–263. [CrossRef] [PubMed]
24. Italian National Institute of Statistics (ISTAT). *Le Sezioni Di Censimento: Definizione, Metodologia E Utilizzo; Atti-CIS*; Fascicolo 5; ISTAT: Rome, Italy, 2016. Available online: https://www.istat.it/it/files/2016/02/Atti-CIS_Fascicolo_5.pdf (accessed on 11 June 2025).
25. Lelo, K.; Monni, S.; Tomassi, F. Socio-spatial inequalities and urban transformation. The case of Rome districts. *Socio-Econ. Plan. Sci.* **2019**, *68*, 66–78. [CrossRef]
26. Salvatori, L.M.; Marceca, M.; Paglione, L.; Angelozzi, A.; Caminada, S.; Errigo, M.; Greco, A.D.; Iorio, S. Methodological Comparison of Different Projects. In *Green Energy and Technology*; Springer: Cham, Switzerland, 2020; pp. 225–260. [CrossRef]
27. Comune di Roma Capitale. *Annuario Statistico Di Roma Capitale 2023—Capitolo 2: Popolazione E Territorio*; Roma Capitale: Rome, Italy, 2023; pp. 12–15. Available online: https://www.comune.roma.it/web-resources/cms/documents/02_Popolazione_Annuario_2023_def.pdf (accessed on 11 June 2025).
28. Comune di Roma Capitale. Ufficio Statistico. Zone Urbanistiche Di Roma Capitale: Istituzione, Aggiornamenti E Criteri Di Delimitazione; Roma Capitale: Ufficio Statistico, 1977. Available online: https://www.comune.roma.it/web-resources/cms/documents/ElencoZ_Urbanistiche_rg_A.pdf (accessed on 15 June 2025).
29. Lelo, K.; Monni, S.; Reynaud, C.; Tomassi, F. Demographic Trends in the Rome Censuses 1981–2021: The 155 Urban Zones. Econ. Politica 2024, (Oct–Nov), 1–18. Available online: https://www.economiaepolitica.it/lavoro-e-diritti/i-cambiamenti-demografici-nei-censimenti-dal-1981-al-2021-le-155-zone-urbanistiche-di-roma/ (accessed on 11 June 2025).
30. Comune di Roma Capitale. Ufficio Toponomastica. I Rioni Di Roma: Suddivisioni Storiche E Amministrative; Roma Capitale: Ufficio Toponomastica, 2021. Available online: https://www.comune.roma.it/web-resources/cms/documents/01_Territorio_Annuario_2021_agg30062022.pdf (accessed on 11 June 2025).
31. Comune di Roma Capitale. Ufficio Statistico. I Quartieri Di Roma: Definizione, Origine E Sviluppo; Roma Capitale: Ufficio Statistico, 2022. Available online: https://www.comune.roma.it/web-resources/cms/documents/Annuario_2022_completo.pdf (accessed on 11 June 2025).
32. Insolera, I. *Modern Rome: From Napoleon to the Twenty-First Century*; Cambridge Scholars Publishing: Newcastle-upon-Tyne, UK, 2018; xxvii+479p, ISBN 978-1-5275-1664-9.
33. Lelo, K.; Chiaradia, F.; Monni, S.; Tomassi, F. The 15-Minute City: An Attempt to Measure Proximity to Urban Services in Rome. *Sustainability* **2024**, *16*, 9432. [CrossRef]
34. Paoletti, F.; Greco, A.D. Participatory Local Welfare in Trieste: A Laboratory for Local Welfare Activities. In *Urban Health*; Battisti, A., Marceca, M., Iorio, S., Eds.; AIMETA 2019. Green Energy and Technology; Springer: Cham, Switzerland, 2020. [CrossRef]
35. Tocci, W. *Avanti C'è Posto. Storie E Progetti Del Trasporto Pubblico a Roma*; Donzelli Editore: Roma, Italy, 2008; pp. 1–288, ISBN 978-88-6036-276-6.
36. Celesti Grapow, L.; Fanelli, G. The Vanishing Landscape of the Campagna Romana. *Landsc. Urban Plan.* **1993**, *24*, 69–76. [CrossRef]
37. Bradley, G. Mobility and Secession in the Early Roman Republic. *Antichthon* **2017**, *51*, 149–171. [CrossRef]

38. Cattarulla, C. "Luoghi di Memoria Ispanoamericani a Roma (Simón Bolívar e il Giuramento sul Monte Sacro)." 1. MARINELLA ROCCA LONGO, MADDALENA PENNACCHIA (a cura di), Turismo Creativo e Identità Culturale, 2015: 107. Available online: https://romatrepress.uniroma3.it/wp-content/uploads/2024/12/6.-Camilla-Cattarulla.pdf (accessed on 15 June 2025).
39. Angelidou, M.; Kanellopoulos, I. Peripheral development and deregulated urban growth in Rome: Regional competitiveness and urban planning challenges. *Land Use Policy* **2015**, *48*, 167–175. [CrossRef]
40. Mannucci, S.; Kwakkel, J.H.; Morganti, M.; Ferrero, M. Exploring potential futures: Evaluating the influence of deep uncertainties in urban planning through scenario planning—A case study in Rome, Italy. *Futures* **2023**, *154*, 103265. [CrossRef]
41. Glorioso, V.; Subramanian, S.V. Equity in access to health care services in Italy. *Health Serv. Res.* **2014**, *49*, 950–970. [CrossRef] [PubMed] [PubMed Central]
42. Comune di Roma Capitale. Portale Open Data—Roma Capitale. Available online: http://www.datiopen.it/it/catalogo-opendata/daticomuneromait (accessed on 11 June 2025).
43. Comune di Roma Capitale. Delibera Consiliare n. 2983 Del 29–30 Luglio 1977: Istituzione Delle Zone Urbanistiche; Roma Capitale: Rome, Italy. Available online: https://it.wikipedia.org/wiki/Zone_urbanistiche_di_Roma (accessed on 11 June 2025).
44. Agenzia delle Entrate. OMI—Osservatorio Del Mercato Immobiliare. Consultazione Zone E Valori Immobiliari; Agenzia Delle Entrate: Rome, Italy. Available online: https://www1.agenziaentrate.gov.it/servizi/geopoi_omi/index.php (accessed on 11 June 2025).
45. Italian Ministry of Infrastructure and Transport. OpenData: ANAS Road Network Graph. Available online: https://dati.mit.gov.it/catalog/dataset/grafo-stradale-anas (accessed on 19 July 2025).
46. Rosano, A.; Pacelli, B.; Zengarini, N.; Costa, G.; Cislaghi, C.; Caranci, N. Aggiornamento e revisione dell'indice di deprivazione italiano 2011 a livello di sezione di censimento [Update and review of the 2011 Italian deprivation index calculated at the census section level]. *Epidemiol Prev.* **2020**, *44*, 162–170. [CrossRef] [PubMed]
47. Open House Roma. Complesso Residenziale Monti Della Breccia. Available online: https://www.openhouseroma.org/sito/complesso-residenziale-monti-della-breccia (accessed on 11 June 2025).
48. Cesaroni, G.; Venturini, G.; Paglione, L.; Angelici, L.; Sorge, C.; Marino, C.; Davoli, M.; Agabiti, N. Differenziali di mortalità a Roma: Il ruolo dell'istruzione e dei prezzi immobiliari del quartiere di residenza [Mortality inequalities in Rome: The role of individual education and neighbourhood real estate market]. *Epidemiol Prev.* **2020**, *44* (Suppl. 1), 31–37. [CrossRef] [PubMed]
49. Coffee, N.T.; Lockwood, T.; Hugo, G.; Paquet, C.; Howard, N.J.; Daniel, M. Relative residential property value as a socio-economic status indicator for health research. *Int. J. Health Geogr.* **2013**, *12*, 22. [CrossRef] [PubMed]
50. Moudon, A.V.; Cook, A.J.; Ulmer, J.; Hurvitz, P.M.; Drewnowski, A. A neighborhood wealth metric for use in health studies. *Am. J. Prev. Med.* **2011**, *41*, 88–97. [CrossRef]
51. Lockwood, T.; Coffee, N.T.; Rossini, P.; Niyonsenga, T.; McGreal, S. Does where you live influence your socio-economic status? *Land Use Policy* **2018**, *72*, 152–160. [CrossRef]
52. Juhn, Y.J.; Beebe, T.J.; Finnie, D.M.; Sloan, J.; Wheeler, P.H.; Yawn, B.; Williams, A.R. Development and Initial Testing of a New Socioeconomic Status Measure Based on Housing Data. *J. Urban Health* **2011**, *88*, 933–944. [CrossRef]
53. Galobardes, B.; Shaw, M.; Lawlor, D.A.; Lynch, J.W.; Smith, G.D. Indicators of socioeconomic position (part 1). *J. Epidemiol. Community Health* **2006**, *60*, 7–12. [CrossRef]
54. Drewnowski, A.; Aggarwal, A.; Tang, W.; Moudon, A.V. Residential property values predict prevalent obesity but do not predict 1 year weight change. *Obesity* **2015**, *23*, 671–676. [CrossRef]
55. Rehm, C.D.; Moudon, A.V.; Hurvitz, P.M.; Drewnowski, A. Residential property values are associated with obesity among women in King County, WA, USA. *Soc. Sci. Med.* **2012**, *75*, 491–495. [CrossRef]
56. Tulumello, S.; Dagkouli-Kyriakoglou, M. Financialization of Housing in Southern Europe: Policy Analysis and Recommendations; European Parliament: 2021. Available online: https://www.researchgate.net/profile/Simone-Tulumello-2/publication/349436154_Financialization_of_housing_in_Southern_Europe_Policy_analysis_and_recommendations_Final_report_-_submitted_28012021/links/602fb529a6fdcc37a8381322/Financialization-of-housing-in-Southern-Europe-Policy-analysis-and-recommendations-Final-report-submitted-28-01-2021.pdf (accessed on 11 June 2025).
57. Fujita, K. Introduction: Residential segregation in context. In *Residential Segregation in Comparative Perspective*; Maloutas, T., Ed.; Routledge: Farnham, UK, 2012; pp. 1–36.
58. Logan, J.R.; Martinez, M. The Spatial Scale and Spatial Configuration of Residential Settlement: Measuring Segregation in the Postbellum South. *Am. J. Sociol.* **2018**, *123*, 1161–1203. [CrossRef] [PubMed] [PubMed Central]
59. Barreca, A.; Curto, R.; Rolando, D. Housing vulnerability and property prices: Spatial analyses in the Turin real estate market. *Sustainability* **2018**, *10*, 3068. [CrossRef]
60. Comune di Roma Capitale. *Bollettino Statistico. Popolazione Residente Nei Municipi E Nelle Zone Urbanistiche Di Roma Capitale—Giugno 2023*; Roma Capitale: Rome, Italy, 2023. Available online: https://www.comune.roma.it/web-resources/cms/documents/Bollettino_statistico_giugno_2023_def.pdf (accessed on 11 June 2025).
61. Monni, S.; Lelo, K.; Tomassi, F. Roma, tra centro e periferie: Come incidono le dinamiche urbanistiche sulle disuguaglianze socio-economiche. *Roma Mod. E Contemp.* **2017**, *25*, 131–146. [CrossRef]

62. Temam, S.; Varraso, R.; Pornet, C.; Sanchez, M.; Affret, A.; Jacquemin, B.; Clavel-Chapelon, F.; Rey, G.; Rican, S.; Le Moual, N. Ability of ecological deprivation indices to measure social inequalities in a French cohort. *BMC Public Health* **2017**, *17*, 956. [CrossRef] [PubMed]
63. Mohammed, S.; Bailey, G.A.; Farr, I.W.; Jones, C.; Rawlings, A.; Rees, S.; Scully, S.; Wang, T.; Evans, H.T. Using the Welsh Index of Multiple Deprivation in research: Estimating the effect of excluding domains on a routine health data study. *BMC Public Health* **2025**, *25*, 1178. [CrossRef] [PubMed]
64. Lelo, K.; Monni, S.; Tomassi, F. *Le Sette Rome*, 1st ed.; Donzelli Editore: Roma, Italy, 2021; p. 37.
65. Khashoggi, B.F.; Murad, A. Issues of Healthcare Planning and GIS: A Review. *ISPRS Int. J. Geo-Inf.* **2020**, *9*, 352. [CrossRef]
66. Biu, P.W.; Nwasike, C.N.; Nwaobia, N.K.; Ezeigweneme, C.A.; Gidiagba, J.O. GIS in Healthcare Facility Planning and Management: A Review. *World J. Adv. Res. Rev.* **2024**, *21*, 12–19. [CrossRef]
67. Ministero della Salute. *Piano Nazionale Della Cronicità*; Ministero della Salute: Rome, Italy, 2016. Available online: https://osservatoriocronicita.it/images/PDF/Piano_Nazionale_cronicit.pdf (accessed on 11 June 2025).
68. Regione Lazio. *La Salute È Equità: Piano Regionale Della Prevenzione 2020–2025. Documento Tecnico*; Regione Lazio: Rome, Italy, 2021; ISBN 978-88-490-0751-0. Available online: https://www.deplazio.net/images/stories/files/Salute-equita.pdf (accessed on 11 June 2025).
69. Whitehead, J.; Pearson, A.L.; Lawrenson, R.; Atatoa-Carr, P. Spatial equity and realised access to healthcare—A geospatial analysis of general practitioner enrolments in Waikato, New Zealand. *Rural. Remote Health* **2019**, *19*, 5349. [CrossRef]
70. Higgs, G. The role of GIS for health utilization studies: Literature review. *Health Serv. Outcomes Res. Methodol.* **2009**, *9*, 84–99. [CrossRef]
71. Turnbull, J.; Martin, D.; Lattimer, V.; Pope, C.; Culliford, D. Does distance matter? Geographical variation in GP out-of-hours service use: An observational study. *Br. J. Gen. Pract.* **2008**, *58*, 471. [CrossRef]
72. Freeman, R.E.; Boggs, K.M.; Sullivan, A.F.; Faridi, M.K.; Freid, R.D.; Camargo, C.A., Jr. Distance From Freestanding Emergency Departments to Nearby Emergency Care. *Ann. Emerg. Med.* **2021**, *77*, 48–56. [CrossRef] [PubMed]
73. Oppio, A.; Bottero, M.; Giordano, G.; Arcidiacono, A. A multi-methodological evaluation approach for assessing the impact of neighbourhood quality on public health. *Epidemiol. Prev.* **2016**, *40*, 249–256. [CrossRef] [PubMed]
74. World Health Organization Regional Office for Europe. *Healthy, Prosperous Lives for All: The European Health Equity Status Report*; WHO Regional Office for Europe: Copenhagen, Denmark, 2019. Available online: https://iris.who.int/bitstream/handle/10665/344116/9789289052498-eng.pdf?sequence=1 (accessed on 13 May 2025).
75. Beute, F.; Davies, Z.G.; De Bell, S.; Houlden, V.; White, M.P.; O'Brien, L.; Lomas, T.; Mavoa, S.; Badal, K.; Thompson, C.W. *Types and Characteristics of Urban and Peri-Urban Green Spaces Having an Impact on Human Mental Health and Wellbeing: A Systematic Review*; An EKLIPSE Expert Working Group Report; EKLIPSE: Wallingford, UK, 2020. Available online: https://iris.uniroma1.it/bitstream/11573/1535901/1/Beute_Types_2020.pdf (accessed on 11 June 2025).
76. LaReaux, J.; Watkins, D. Geospatial analysis for promoting urban green space equity: Case study of Detroit, Michigan, USA. *Urban For. Urban Green.* **2025**, *105*, 128716. [CrossRef]
77. Zeynali, R.; Mandanici, E.; Sohrabi, A.H.; Trevisiol, F.; Bitelli, G. GIS-Based Urban Heat Island Mapping and Analysis: Experiences in the City of Bologna. In Proceedings of the 2024 IEEE International Workshop on Metrology for Living Environment (MetroLivEnv), Chania, Greece, 12–14 June 2024; pp. 230–234. [CrossRef]
78. National Research Council (US). *Chapter 2: Determining the Extent of Food Deserts. The Public Health Effects of Food Deserts: Workshop Summary*; National Academies Press: Washington, DC, USA, 2009. Available online: https://www.ncbi.nlm.nih.gov/books/NBK208011/ (accessed on 11 June 2025).
79. Smooth, W.G. Intersectionality from Theoretical Framework to Policy Intervention. In *The Politics of Intersectionality*; Wilson, A.R., Ed.; Situating Intersectionality. Palgrave Macmillan: New York, NY, USA, 2013. [CrossRef]
80. Bauer, G.R.; Churchill, S.M.; Mahendran, M.; Walwyn, C.; Lizotte, D.; Villa-Rueda, A.A. Intersectionality in quantitative research: A systematic review of its emergence and applications of theory and methods. *SSM—Popul. Health* **2021**, *14*, 100798. [CrossRef]
81. Bauer, G.R. Incorporating intersectionality theory into population health research methodology: Challenges and the potential to advance health equity. *Soc. Sci. Med.* **2014**, *110*, 10–17. [CrossRef]
82. Bono, M.; Lapierre, J.; Morin, P. Caring for People and Territories: Brief Historical Review of the Intersectoral Social Innovation Experience of Trieste and Its Habitat Micro-Area Program. *Can. J. Nonprofit Soc. Econ. Res.* **2024**, *15*, 46–58. [CrossRef]
83. De Vidovich, L. Urban Regeneration through the Territorialisation of Social Policies: Findings from the Microareas Programme in Trieste, Italy. *J. Urban Regen. Renew.* **2020**, *13*, 406–421. [CrossRef]
84. Costa, G.; Stroscia, M.; Zengarini, N.; Demaria, M. *40 Anni Di Salute a Torino: Spunti per Leggere i Bisogni E I Risultati Delle Politiche*; Inferenze: Milano, Italy, 2017; ISBN 978-88-942239-0-3.
85. Gentilini, V.; Bodini, C.; Di Girolamo, C.; Campione, I.; Cavazza, G.; Marzaroli, P.; Musti, M.A.; Perlangeli, V.; Pandolfi, P.; Pizzi, L.; et al. An ecological study on health inequalities in the city of Bologna (Emilia-Romagna Region, Northern Italy): Bridging knowledge and action. *Epidemiol. E Prev.* **2020**, *44* (Suppl. 1), 45–53. [CrossRef]

86. Consolazio, D.; AlSayed, A.; Serini, M.; Benassi, D.; Sarti, S.; Terraneo, M.; Celata, C.; Russo, A.G. Social inequalities in health within the City of Milan (Lombardy Region, Northern Italy): An ecological assessment. *Epidemiol. E Prev.* **2024**, *48*, 298–308. [CrossRef]
87. Lindberg, M.H.; Chen, G.; Olsen, J.A.; Abelsen, B. Combining education and income into a socioeconomic position score for use in studies of health inequalities. *BMC Public Health* **2022**, *22*, 969. [CrossRef]
88. Moss, J.L.; Johnson, N.J.; Yu, M.; Altekruse, S.F.; Cronin, K.A. Comparisons of individual- and area-level socioeconomic status as proxies for individual-level measures: Evidence from the Mortality Disparities in American Communities study. *Popul. Health Metr.* **2021**, *19*, 1. [CrossRef]

Disclaimer/Publisher's Note: The statements, opinions and data contained in all publications are solely those of the individual author(s) and contributor(s) and not of MDPI and/or the editor(s). MDPI and/or the editor(s) disclaim responsibility for any injury to people or property resulting from any ideas, methods, instructions or products referred to in the content.

MDPI AG
Grosspeteranlage 5
4052 Basel
Switzerland
Tel.: +41 61 683 77 34

Land Editorial Office
E-mail: land@mdpi.com
www.mdpi.com/journal/land

Disclaimer/Publisher's Note: The title and front matter of this reprint are at the discretion of the Guest Editors. The publisher is not responsible for their content or any associated concerns. The statements, opinions and data contained in all individual articles are solely those of the individual Editors and contributors and not of MDPI. MDPI disclaims responsibility for any injury to people or property resulting from any ideas, methods, instructions or products referred to in the content.

www.ingramcontent.com/pod-product-compliance
Lightning Source LLC
Chambersburg PA
CBHW040826250926
51850CB00007BA/1269
9783725858279